AF324835

Book Series on
Complex Metallic Alloys – Vol. 4

MECHANICAL PROPERTIES OF COMPLEX INTERMETALLICS

Series on Complex Metallic Alloys

Series Editor: Jean Marie Dubois

Book Series on
Complex Metallic Alloys – Vol. 4

MECHANICAL PROPERTIES OF COMPLEX INTERMETALLICS

edited by

Esther Belin-Ferré

Centre National de la Recherche Scientifique, Université Pierre et Marie Curie, France

NEW JERSEY · LONDON · SINGAPORE · BEIJING · SHANGHAI · HONG KONG · TAIPEI · CHENNAI

Published by

World Scientific Publishing Co. Pte. Ltd.

5 Toh Tuck Link, Singapore 596224

USA office: 27 Warren Street, Suite 401-402, Hackensack, NJ 07601

UK office: 57 Shelton Street, Covent Garden, London WC2H 9HE

British Library Cataloguing-in-Publication Data
A catalogue record for this book is available from the British Library.

ISBN-13 978-981-4322-16-4
ISBN-10 981-4322-16-4

Printed in Singapore by World Scientific Printers.

FOREWORD

This is the fourth and last volume of a series of books issued from the lectures delivered at the Euro-School on Materials Science, organised yearly in Ljubljana, by the European Network of Excellence Complex Metallic Alloys (CMA) under contract NMP3-CT-2005-500145.

The last session of the CMA Euro-School was held in May 2009, it was mostly dedicated to mechanical properties and plasticity of complex metallic alloys. Nine chapters of this volume are dedicated to these matters. Another chapter refers to mechanical properties of nanotubes containing transition metals. The policy of the Euro-School has always been to invite outstanding specialists on subjects not directly in line with the main theme of the session. This is why two chapters of this volume are dedicated to introduce the reader to spintronics. Finally, the last chapter is the corrigendum of chapter 4 published in volume 2 of the series in which many typing mistakes remained; they are accounted for in the present book.

The European Commission is once more gratefully acknowledged for its financial support. Again, warm thanks go to all the authors whose contributions have made possible editing of all the volumes of the books series on Complex Metallic Alloys.

Esther Belin-Ferré
Paris, February 2010

CONTENTS

CHAPTER 1

THE PLASTICITY OF METALS: BASIC CONCEPTS

Mick Feuerbacher

Forschungszentrum Jülich GmbH
52425 Jülich, Germany
Email: M.Feuerbacher@fz-juelich.de

An introduction to the mechanisms of plastic deformation in metals and alloys is presented. After a brief overview over the basic terms and observables, the microstructural aspects of deformation are addressed. The concept of dislocations, a class of one-dimensional crystal defects, which play a pivotal role in the plasticity of most metals, is introduced. We describe their characteristic parameters and the mechanisms of dislocation construction and movement. Macroscopic experimental approaches for the investigation and characterization of the deformation behaviour are described. For the interpretation of deformation experiments we will introduce a thermodynamic approach including the concept of thermal activation. On this basis, we describe the relations between the thermodynamic activation parameters and the experimentally accessible observables.

1. Basic Terms

1.1. *Definition*

The term plastic deformation, or plasticity in short, describes the ability of a piece of matter to undergo a shape change under mechanical stress without fracture, if the following conditions are fulfilled. i) the strains occurring due to the shape change are relatively high, typically larger than about 1 %. ii) The shape change is irreversible. iii) The relation of the stress (leading to the deformation) and the strain is nonlinear.

This practical definition discriminates plasticity from other types of mechanical deformation. Elastic deformation (or elasticity) describes small, reversible, and linear deformations of matter. Elastic deformation can be described by of Hooke's law and, in terms of a spring-ball model of a solid, be understood as a compression of the springs (representing the interatomic forces) between the balls (representing the atoms). Accordingly, elastic deformation is connected with a volume change of the material. Anelastic deformation (or anelasticity) also concerns small and reversible i.e. elastic, deformations but takes into account the corresponding relaxation times: Elastic deformations in practice do not take place instantaneously but finite relaxation times of the atom positions are involved.

1.2. *Deformation geometry*

Plastic deformation experiments can be carried out under several different geometrical conditions, which determine the nature of the stress field the sample is subjected to. The simplest variation is uniaxial testing, in which a rod-shaped sample is compressed (uniaxial compression test) or dilated (uniaxial tensile test) along its long axis. Uniaxial testing has the advantage that the stress field is cylindrical symmetric and constant over the length of the sample. In a bending test a usually cuboïd-shaped sample is resting on two distant supports and subjected to an opposing load, applied in one (three-point bending test) or two (four-point bending test) points. Further, less common deformation geometries, realized by dedicated specimen shapes and loading devices are e.g. the shear test, in which the specimen is subjected to pure shear stress, or torsion testing, in which a rod shaped sample is twisted along its long axis.

1.3. *Dynamics and observables*

Let us now and for the remainder of this chapter restrict the discussion to uniaxial compression tests. Consider a sample of initial length l_0 and an area A_0 (perpendicular to the compression direction). Compression of this sample along its long axis can be performed in two different modes.

First, one can subject the sample to a constant load, F, and measure the resulting compression of the sample, Δl, as a function of time t. From these basic observables, the strain $\varepsilon = \Delta l / l_0$ and the strain rate $\dot{\varepsilon} = \partial \varepsilon / \partial t$ are calculated. The parameters of this type of experiment are the stress $\sigma = F/A_0$ and the temperature T at which the test is carried out. The corresponding basic diagram representing the test is $\varepsilon(t)\big|_{\sigma,T}$, or, frequently used as different deformation stages can more clearly be discerned, $\varepsilon(\dot{\varepsilon})\big|_{\sigma,T}$. This type of experiment is referred to as a *creep test*.

Second, one can deform the sample at a constant velocity and measure the required force. From these observables the σ and ε are calculated (as described above) and the parameters of the experiment are $\dot{\varepsilon}$ and T. The diagram representing the test is $\sigma(\varepsilon)\big|_{\dot{\varepsilon},T}$, the so-called stress-strain curve. This type of experiment is frequently referred to as a *dynamical test*.

Note that pure plastic deformation, in contrast to elastic deformation, takes place at constant volume and therefore a change in specimen length results in a change of area. Accordingly, the values of strain and stress change during the course of the experiment.

The values referencing to the initial values length l_0 and A_0 are referred to as "technical strain" and "technical stress" as opposed to the "true strain" and "true stress", referencing to the current value at a given degree of deformation. For the sake of simplicity we do not discern between technical and true quantities in this chapter.

2. Dislocations

2.1. *Theoretical shear stress*

Plastic deformation of a crystalline solid necessarily involves relative movement of atomic planes. The shear stress required for sliding of atomic planes in the complete absence of crystal defects can be calculated assuming a sinusoidal potential and, in the small-strain limit, reduces to

$$\tau_{th} = \frac{b}{a}\frac{\mu}{2\pi}$$

where b is the spacing of the atoms in the direction of the shear stress, a the spacing of the atom rows, and μ the shear modulus.[1] Since $b \approx a$, this so-called *theoretical shear stress* assumes values of the order of $\mu/10$, i.e. values around 10^9 N/m^2. Although these are merely rough estimates, it is obvious that τ_{th} is many orders of magnitude larger than the observed shear stresses, which are of the order of 10^5 to 10^6 N/m^2 for normal metals.

This striking discrepancy, in the 1930s lead to the presumption that crystal imperfections are responsible for the low values of shear stress observed experimentally in real materials, which eventually culminated in the prediction of dislocations by three independent researchers.[2-4] In this context it is worth mentioning that in recent years, for small, essentially dislocation-free fibres, so-called whiskers, stresses close to τ_{th} have been observed.

2.2. *The Volterra process*

Let us introduce the concept of dislocations by describing their construction in a simple model crystal according to a procedure, which is referred to as the Volterra process. In Fig. 1a a crystalline lattice is schematically shown. A half plane in the crystal is marked by white circles. This half plane is removed from the crystal (Fig. 1b). Then we close the gap, i.e. we connect the lips of the upper half crystal, and relax the atom positions around the end of the missing half plane (Fig. 1c).

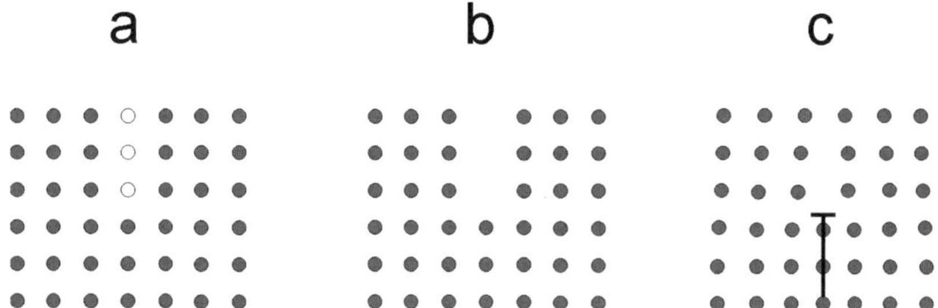

Fig. 1. Volterra construction of an edge dislocation.

What remains is a defect in a crystal that is otherwise perfect (assuming that the crystal has infinite extension). The defect is a line defect (i.e. a one-dimensional defect) due to the extension of the crystal along the viewing direction, and the lattice displacements are localized at the core of the defect. The *line direction* of a dislocation is defined as the direction along which the strain field is constant. For the dislocation in Fig. 1c this is along the viewing direction.

The so-constructed defect is the simplest form of an edge dislocation, and it is represented schematically by the symbol ⊥. Note that the dislocation can be regarded as a removed half plane as shown above, or, equivalently, as an inserted half plane.

2.3. *The Burgers vector*

The Burgers vector is the primary parameter describing a dislocation. It is constructed as follows (Fig. 2): Take an atom-to-atom path, by convention in clockwise direction, forming a closed circuit around a dislocation core (Fig. 2 left). Then follow the same atom-to-atom sequence in an ideal part of the crystal (Fig. 2 right). This circuit will not close, and the vector required to close it is the Burgers vector.

The Burgers vector describes the major elastic strain-field component of a dislocation and, as we will see below, defines the atomic displacement produced as the dislocation moves across its slip plane.

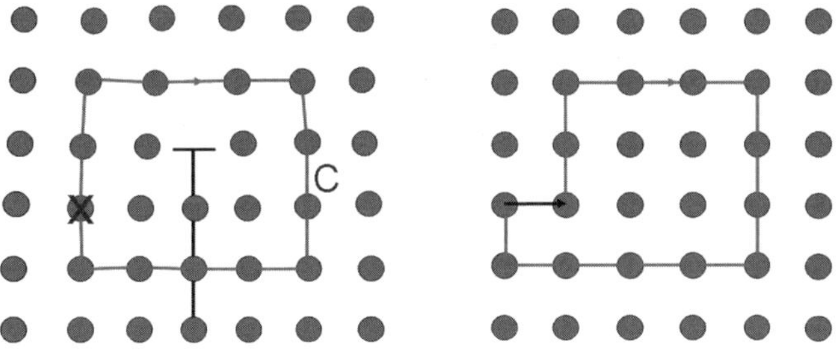

Fig. 2. Burgers circuit around an edge dislocation.

Fig. 3. High resolution TEM micrograph of an edge dislocation in $BaTiO_3$[5] imaged with the viewing direction parallel to the line direction.

Figure 3 is a high resolution transmission electron microscopy (TEM) image of an edge dislocation in $BaTiO_3$[5] , imaged with the viewing direction parallel to the line direction (end-on orientation). The dislocation core is in the centre of the micrograph (arrow).

2.4. *Types of dislocation*

The dislocation constructed in 2.2 represents a special case where the Burgers vector is perpendicular to the line direction. This is the characteristic of an *edge dislocation*. Generally, the Burgers vector can have any direction relative to the line direction, and for example in the case of a curved dislocation line, the angle between the Burgers vector and the line direction changes change along its length. Figure 4 shows, in a three-dimensional sketch, a curved dislocation segment (dashed line) in a cube of matter.

At the intersection with the left front surface, the dislocation has pure edge character (cf. Fig. 2). At the intersection with the right front surface, the Burgers vector is parallel to the line direction. This is the characteristic of a *screw dislocation*. The corresponding Burgers circuit around the dislocation core is seen. The intermediate segment of the dislocation has mixed character.

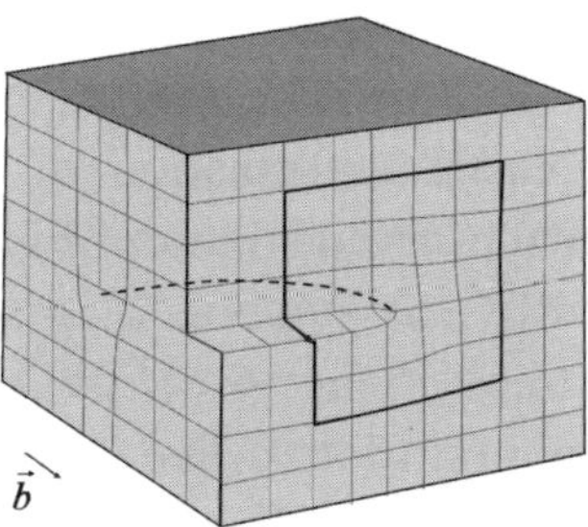

Fig. 4. Three dimensional schematic of a curved dislocation segment in a cube of matter.

2.5. *Dislocation loops*

Dislocations are one-dimensional structural defects representing the boundary of an area over which a given displacement has occurred. Hence, a dislocation line evidently cannot end within an otherwise perfect region of crystal. It must terminate for instance at a free surface, another dislocation line, or a grain boundary.[6] Generally, a dislocation in a perfect and infinite crystal is a closed loop. Isolated segments observed experimentally are parts of loops arbitrarily cut out by the specimen surfaces. A schematic example is shown in Fig. 4. The segment shown is to be understood as a part of a loop, which is terminated by the cube surfaces.

Figure 5 is a TEM bright-field micrograph of dislocations in decagonal Al-Ni-Co.[7] On the right hand side, a complete dislocation loop is seen. Further dislocations, single short segments and connected ones forming a network, are seen on the left.

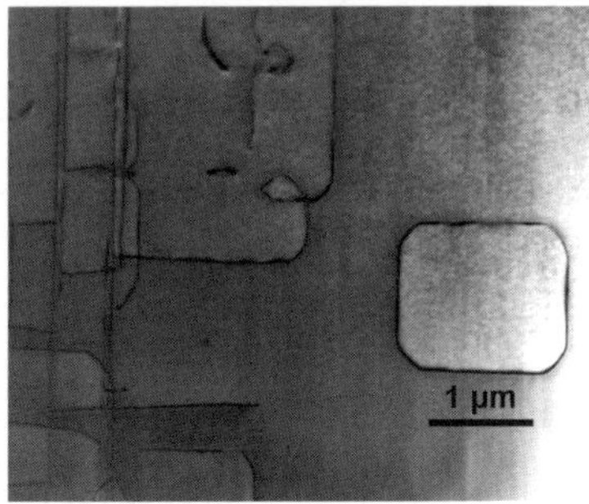

Fig. 5. TEM bright-field micrograph of dislocations in decagonal Al-Ni-Co.[7]

2.6. *Dislocation motion*

Dislocations mediate plastic deformation of a crystal by movement under a stress field in the crystal lattice. Figure 6 a shows an edge dislocation, represented by the symbol ⊥, in an initial stable position. Subjected to a shear stress τ (Fig. 6 b), the dislocation opens atomic bonds in direction of motion (dotted line) and closes bonds in its back (solid line), thereby moving ahead by one Burgers vector length (Fig. 6 c). This process can be repeated sequentially, leading to movement over macroscopic distances. In Fig. 6 the dislocation moves in a plane that also contains the Burgers vector. This mode of motion is referred to as dislocation *glide* or *slip*, and is most frequently observed. The plane in which the dislocation moves is the glide plane. On the other hand, if the Burgers vector of the moving dislocation has a component pointing out of the plane of motion, it is referred to as dislocation *climb*. The special case of movement perpendicular to the Burgers vector is called *pure climb*. In contrast to dislocation glide, climb motion generally requires transport of matter in the crystal, from or to the dislocation core, i.e. additional diffusion is necessary. Thus, this type of movement is referred to as non-conservative and it generally takes place at relatively high temperatures only.

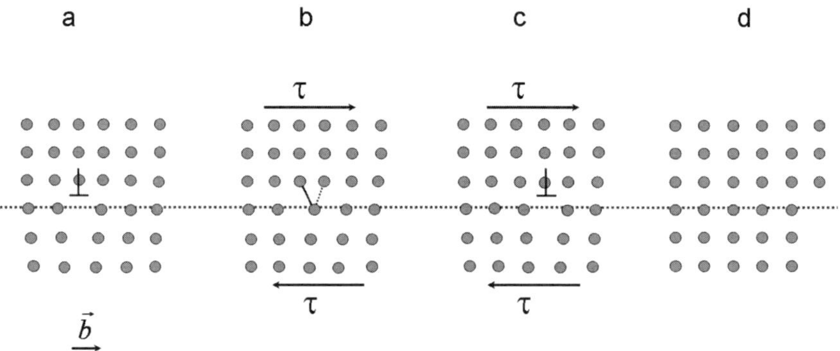

Fig. 6. Glide movement of an edge dislocation.

2.7. *Partial dislocations*

Dislocation movement leaves the structure of the crystal unaltered if and only if the Burgers vector corresponds to a translation vector of the crystal lattice. In that case, the dislocation is called a *perfect dislocation*. In contrast, a *partial dislocation*, possesses a Burgers vector which is *not* a translation vector of the lattice. As a direct consequence, a partial dislocation is necessarily connected to a planar fault, i.e. a stacking fault or antiphase boundary, in the lattice (see e.g. reference 7). An example is shown in Fig. 7. Consider a cubic crystal lattice of two atom species (Fig. 7a). In this lattice, the vector connecting neighboured atoms (grey arrow) is not a translation vector (black arrow). A dislocation having the short vector as Burgers vector will terminate a planar fault in the lattice (Fig. 7b). The fault connects two crystal parts such that equal atoms are on neighboured sites, which is not the case in the ideal lattice.

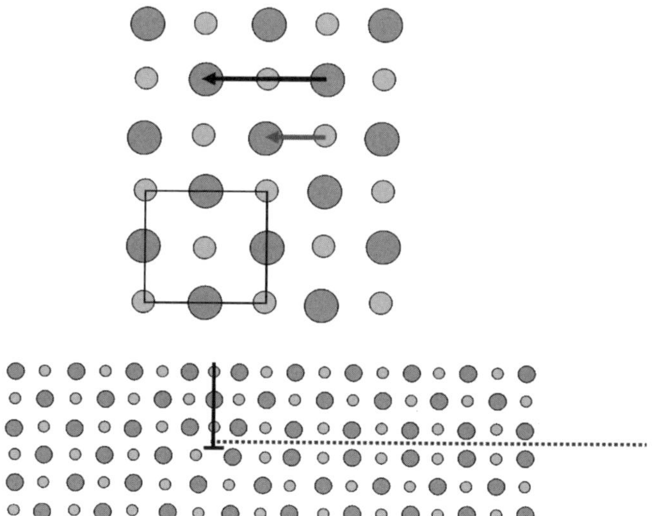

Fig. 7. Partial dislocation in an ordered cubic crystal of two atom species. A dislocation having the short vector as Burgers vector will terminate a planar fault in the lattice.

3. Macroscopic Deformation

3.1. *Stress-strain curves*

The stress-strain curve describes the response of a sample to a dynamical deformation test (see 1.3). Fig. 8a shows a schematic stress-strain curve for strains in the range from 0 to about 10 %. At low strains, the curve exhibits a linear regime, which is due to elastic deformation of the sample (A). The slope (dashed line) corresponds to the modulus of elasticity (or Young's modulus). At the onset of plastic deformation, i.e. when dislocations become active in the deformation process, the sample yields and the curve deviates from the linear behaviour. The curve then develops into a maximum (B), which is referred to as the *upper yield point*. The drop after the upper yield point is due to massive dislocation production and is referred to as yield drop. The stress drops until the curve reaches a local minimum (C), the *lower yield point*, at which the microstructure of the sample enters a dynamic equilibrium, in which dislocation multiplication and annihilation processes balance, leading to a basically flat stress-strain curve (D). This stage, however, is in most materials superposed by hardening processes, which may take place in several successive stages, as described in the next section

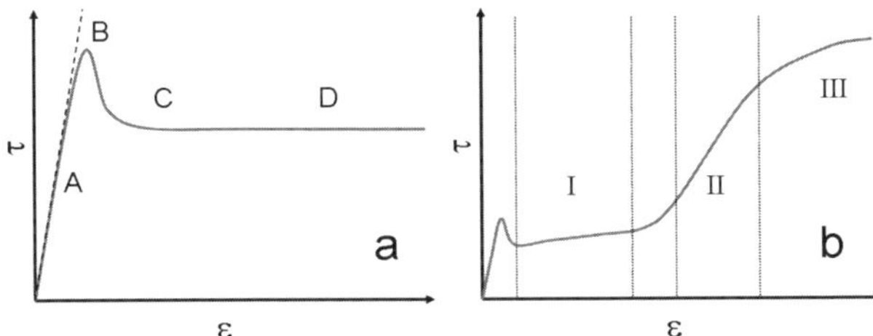

Fig. 8. Schematic stress-strain curve for strains (a) in the range from 0 to about 10 % and (b) for large strains (30 – 50%) (see text).

3.2. *Deformation stages*

At higher strains, typical metals will show hardening stages after initial yielding. Fig. 8b schematically shows a stress-strain curve representative of the behaviour of a single-crystalline material in, for strains in the range from 0 to 30–50 % and initial single-slip orientation. The hardening stages are typically referred to as stage I to III, and occur due to different dislocation interaction processes. Stage I is the *easy glide region*. The dislocations slip on the primary slip plane at high velocities and with a large path length. The hardening coefficient $\Delta\sigma/\Delta\varepsilon$ is low and amounts to about 10^{-4} to 10^{-3} μ, where μ is the shear modulus of the material. After a certain amount of strain, the sample turns into an orientation in which double slip can occur, leading to stage II. Double slip, i.e. movement of dislocations on an additional slip plane causes increased dislocation intersection processes. This leads to demobilization of dislocations which is compensated by dislocation multiplication. The dislocation density increases and the hardening coefficient increases typically by a factor 2 to 10. Stage III, is an approximately parabolic hardening stage in which the hardening rate steadily decreases due dislocation annihilation processes. The extensions of the different stages can vary considerably, depending on external variables such as temperature and strain-rate, and internal variables such as the stacking-fault energy.

4. Activation Thermodynamics

4.1. *The problem*

In this chapter we will introduce a brief formalism for the description of macroscopic phenomena of plastic deformation. For more detailed reading please consult e.g. Haasen[8], Evans and Rawlings[9] and Kocks, Argon, and Ashby.[10] Macroscopic deformation involves large numbers of dislocation processes on the microscopic scale – the dislocation densities, i.e. the dislocation line length per unit volume, involved in the deformation of a real crystal are of the order of 10^7 to 10^{10} cm^{-2}. The description of the macroscopic phenomena can therefore only be carried out considering averages over large numbers of contributing dislocations, in terms of a thermodynamic formalism.

4.2. *Rate theory*

Assume a crystal containing dislocations at a density ρ possessing a Burgers vector length b. Movement of these dislocations over an average distance x leads to a plastic strain

$$\varepsilon_{plast} = \rho b x \tag{1}$$

The time derivation of Eq. (1) yields the Orowan equation

$$\dot{\varepsilon}_{plast} = \rho b v \tag{2}$$

where v is the average dislocation velocity and $\dot{\varepsilon}_{plast}$ is the plastic strain rate. The Orowan equation is of central importance, as it relates the microscopic parameters ρ, b and v to the macroscopic parameter $\dot{\varepsilon}_{plast}$.
An external stress σ which is applied to a sample in tensile direction causes a shear stress τ on the glide plane of

$$\tau = m_s \sigma \tag{3}$$

where the *Schmid factor* $m_s = cos\phi \, cos\lambda$ (ϕ and λ denote the angles between compression direction and slip plane and compression direction and slip direction, respectively) is a geometric factor assuming values between 0 and 0.5. A dislocation driven by the shear stress τ is simultaneously subjected to the oppositional internal stress τ_i, originating from long-range stress fields of other dislocations in the sample, which counteract its movement. The effective stress acting on a moving dislocation τ_{eff} is then given by

$$\tau_{eff} = \tau - |\tau_i| \tag{4}$$

The velocity, v, of a dislocation is limited by energetical obstacles to be overcome during movement. The friction stress on the dislocation originating from this process is τ_f. If $\tau_{eff} > \tau_f$, the dislocation can overcome the obstacles and continuously move through the crystal. If, on the other hand, $\tau_{eff} < \tau_f$ at an obstacle, dislocation motion is halted. However, even in the case $\tau_{eff} < \tau_f$, an obstacle can be overcome by *thermal activation.* This means that dislocation motion is aided by thermal fluctuations leading to a non-zero probability to overcome the obstacle.

The energy barrier overcome by thermal fluctuations is then given by

$$\Delta G = \int_{x_1}^{x_2} (\tau_f - \tau_{eff})lbdx \tag{5}$$

where ΔG is the Gibbs free energy and l the length of the dislocation line at dislocation line at the obstacle. The probability of thermal activation to overcome the obstacle is

$$P = \exp\frac{-\Delta G}{kT} \tag{6}$$

where k is Boltzmann´s constant and T is the absolute temperature. The total free energy to overcome the obstacle is $\Delta F = \Delta G + \Delta W$, where ΔW is the work-term, given by the energy contribution of the effective stress, i.e.

$$\Delta W = \int_{x_1}^{x_2} \tau_{eff}lbdx \tag{7}$$

The corresponding energy contributions are shown in Fig. 9 in a schematic force-distance diagram. The respective forces are related to the friction stresses by $lb\tau$. The dislocation, assisted by thermal fluctuations, follows the reaction path from the stable equilibrium position x_1 to the unstable position x_2 overcoming the activation distance $\Delta x = x_2 - x_1$.

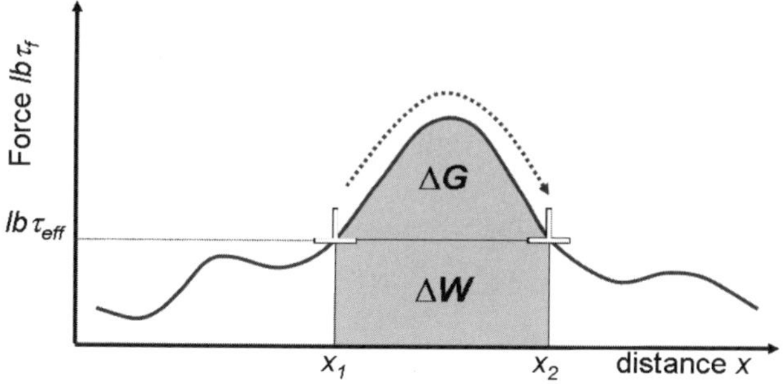

Fig. 9. Schematic force-distance diagram. A dislocation follows the reaction path from the stable equilibrium position x_1 to the unstable position x_2.

If thermal activation is the rate-controlling process, the dislocation velocity is given by $v = v_0 \Delta x P$, where v_0 is the attempt frequency. Hence we obtain, using equations (2) and (6),

$$\dot{\varepsilon}_{plast} = \dot{\varepsilon}_0 \exp\frac{-\Delta G}{kT} \tag{8}$$

with $\dot{\varepsilon}_0 = \rho b \Delta x v_0$.

The Gibbs free energy is a thermodynamic variable of state which depends on the temperature and the stress. The differential is $d(\Delta G) = -\Delta S\, dT - V\, d\tau$ with the definitions

$$\Delta S \equiv -\left.\frac{\partial(\Delta G)}{\partial T}\right|_{\tau_{eff}} \tag{9}$$

and

$$V \equiv -\left.\frac{\partial(\Delta G)}{\partial \tau_{eff}}\right|_{T} \tag{10}$$

ΔS is the activation entropy. V is called the activation volume and can be written as

$$V = lb\Delta x = b\Delta A \tag{11}$$

The *activation area* ΔA can be interpreted as the area that is passed by a dislocation line during the thermally activated overcoming of an obstacle. This is illustrated in Fig. 10. Before activation (solid line) the dislocation is in a stable position corresponding to x_1 in Fig. 9. Upon activation the dislocation moves by Δx to the position given by the dashed line. The area between both lines is the activation area ΔA.

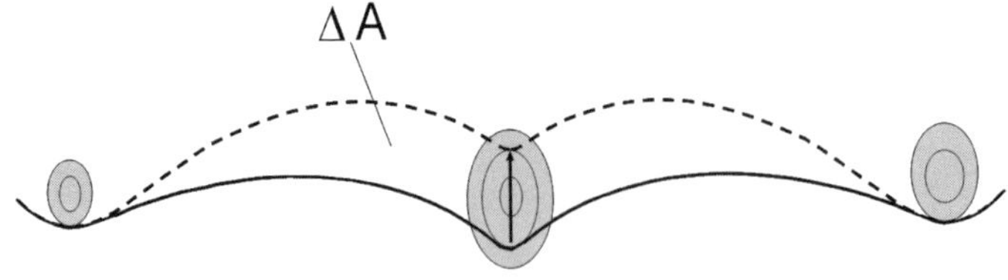

Fig. 10. Activation Area A of a dislocation overcoming an energetic obstacle.

4.3. *Connection to experiment*

The experimental determination of thermodynamic activation parameters is performed by incremental tests during plastic deformation. Most important are stress-relaxation and temperature-change experiments.

4.3.1. *Stress-relaxation experiments*

The experimental determination of the activation volume is possible according to

$$V = -\left.\frac{\partial(\Delta G)}{\partial \tau_{eff}}\right|_T = \frac{kT}{m_S}\left.\frac{\partial(\ln \dot{\varepsilon}_{plast})}{\partial \sigma}\right|_T \tag{12}$$

where Eq. (8) was used. This relation contains the experimentally accessible parameters $\dot{\varepsilon}_{plast}$, the plastic strain rate, and σ, the applied stress. In a stress-relaxation experiment the deformation machine is suddenly halted during plastic deformation and the stress is measured as a function of time. As the total strain of the sample remains constant, the total strain rate is given by $\dot{\varepsilon} = \dot{\varepsilon}_{plast} + \dot{\varepsilon}_{elast} = 0$, i.e. $\dot{\varepsilon}_{plast} = -\dot{\varepsilon}_{elast}$. For the elastic strain, $\varepsilon_{elast} \propto \sigma$ holds according to Hooke´s law. Hence, during a stress relaxation experiment we have $\dot{\varepsilon}_{plast} \propto -\dot{\sigma}$. With Eq. (12) this yields a relation for the experimental determination of the activation volume

$$V = \frac{kT}{m_S}\left.\frac{\partial(\ln-\dot{\sigma})}{\partial \sigma}\right|_T \tag{13}$$

The comparison of the experimental activation volume V with the atomic volume V_a, i.e. the average volume per atom in the unit cell, of a given crystal structure permits inference on the mode of plastic deformation. For the case of a diffusion-controlled mode of plastic deformation the experimental activation volume is expected to be of the same order of magnitude as V_a. If dislocation motion, on the other hand, is controlled by the thermally activated overcoming of obstacles the experimental activation volume is rather of the order of the obstacle volume.

4.3.2. *Temperature changes*

The Gibbs free energy of activation is not directly accessible in deformation experiments. However, the activation enthalpy ΔH can be determined via

$$\Delta H = \Delta G + T\Delta S = -kT^2 \frac{\partial \ln \dot{\varepsilon}_{plast}}{\partial \sigma}\bigg|_T \frac{\partial \sigma}{\partial T}\bigg|_{\dot{\varepsilon}_{plast}} \tag{14}$$

During a constant strain-rate test, a change of deformation temperature is used to determine $\Delta\sigma/\Delta T \approx \partial\sigma/\partial T$. In practice, the temperature is changed by an amount ΔT after unloading the sample. After thermal equilibrium is attained in the deformation machine the sample is reloaded. The temperature dependence of the stress $\Delta\sigma/\Delta T$ together with the result of the stress relaxation experiments (Eq. 13) yields the activation enthalpy ΔH.

Acknowledgments

The author thanks Dr. M. Heggen for continuous scientific cooperation and stimulating discussions. This work was supported by the 6th Framework EU Network of Excellence "Complex Metallic Alloys" (Contract No. NMP3-CT-2005-500140).

References

1. J. P. Hirth and J. Lothe, *Theory of Dislocations* (Wiley, New York, 1982)
2. E. Orowan, *Z. Phys.* **89** 634 (1934).
3. M. Polanyi, *Z Phys.* **89** 660 (1934).
4. G. I. Taylor, *Proc. R. Soc.* **A 145** 362 (1934)
5. C. L. Jia, unpublished result, 2008.
6. R. E. Smallman and R. J. Bishop, *Modern Physical Metallurgy & Materials Engineering*, (Butterworth-Heinemann, Oxford, 1999).
7. P. Schall, private communication (2002).
8. P. Haasen, *Physical Metallurgy* (Cambridge University Press, Cambridge, 1986)
9. A. G. Evans and R. D. Rawlings, *Phys. Stat. Sol.* **34** 9 (1969).
10. U. F. Kocks, A. S. Argon and M. F. Ashby, *Thermodynamics and kinetics of slip* (Pergamon Press, Oxford, 1975).

CHAPTER 2

BASICS OF MECHANICAL PROPERTIES OF METALS

Jean Philippe Chateau

Institut Jean Lamour, Ecole des Mines
Parc de Saurupt, 54042 Nancy cedex, France
E-mail: Jean-Philippe.Chateau@mines.inpl-nancy.fr

Metallic bondings promote a good compromise between metals' stiffness, resistance and deformability. They have a linear elastic behaviour below a critical stress and deform plastically by dislocation glide above. Dislocations produce glide along the denser planes of the crystal in the dense directions, constituting slip systems. The critical shear stress depends on the lattice friction, the presence of solute atoms and strain hardening due to dislocation interactions. The three stages of the tensile curve of a single crystal are characteristic of these three contributions. Plasticity can also involve grain boundary sliding, mechanical twinning or martensitic transformations, which lead to outstanding mechanical properties. At the macroscopic scale, tensile tests reveal that the yield stress inversely scales like the square root of the grain size in polycrystals. Their homogeneous elongation is limited by necking, favoured by low strain hardening rates. Plasticity is thermally activated and generally enhanced at high temperature and low strain rates. Superplasticity can be observed when the strain rate sensitivity is high, while a negative one produces plastic instabilities. Dynamic recovery and recrystallisation mechanisms intervene in plasticity and reduce strain hardening. The numerous thermally activated mechanisms involved in plastic deformation lead to creep, even below the yield stress.

1. Introduction

Metals and metallic alloys are widely used for structural applications thanks to their mechanical properties. They offer a good compromise between mechanical resistance and deformability. A metallic crystal is composed of cations sealed in a sea of free electrons. The attraction between the cations and the electron cloud leads to high cohesion of the crystal but the non localised nature of the interatomic bondings also allows large deformations by glide along crystallographic planes. In comparison, ceramics have strong interatomic bondings: covalent, iono-covalent or ionic. As a result, they have a higher mechanical resistance than metals, but are brittle at room temperature. In polymers, weak bondings between the macromolecules (hydrogen or Van der Walls) lead to higher deformability but much lower mechanical resistance.

The basics of deformation mechanisms will be presented at the microscopic scale first. At small strains, a metal has a linear elastic behaviour, before yielding by glide of dislocations at a certain stress threshold. The tensile curve of a single crystal and the basic hardening mechanisms will be presented. At the macroscopic scale, the tensile curve of a polycrystalline will be analysed and other plasticity mechanisms will be introduced. Plasticity is thermally activated. We will present the influence of temperature and strain rate and introduce the concepts of dynamic recovery and re-crystallisation and of creep.

2. Lattice Deformation

2.1. *Elasticity*

2.1.1. *Entropic elasticity*

Solids like elastomers and amorphous polymers are composed of long macromolecular chains randomly distributed when the material is unstressed (Fig. 1a). Due to the random distribution of each chain, the number of configurations Ω_0 and the entropy $S_0 = \ln \Omega_0$ are high. Applying a tensile force tends to align the chains in the tensile direction by free rotations within the macromolecules and does not require any stretching of the covalent bondings between atoms. Thus the number of configurations and the entropy S decrease, while the bonding enthalpy H remains hardly equal to the initial state H_0 (Fig. 1b). When the material is unloaded, the system tends to minimise the thermodynamic potential $G = H - TS$ by maximising the entropy, which is the driving force of the retraction. Such materials have a large elastic domain (up to 800% for rubber) and there is no linear relation between tensile stress and elongation.

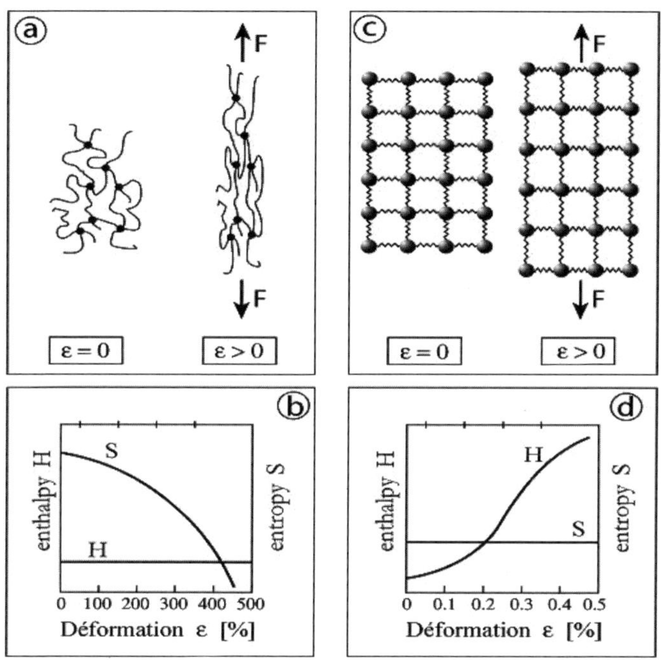

Fig. 1. Microscopic elasticity mechanisms and evolution of enthalpy and entropy in the case of entropic (a and b) and enthalpic (c and d) elasticity.

2.1.2. *Enthalpic elasticity*

In metals or ceramics, a repel force appears between atoms when they are spread from their equilibrium distance at the considered temperature (Fig. 1c). There is no displacement of the atoms on large distances so that the number of configurations and the entropy remain hardly constant, while the bonding enthalpy increases (Fig. 1d). When the material is unloaded, the system tends to minimise the enthalpy, which is the driving force for the retraction. The elastic domain of such materials is small (<< 1%) and limited by the onset of plasticity or fracture.

If we consider two neighbouring atoms in the solid, the bonding potential of the system can be approached by the relation:

$$V(r) = -\frac{A}{r^n} + \frac{B}{r^m} \qquad (1)$$

(Fig. 2.a) where r is the distance between the atoms. A, B, n and m are positive constants and $n > m$, so that the first term is a long range attraction term (electrostatic force between an anion and a cation in the case of ionic bonding, or between the cations and the electron cloud in metals, for instance) and the second term is a short range repulsion term due to the electron cloud of each atom (exclusion principle of Pauli).

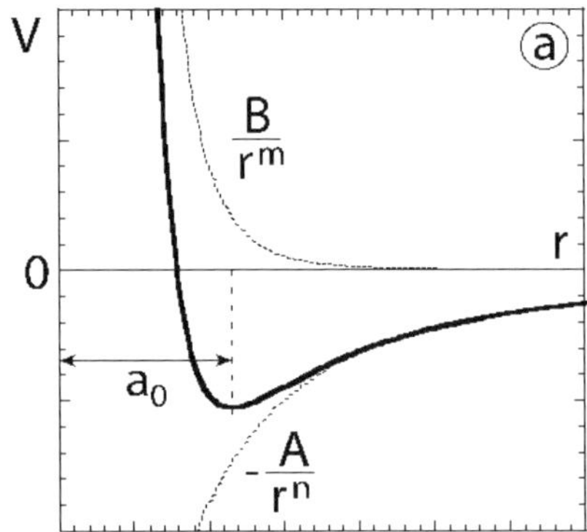

Fig. 2. a) Potential interaction energy and b) interaction force between two atoms.

The interaction force between the atoms derives from the potential:

$$F = -\frac{\partial V}{\partial r} = -\frac{nA}{r^{n+1}} + \frac{mB}{r^{m+1}} \qquad (2)$$

The dependence of F with r is shown in Fig. 2b (Condon-Morse curve). The value of r corresponding to the minimum energy is equal to the interatomic equilibrium distance a_0. At $r = a_0$, the interaction force vanishes and for a given displacement around this position, it repels the atoms to the equilibrium distance. For small displacements in the elastic domain, the curve can be approached by its slope by a first order Taylor development. At the scale of the entire crystal, this leads to a linear relation between applied tensile stress and elongation, where the proportionality constant is the Young modulus E.

In materials with enthalpic elasticity, the superposition principle of small displacements, strains and stresses applies.

2.1.3. *Elastic constants*

In an isotropic material, such as a polycrystal with no texture, two elastic constants are required to express the elasticity law, given by Hooke law: E and the Poisson ratio v, or λ and μ, the Lamé parameters. $\mu = E/2(1+v)$ is the shear modulus and is the proportionality constant between the non diagonal components of the stress tensor (shear stresses) and of twice the strain tensor (glides). In a crystal, which is anisotropic, more constants are required, depending on the symmetries of the crystal; for example, 5 for the hexagonal structure, and 3 for the cubic structures. It cannot be defined a given Young modulus in a crystal, as it depends on the crystallographic tensile direction. In cubic structures, E is generally minimal in the <001> directions and maximal in the <111> directions (e.g. in Cu $E_{100} = 68$ GPa, $E_{111} = 190$ GPa, in Fe $E_{100} = 130$ GPa, $E_{111} = 277$ GPa). Most models of plastic behaviour at the dislocations scale neglect this anisotropy and the results are functions of μ.

Table 1 shows values of the Young modulus for materials of different classes at room temperature. Metals have intermediate values of E (tens to hundreds of GPa) between ceramics (hundreds to 1000 GPa for diamond) and polymers (several down to tenths of GPa).

Table 1. Young modulus for various materials.

Material	E [GPa]	Material	E [GPa]	Material	E [GPa]
Diamond, nanotubes	1000	Ni and alloys	130-234	NaCl, LiF	15-68
WC	450-650	CFRP	70-200	Concrete	45-50
Cermets Co/WC	400-530	Fer, α-steels	196-207	GFRP	35-45
(Ti, Zr, Hf)-Bo,	500	γ-steels	190-200	Mg and alloys	41-45
SiC	450	Pt	172	Pb and alloys	14
TiN	450	Cu and alloys	120-150	Polyesters	1-5
Tungstène	406	Ti and alloys	80-130	PMMA	3.4
Al_2O_3	390	Si	107	Polystyren	3-3.4
TiC	379	Zr and alloys	96	Polycarbonate	2.6
Mo and alloys	320-365	SiO_2	94	Polypropylen	0.9
Cr	289	Zn and alloys	43-96	Polyethylen	0.2 - 0.7
MgO	250	Au	82	Rubber	0.01-0.1
Co and alloys	200-248	Al and alloys	69-79	PVC	0.003-0.01

2.1.4. *Temperature dependence*

In most materials, the Young and shear moduli linearly decrease when temperature increases, as long as no phase transition is observed. For example, Fig. 3 shows the general empirical law for austenitic steels (fcc).[1] Figure 3 also shows the case of a Fe-Mn-C steel which exhibits an anomaly of the elastic constants (an increase with temperature in a given domain), whereas it remains austenitic in the entire domain. This anomaly is due to a phase transition of magnetic nature, as it undergoes a Néel transition at $T_N \sim 50°C$: it is paramagnetic at high temperature (randomly oriented magnetic moments) and anti ferromagnetic at low temperature (parallel magnetic moments with alternate signs). The anti ferromagnetic phase has a much lower shear modulus, as measured by DMTA, so that it rapidly decreases at T_N when temperature decreases. The measured Young modulus at room temperature is 160 GPa, which is rather low for an austenitic steel (190 GPa) and corresponds to values usually observed at 400°C.

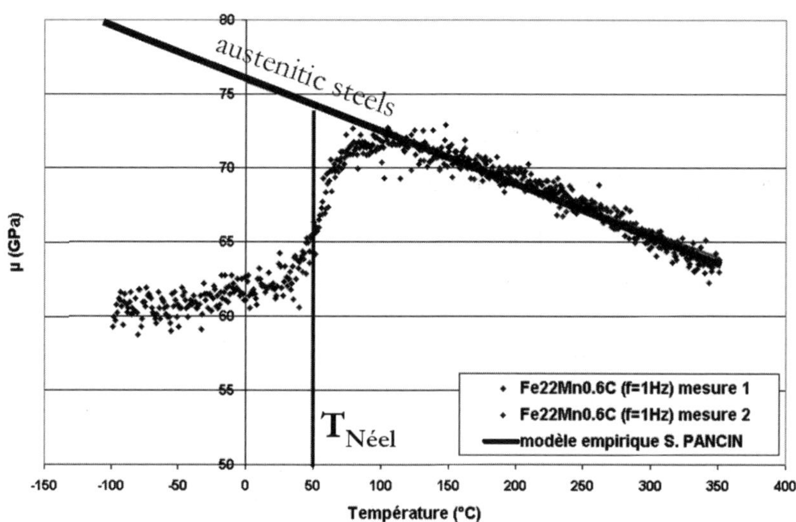

Fig. 3. Empirical law for the evolution of the shear modulus with temperature in austenitic steels and dynamical mechanical thermal analysis (DMTA) measurement for Fe-22Mn-0.6C: an elastic flexion is applied to a bar with a given frequency at each temperature and the force is measured.

2.3. *Elastic limit - plasticity*

2.3.1. *Lattice friction*

In metals, after a small elastic deformation, a part of the deformation becomes permanent and irreversible, which corresponds to the onset of plasticity. The observation of the initially polished surface of a crystal then deformed plastically shows step lines. Plasticity is achieved by glide along crystallographic planes when the shear stress along these planes in a given direction reaches a critical value τ_c. The estimation of the stress necessary to shear a perfect crystal rises to a theoretical value of the order of 0.1μ, which is 2 to 4 orders of magnitude higher than the experimental measurements. This emphasises the existence of linear defects called dislocations which displacement creates an elementary glide (Fig. 4.a).

                                                    J. P. Chateau

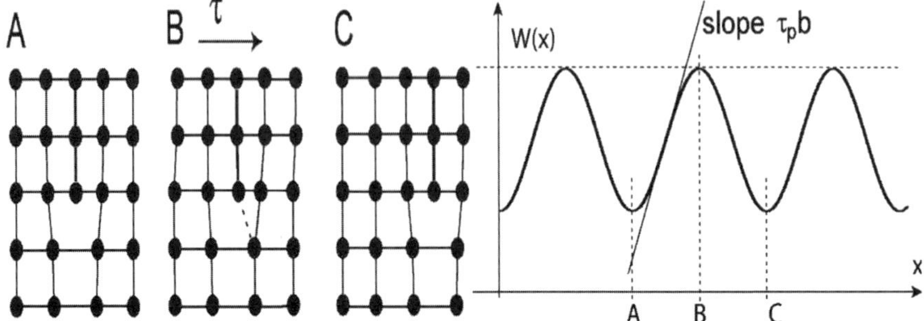

Fig. 4. a) Dislocation glide from an equilibrium position to the next one. b) Corresponding potential energy. The maximum shear stress is $\tau_p =(1/b)$max (dW/dx).

Dislocations are characterised by their glide plane and Burgers vector $\vec{b}$. When a dislocation moves through a crystal along its plane it creates a glide step of length b, which requires a shear stress far lower than the theoretical stress. During glide, a dislocation moves through a periodic potential W corresponding to atomic rows. In Fig. 4.b, the dislocation is in equilibrium in positions A and C, but has to overcome a potential barrier in B. The applied stress has to reach a critical value τ_p, called Peierls stress, to move the dislocation from A to C at 0 K. At a higher temperature, the thermal fluctuations of atoms help the movement of the dislocation. A lower stress τ_f is required, called the lattice friction.

The glide plane of the dislocation and its Burgers vector (glide direction) define a slip system. The plastic deformation of a crystal will result of the combination of elementary glides along different slip systems and is highly dependent of the crystallography.

2.3.2. *Slip systems*

Due to the crystalline nature of metals and its anisotropy, τ_f varies from a family of crystallopgraphic planes to another. Experiments show that the lattice friction is the lowest in the dense planes, as they have also the highest inter reticular distance and are the most able to glide. The ideally dense planes are the {111} planes in the fcc structure and {0001} in the hcp structure. These planes have the same compact packing of atoms in the two structures. In fcc. metals, an estimation of the lattice friction is about 0.01 MPa and can be neglected in comparison with any other plane family. Dislocations in this structure glide hardly exclusively in these planes, with no lattice friction and without aligning along the Peierls potential valleys. In hcp metals, the situation is similar, but where there are 4 compact {111} planes in the fcc structure, there is only one {0001} plane. This is insufficient to produce any imposed deformation by glide (e.g. an elongation along <0001>) so that other planes are often observed. On the contrary, in bcc metals, the lattice friction is not negligible in the denser planes {110}, {112} and {123} and there is a smaller difference between them so that dislocations are observed in these 3 plane families.

The slip direction is determined by the Burgers vector of the dislocations. As a linear defect, a dislocation is characterised by an energy per unit length which is shown to be of the order of μb^2. The dislocations observed in a crystal are the ones with the minimum line energy. The Burgers vector is thus the smallest distance between two atoms and the observed slip directions are the dense directions of the crystal: <110> in fcc, <111> in bcc and <11$\bar{2}$0> in hcp. The observed slip systems in the three main structures are summarised in Fig. 5.

The main differences between the two cubic structures are that there are more possible slip systems in bcc (48) than in fcc (12), but the lattice friction is higher in bcc and strongly depends on temperature, which has important consequences on the mechanical properties. In hcp, pyramidal slip planes of second order are observed which do not contain any <11$\bar{2}$0> direction. In this case, higher line energy dislocations are observed and the slip direction is <11$\bar{2}$3>.

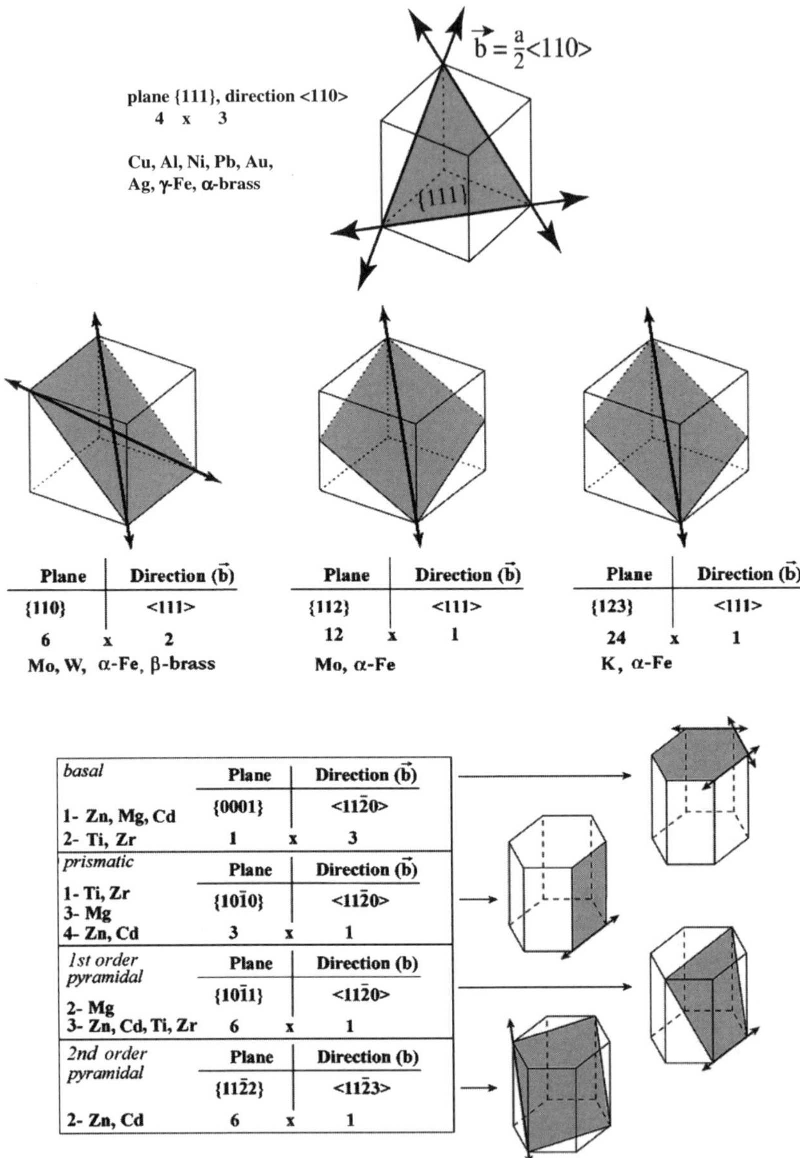

Plane	Direction ($\vec{b}$)
{110}	<111>
6 x	2

Mo, W, α-Fe, β-brass

Plane	Direction ($\vec{b}$)
{112}	<111>
12 x	1

Mo, α-Fe

Plane	Direction ($\vec{b}$)
{123}	<111>
24 x	1

K, α-Fe

basal	Plane		Direction ($\vec{b}$)
1- Zn, Mg, Cd	{0001}		<11$\bar{2}$0>
2- Ti, Zr	1	x	3
prismatic	Plane		Direction ($\vec{b}$)
1- Ti, Zr 3- Mg	{10$\bar{1}$0}		<11$\bar{2}$0>
4- Zn, Cd	3	x	1
1st order pyramidal	Plane		Direction (b)
2- Mg	{10$\bar{1}$1}		<11$\bar{2}$0>
3- Zn, Cd, Ti, Zr	6	x	1
2nd order pyramidal	Plane		Direction (b)
	{11$\bar{2}$2}		<11$\bar{2}$3>
2- Zn, Cd	6	x	1

Fig. 5. Slip systems in the principal metallic structures. a) fcc., b) bcc, c) hcp.

2.4. *Tensile test on a single crystal*

2.4.1. *Resolved shear stress*

To study the plastic deformation of metals, many experiments have been conducted by tensile tests on single crystals to determine the crystallography of slip and the lattice friction. When a crystal is subjected to a given applied stress tensor $\overline{\overline{\sigma}}$, the force exerted on a dislocation present in the crystal is given by the Peach & Köhler relation:

$$\vec{F}_{PK} = \left(\overline{\overline{\sigma}} \cdot \vec{b}\right) \times \vec{L} \tag{3}$$

where $\vec{L}$ is the line vector of the dislocation. The force is perpendicular to the dislocation line at any point and its glide component reduces to:

$$F_g = \left(\overline{\overline{\sigma}} \cdot \vec{n}\right) \cdot \vec{b} = \tau b \tag{4}$$

where $\vec{n}$ is the normal to the slip plane. τ is the component of the shear stress in the slip plane parallel to $\vec{b}$ and is called resolved shear stress.

In the case of a uniaxial tensile force $\vec{F}$, the resolved shear stress on a slip system can be determined knowing the angles λ between $\vec{F}$ and $\vec{b}$ and ϕ between $\vec{F}$ and $\vec{n}$. The projection of the applied force in the slip direction is:

$$F_b = F \cos \lambda \tag{5}$$

and the resolved shear stress is:

$$\tau = \frac{F_b}{A} = \frac{F \cos \lambda}{A/\cos \phi} = \frac{F}{A_0} \cos \lambda \cos \phi \tag{6}$$

The relation between the tensile stress $\sigma = F/A_0$ and τ is:

$$\tau = m\sigma \tag{7}$$

Where $m = \cos \lambda \cos \phi \leq 0.5$ is called Schmid factor. The maximum value of m is 0.5, which is the ratio between the maximum shear and the tensile stress in an isotropic solid. The contribution of an increment of glide along the considered slip system $d\gamma$ to the deformation in the tensile direction ε is:

$$d\varepsilon = m \cdot d\gamma \tag{8}$$

The Schmid factors vary from a slip system to another and when the applied force increases, the slip system with the highest m will glide first (in the case of a critical resolved shear stress equal for each system) and is called the primary slip system. The number of activated slip systems depends on the crystallographic tensile direction. In the cubic structures, space can be decomposed into equivalent crystallographic zones materialised by the standard triangles on the stereographic projection, delimited by the traces of the {100} and {110} planes. In each triangle, there is one and only one given direction [ijk] of each crystallographic family <ijk>.

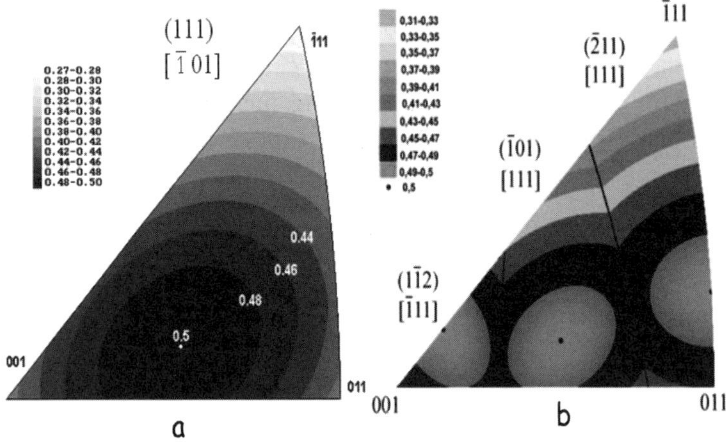

Fig. 6. Schmid factor and primary slip system in the standard triangle. a) fcc, b) bcc.

In the fcc structure, a tensile axis in one triangle corresponds to one given primary slip system. The crystal deforms by single slip. The value of m in the standard triangle defined by the poles [100], [011] and [$\bar{1}$11] is shown in Fig. 6a. In this triangle, the primary system is (111)[$\bar{1}$01]. m

is equal to 0.5 for a tensile direction close to [$\bar{1}$ 49] (easiest deformable orientation) and has a minimal value equal to 0.272 for [$\bar{1}$ 11] (most resistant orientation). A tensile axis on one of the triangle's frontiers or summits (i.e. common to several triangles) corresponds to several slip systems with equal maximum Schmid factors and the crystal deforms by multiple slip: 8 systems for the [100] tensile axis, 4 for [011] and 6 for [$\bar{1}$ 11]. In the b.c.c. structure, several primary slip systems are found in each standard triangle due to the higher number of slip systems in this structure (Fig. 6b).

2.4.2. *Tensile curve in single slip orientation*

Practically, the chosen orientation for single slip is generally the [$\bar{1}$ 23] tensile direction, which is the furthest direction to any side of the standard triangle, even if the Schmid factor of the primary system is equal to 0.467 lower than for the [$\bar{1}$ 49] orientation. When the critical resolved shear stress is reached, the crystal deforms along the primary (111) planes, which leads to an elongation in the tensile direction, but would also lead to a translation in the transverse direction (Fig. 7b). The heads of the sample must then rotate to stay aligned with the tensile direction, inducing a rotation of the crystal (Fig. 7c). As a consequence, the crystallographic tensile direction evolves and moves towards the slip direction [$\bar{1}$ 01] (Fig. 7d).

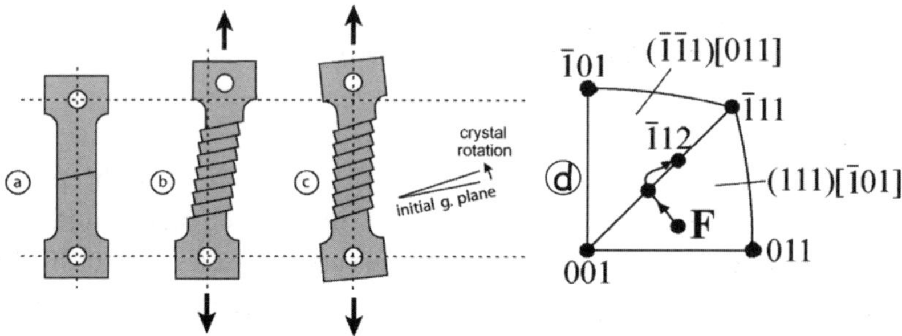

Fig. 7. a) to c) Rotation of a crystal in single slip orientation during tensile test. d) Evolution of the tensile direction F in the standard triangle in single slip orientation.

The Schmid factor of the primary slip system decreases, until the tensile axis reaches the side of the standard triangle, where the Schmid factor on the primary systems of two triangles are equal and a secondary (or conjugate) activates: $(11\bar{1})[011]$ (Fig. 7d). This system induces a different rotation of the crystal, towards the [011] direction and the tensile axis moves towards $[\bar{1}01]+[011]=[\bar{1}12]$. Once reached the $[\bar{1}12]$ tensile direction, the crystal rotation stops.

The typical tensile test curve, expressed in terms of resolved stress and glide, obtained in single slip is schemed and examples for Cu-Al[2] alloys are presented in Fig. 8. Three stages of the plastic deformation are observed[3-5], characterised by different hardening rates $\theta = \mathrm{d}\tau/\mathrm{d}\gamma$:

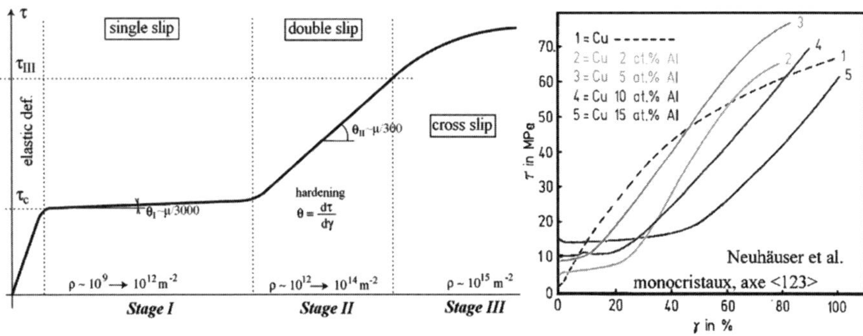

Fig. 8. a) Scheme of the resolved stress- resolved glide curve for a single crystal oriented for single slip, showing the 3 stages of deformation. b) Tensile curves of single crystals of Cu-Al alloys with <123> orientation. The increase of Al concentration increases the critical resolved shear stress for the beginning of stage I.

- elastic domain; $\theta = \mu$,

- Stage I or easy glide: above the critical stress τ_c, only the primary system is active. Sources emit dislocations in parallel planes and a low but positive hardening $\theta_I \sim \mu/3000$ is observed due to the rotation of the crystal and to middle and long range elastic interactions between the dislocations,

- Stage II: the conjugate system activates. Two families of dislocations glide in secant planes and the hardening $\theta_{II} \sim \mu/300$ is much higher because of short range interactions between secant dislocations and formation of junctions, called work or strain hardening. The density of dislocations increases rapidly and their distribution in the crystal is homogeneous.

    - Stage III: the hardening decreases because of cross slip of screw parts of the dislocations which allows annihilation of dislocations of opposite signs and the dislocations density increases less rapidly. The stress τ_{III} and hardening depend on temperature and strain rate as will be developed in part 4. The distribution of dislocations becomes inhomogeneous to minimise the total elastic energy stored and a cell structure forms: walls of high dislocation density separating cells of low dislocation densities.

2.4.3. *Lattice strengthening*

The tensile curves obtained in Cu-Al alloys presented in Fig. 8b show different origins of the lattice strength. In addition to the intrinsic friction of the lattice τ_f, which is very low in pure copper (fcc structure), the presence of aluminium solute atoms in the copper lattice increases the critical resolved shear stress for plasticity onset. Solute atoms interact with dislocations and exert additional forces on them, due to local lattice deformation. In the case of a purely spherical deformation around a solute atom, an elastic analysis shows that the force exerted on an edge dislocation is:

$$F = \frac{\mu b\,\Delta V}{\pi z^2}\,\frac{2(x/z)}{\left[1+(x/z)^2\right]^2} \tag{9}$$

where ΔV is the difference in atomic volume between the solute and the host atoms in the case of a substitutional solid solution and between the atomic volume of the solute and the radius of the interstitial sites in the case of an interstitial solid solution. z is the distance between the solute atom and the slip plane of the dislocation and x is the position of the dislocation in its slip plane (origin at the vertical of the atom). In addition, the valence of the solute atom also plays a role in the interaction with dislocations.

The increase of the critical shear stress due to an atomic concentration c of solute atoms is given by:

$$\Delta\tau_{sol} = Kc^n \qquad \text{with} \qquad 1/3 \leq n \leq 2/3$$

Interstitial solute atoms create a higher hardening than substitutional ones and this effect is stronger in bcc than in fcc.

Interactions between dislocations produce a hardening increasing along with strain. They have two origins: middle and long range elastic interactions which produce hardening in stage I and short range crossing of secant dislocations which produce the main part of hardening in stage II. The second term is isotropic: the stress required to move the dislocations backwards is equal to the stress that was necessary to move them forwards. Elastic interactions produce an oriented hardening, called kinematic hardening. It is at the origin of the Baushinger effect: when a metal is deformed to a given stress level τ_t in tension and then relaxed, it requires to reach τ_t again to deform plastically when reloaded in tension, while when reloaded in compression, plasticity begins at a stress τ_c lower in absolute value than τ_t, due to relaxation of pile-ups on obstacles. This effect is higher when strong obstacles are present (precipitates, interfaces).

In both cases, the strain hardening depends on the average distance between dislocations $1/\sqrt{\rho}$ with ρ the dislocation density and can be written as:

$$\Delta\tau_{work} = \alpha\mu b\sqrt{\rho} \qquad \text{with} \qquad 0.1 \leq \alpha \leq 0.4$$

α is a parameter relevant of the average strength of the interaction between dislocations. The formation of jogs along secant dislocations when they cross each other contributes to hardening in stage II. To summarize, the stress at which plastic deformation occurs is mainly the sum of three contributions in single crystals of metallic alloys:

$$\tau_c = \tau_f + Kc^n + \alpha\mu b\sqrt{\rho} \tag{10}$$

The lattice plasticity leads to specific deformation properties at the microscopic scale. Most metallic alloys are elaborated as polycrystals. At the macroscopic scale, the collective behaviour of the grains plays a role of first order in the plastic behaviour.

3. Macroscopic Behaviour

3.1. *Tensile test*

The tensile test is used to characterise the mechanical properties of materials: stiffness, strength and ductility. A sample is composed of a reduced part with a constant cross section and heads fixed in holding grips. A testing machine applies an elongation to the sample, by displacing the holding grips, most frequently at an imposed constant speed. The force necessary to achieve the imposed displacement is measured with a load cell as the response of the tested material. The measurement of the elongation of the sample requires caution, as the displacement of the crossheads is the result of the deformation of the sample but also of the elastic deformation of the testing machine and mounting. Extensometers fixed in the reduced part of the sample are usually used to measure the distance between two points. The test can be performed at a constant spreading speed of the extensometer and the results can differ from tests at a constant traverse speed, especially if plastic instabilities occur.

3.1.1. *Engineering stress-strain curve*

The force F as a function of the displacement ΔL raw curve (Fig. 9a) is not characteristic of the material as it also depends on the dimensions of the sample. The easiest normalization of F and ΔL is to introduce the engineering stress s and engineering strain e:

$$s = \frac{F}{A_0} \text{ (MPa)} \qquad\qquad e = \frac{\Delta L}{L_0} \text{ (\%)} \tag{11}$$

where A_0 is the initial cross-section area in the reduced part and L_0 the initial gauge length.

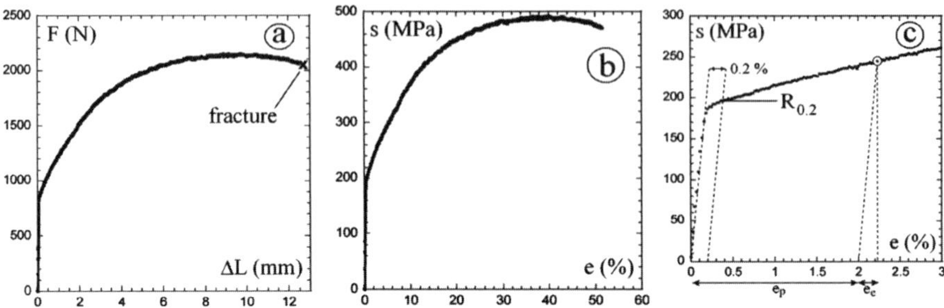

Fig. 9. Tensile test result on Fe-22Mn-0.6C at 400°C, L_0 = 24.3 mm, S_0 = 4.4 mm^2. a) Force-elongation curve, b) engineering stress-strain curve, c) magnification of b for the 3 first percent of elongation. e_e: elastic strain s/E, e_p: plastic strain. $e = e_e + e_p$. The yield stress is measured at 0.2% of plastic strain.

Several mechanical characteristics are determined from the engineering stress-strain curve (Fig. 9b):

- <u>Young's modulus *E*</u>: the slope of the curve in the elastic domain for materials with enthalpic elasticity (Fig 9c). Since deformation is small, it can be determined on the *s(e)* curve. It requires a direct measurement of the elongation and anelastic phenomena may induce an experimental error up to 10%. A more precise assessment can be made by measuring the propagation velocity of elastic waves. For small displacements, the total strain is the sum of elastic and plastic strains:

$$e = e_e + e_p = \frac{s}{E} + e_p \qquad (12)$$

- <u>Yield stress σ_y or $R_{0.2}$</u>: the stress level at which plasticity begins, characterised by a change in the slope of the curve. When a continuous transition is observed the yield stress is defined as the stress at 0.2 % of plastic deformation (Fig. 9c).

- <u>Ultimate tensile strength s_u</u>: most materials harden during plastic deformation and the engineering stress-strain curves show a maximum followed by a decrease of the stress in ductile materials before fracture. The difference between s_u and σ_y is an indication of the hardening ability of the material.

- <u>Elongation to fracture e_F</u>: the elongation of the sample at fracture. It indicates the deformability of the material.

- <u>Area reduction q</u>: not measured on the curve, it is defined as:

$$q = \frac{A_0 - A_F}{A_0} \tag{13}$$

where A_F is the final cross section area measured at the place where fracture occurred. It indicates the ductility of the material. $q = 0$ is a characteristic of brittle materials.

Figure 10 schemes typical curves for different classes of materials at room temperature and Table 2 gives values of σ_y, s_u and e_F for different materials:

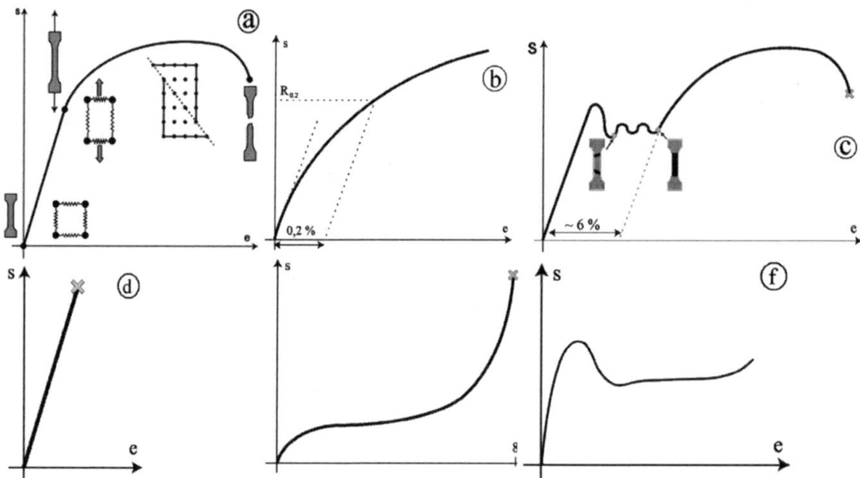

Fig. 10. Schemes of typical engineering stress-strain curves at room temperature. a) metallic alloy, b) annealed pure metal, c) yield point, d) ceramic, e) elastomer, f) thermoplastic polymer.

• Figure 10a and Fig. 9b: typical curve of a ductile metallic alloys. Fracture occurs at large plastic strains, after a neck has appeared and developed beyond the maximum of the curve. For very pure metals, q can be near 100% and the fracture force can be close to zero.

• Figure 10b: in some cases (e.g. annealed copper), no appreciable elastic deformation can be observed at the onset of plasticity. The yield stress must be defined as $R_{0.2}$.

• Figure 10c: yield point and plateau, e.g. low carbon steel after aging. At the onset of plasticity, a rapid decrease of the stress is observed. During aging, solute atoms diffuse and segregate on the dislocations. The pinning of dislocations by solute atoms requires an additional amount of stress to initiate plasticity which vanishes as soon as the dislocations move. The upper yield stress R_{eH} is very dependent on the strain rate and is not an intrinsic characteristic of the material. The lower yield stress R_{eL}, defined as the average stress at the plateau, is considered as the yield stress.

• Figure 10d: brittle material, e.g. glass, ceramics, semiconductors, quenched steel and not tempered. The yield stress is equal to the fracture stress σ_F. It is not a characteristic of the material but also depends on the present defects: geometry and length of microcracks and cavities, rugosity. Because of random distribution of such defects, the dispersion of σ_F can be quite large and requires statistical treatments.

• Figure 10e: elastomer (rubber) with entropic elasticity. The Young modulus E varies with deformation in the large elastic domain. It is worth to note that the stiffness increases with temperature, contrary to the other materials.

• Figure 10f: thermoplastic polymers. At the onset of plasticity, the stress decreases and the entire plastic deformation is achieved by formation and extension of a neck. Plasticity is achieved in the neck (see Fig. 13) by glide of the macromolecules and formation of cavities.

Table 2. Yield stress Re, tensile strength Rm and elongation to fracture e_F of various materials.

Material	Re (MPa)	Rm (MPa)	e_F (%)	Material	Re (MPa)	Rm (MPa)	e_F (%)
diamond	50000	50000	0	Cu Alloys	60-960	250-1000	1-55
SiC	10000	10000	0	Al alloys	100-627	300-700	5-30
Si_3N_4	8000	8000	0	Ni alloys	200-1600	400-2000	1-60
WC	6000	6000	0	Mild steel	220	430	18-25
Al_2O_3	5000	5000	0	γ-inox steels	286-500	760-1280	45-65
glass	3600	3600	0	Co and alloys	180-2000	500-2500	1-60
Cu	60	400	55	Mo and alloys	560-1450	665-1650	1-36
Al	40	200	50	Ti and alloys	180-1320	300-1400	6-30
Ni	70	400	65	PMMA	60-110	110	0
Fe	50	200	30	rubber	30	30	500

In Figs. 10a, c and f, the curves reach a maximum before decreasing. This apparent softening is always associated with an inhomogeneous plastic deformation. A strain localisation induces a local decrease of the section and an increase of the local stress compared to the rest of the sample which favours further deformation in the localised zone, unless the material hardens with strain. In the case of low carbon steel, at the onset of plasticity, localisation bands appear, called Piobert-Lüders bands (Fig. 11). They form and spread along the sample until the end of the stress plateau where the material begins to deform and harden homogeneously.

The engineering stress-strain curve does not take into account the variations of the cross section and the length during the test and therefore *s* and *e* are not the actual true stress and true deformation.

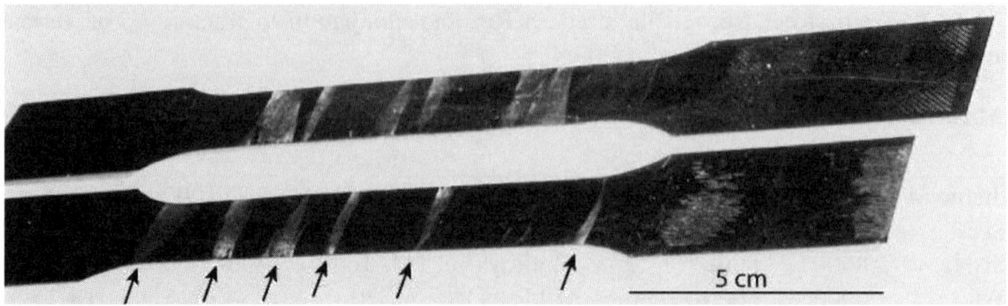

Fig. 11. Piobert-Lüders bands (arrows) in sheets of aged low carbon steels after a small amount of plastic strain.

3.1.2. *True stress-strain curve*

Being achieved by glide in metals, plastic deformation does not change the volume of the sample. By neglecting elastic with regards to plastic deformation, we can assume that for the major part of the test:

$$A_0 L_0 = AL \quad \text{where} \quad L = L_0 + \Delta L$$

and the true stress is equal to:

$$\sigma = \frac{F}{A} = \frac{F}{A_0} \cdot \frac{A_0}{A} = s \cdot \frac{L}{L_0} = s(1+e) \tag{14}$$

which is the actual stress undergone by the material as long as the deformation is homogeneous along the sample. Since the gauge length L increases, an increment of strain $d\varepsilon$ should be calculated using the instantaneous L value and the true strain is defined by:

$$\varepsilon = \int_{t_0}^{t} d\varepsilon(t) = \int_{L_0}^{L} \frac{dL}{L} = \ln\frac{L}{L_0} = \ln(1+e) \tag{15}$$

The true stress-strain curve of the previous steel is presented in Fig. 12. The curve is only valid before point N (necking), as it is based on a homogeneous deformation to estimate the actual cross section. On the curve is measured the homogeneous strain ε_u at point N. This curve gives the constitutive law of plastic deformation $\sigma = f(\varepsilon)$ to be used in structure calculations (and not $s = f(e)$).

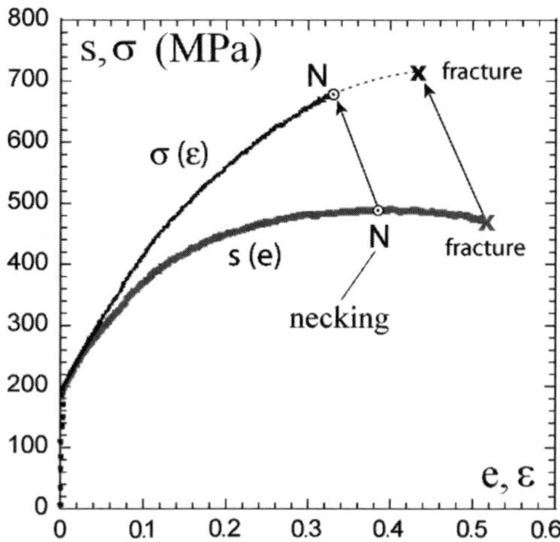

Fig. 12. True stress-strain curve associated with Fig. 9.

The simpler constitutive laws for plastic behaviour usually used are:

• Hollomon law

$$\sigma = K\varepsilon^n \tag{16}$$

n = (dlnσ)/dlnε)< 1 is called the strain hardening coefficient. It has the best fit for materials of Fig. 10.b.

• Ludwik law

$$\sigma = \sigma_y + K_1\varepsilon^{n_1} \tag{17}$$

which is a shift of the previous one along the stress axis to introduce the yield stress σ_y.

• Swift law

$$\sigma = K_2\left(\varepsilon_0 + \varepsilon\right)^{n_2} \tag{18}$$

which introduce the yield stress $\sigma_y = K_2\varepsilon_0^{n_2}$ by a negative shift along the strain axis. Note that n_1 and n_2 are not equal to the strain hardening coefficient n.

                     J. P. Chateau

3.1.3. *Necking*

In ductile materials, after the maximum of the curve, a plastic strain localisation occurs, called necking. The stress in no longer uniform and the stress tensor becomes triaxial in the neck. The final fracture occurs by local damaging in the neck (Fig. 13). Most materials have a positive work hardening rate θ, defined as the slope of the true stress-strain curve in the plastic domain:

$$\theta = \frac{d\sigma}{d\varepsilon}\bigg|_{\text{plastic}} \tag{19}$$

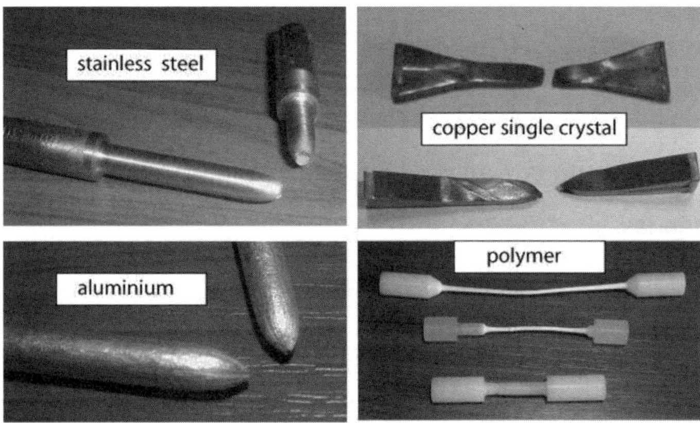

Fig. 13. Examples of necking in ductile materials

When a strain localisation begins to occur due to geometric or materials heterogeneity of the sample, two mechanisms compete: i) because the cross section is smaller in the localised zone, deformation is savoured, ii) since the material strengthens with deformation, the more deformed material in the localised zone becomes more resistant to deformation with respect to the material in the rest of the sample. In the early stages of plastic strain, θ is high and the second mechanism prevents a neck from developing. The specimen is mechanically stable and the deformation is homogeneous. θ decreases with ε and when the first mechanism is dominant, plasticity becomes unstable and the neck forms.

A criterion for the onset of necking can be deduced from the fact that it corresponds to the maximum of the applied force, called the Considère criterion. At *F* maximal:

$$dF = \sigma \cdot dA + A.d\sigma = 0 \tag{20}$$

The volume conservation gives:

$$dV = L \cdot dA + A.dL = 0 \tag{21}$$

and the criterion becomes:

$$\frac{d\sigma}{\sigma} = -\frac{dA}{A} = \frac{dL}{L} = d\varepsilon \tag{22}$$

rewritten as: $\theta = \sigma$ or $(d\ln\sigma)/(d\ln\varepsilon) = \varepsilon$

Note that if the material follows the Hollomon law, the maximum uniform strain ε_u is equal to the strain hardening coefficient *n*. In metallic sheets, a high hardening ability delays the formation of localisation bands, which is of great importance for forming applications by deep drawing. Figure 14 shows the evolution of $d\ln\sigma/d\ln\varepsilon$ with strain for the Fe-22Mn-0.6C steel at different temperatures: ductile fracture (localised necking) or diffuse necking occur at the Considère criterion for each curve.

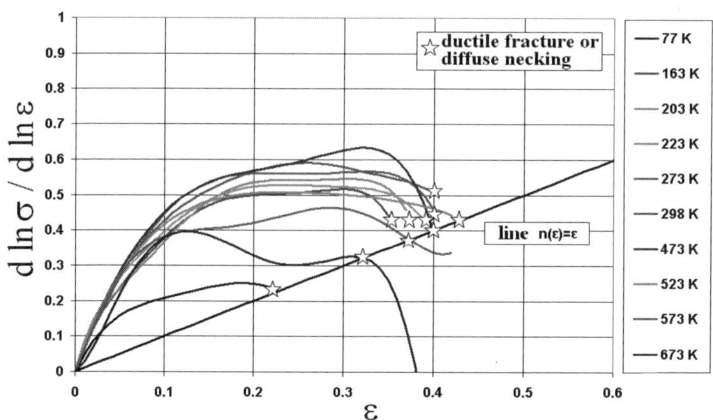

Fig. 14. Actual strain hardening coefficient vs. true strain for Fe-22Mn-0.6C with a grain size of 2.5 μm at various temperatures. The Considère criterion is well satisfied.

This steel has a hardening coefficient reaching a high value of 0.65 at room temperature, hence his excellent deformability ($e_F \sim 50\%$) compared to other high resistance steel ($s_u \sim 1$ GPa).

3.2. Polycrystal

3.2.1. *Microscopic and macroscopic yield stresses*

Industrial materials are mainly used in a polycrystalline state, except e.g. semiconductors for microelectronic applications and superalloys for turbine blades of aircraft reactors. When the stress increases during a tensile test, plasticity will begin in the grains with the most favourable orientation, i.e. the grains where a slip system has a Schmid factor equal to 0.5, when the resolved shear stress overcomes the critical value τ_c. But dislocations are rapidly blocked at a grain boundary, due to misorientation of the lattice in the neighbouring grain. The strain produced inside the deformed grain is low and is not detected on the tensile curve. This corresponds to the so-called microscopic yield stress. The macroscopic yield stress σ_y is reached when plastic deformation propagates through the entire sample, which cannot be achieved simply by crossing the grain boundaries, as they are strong obstacles.

To link the microscopic and macroscopic yield stresses, Sachs supposed that in each grain, plasticity is mainly achieved by the primary slip system, neglecting the secondary systems. In the case of a fully random distribution of the grain orientations (no texture), the average Schmid factor on the primary systems in the fcc structure is 0.446 (average of the values on Fig. 6.a). Then the relation would be:

$$\frac{\sigma_y}{\tau_c} = \frac{1}{m} = M_{Sachs} = 2.24 \tag{23}$$

But, when a grain deforms plastically, the compatibility of strain at its boundaries implies that the neighbouring grains adapt to the imposed deformation, otherwise cracks would form at the interfaces. A given strain tensor is symmetrical and has 6 independent components. But as plastic deformation occurs at constant volume, the trace of the tensor is zero. The number of independent components is then 5 and Von Mises stated that at least 5 independent slip systems were required in each grain to achieve a deformation compatible with the entire medium. When a material does not have 5 independent slip systems (case of hcp metals when glide is mainly achieved in basal planes), other plasticity mechanisms are necessary, such as twinning or intergranular glide. The dislocations emitted to ensure compatibility are called geometrically necessary dislocations, in opposition to the dislocations emitted inside the grains in response to the macroscopic applied stress, which are called statistically stored dislocations.

Taylor used this condition to select 5 systems out of 12 on an energetic criterion over a large number of randomly oriented grains and found[6]:

$$\frac{\sigma_y}{\tau_c} = M_{Taylor} = 3.067 \tag{24}$$

3.2.2. *Hall and Pecth law*

The poor agreement of the two previous approaches with experiments confirms that polycrystals do not behave like a group of isolated single crystals. The deformation compatibility and the blocking of dislocations at the grain boundaries play a major role. The macroscopic yield stress is seen to follow the universal Hall and Petch law:

$$\sigma_y = A + \frac{k}{\sqrt{d}} \tag{25}$$

where d is the mean grain size. It states that the yield stress increases with the volume density of grain boundaries. A and k are characteristic parameters of the material for a given temperature and strain rate and are determined experimentally (e.g. on Fig. 15). A is close to the microscopic yield stress: $A \sim M_{Taylor}\tau_c$. The initial model of Hall and Petch considers dislocation pile-ups at the grain boundaries. A pile-up of N dislocations formed under a resolved shear stress τ in a grain induces a stress $N\tau$ in the neighbouring grain. The smaller is the grain, the fewer dislocations are in the pile-up. They showed that the stress in the neighbouring grain was proportional to $\sqrt{d}$. Besides, this law is also verified for materials where no pile-ups are observed, like metals with high stacking fault energy. Even if the mechanisms are not the same, they have the same scale dependence with d.

This law predicts a very high yield stress for very small grains. For example, the extrapolation of Fig. 15 at 0.5% strain leads to a yield stress of 1.3 GPa for a grain size of 20 nm. This is due to the fact that it is based on intragranular plastic deformation which can require a higher stress than intergranular deformation when the grains are too small.

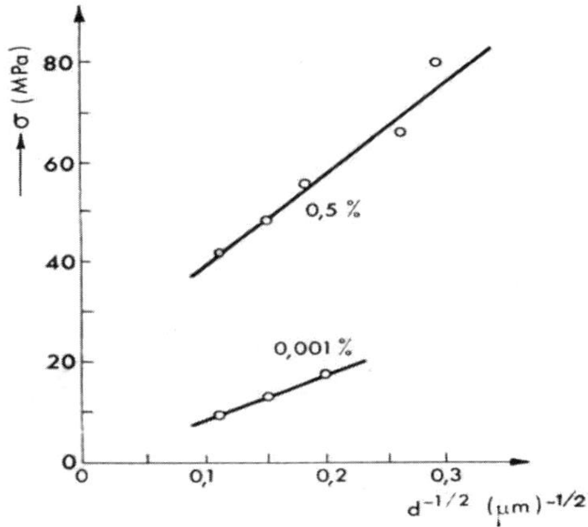

Fig. 15. Yield stress of copper polycrystals (0.001% and 0.5% pre-strained) as a function of the grain size.

3.3. *Other plasticity mechanisms*

3.3.1. *Intergranular glide*

The interatomic bondings between the grains which ensure the cohesion of the polycrystal have their own resistance and it is possible to make them glide on each other with few deformation of them. By analogy, shaping wet sand does not require to deform the sand grains. Besides, glide in the grain boundary requires a high shear stress, usually higher than τ_c. Hence, plasticity occurs in the major cases in the grains and the Hall and Petch law applies. If the grain size of a given material is decreased one observes a critical value d_c below which:

i) the Hall and Petch law ceases to be valid,

ii) the material becomes very ductile, being able to reach elongations up to 1000% in some cases without necking, which behaviour is called super plastic,

iii) plasticity is achieved by glide in the grain boundaries associated with local deformation in the vicinity of the boundary, mainly by diffusion of atoms, to accommodate the compatibility of the grains (Fig. 16).

Intergranular glide is favoured by high temperatures and low deformation rates. Typically, d_c is of the order of 1 μm in materials exhibiting super plasticity at high temperature. At room temperature, the validity limit of the Hall and Petch law is estimated at a few tens of nm in nanostructured materials.

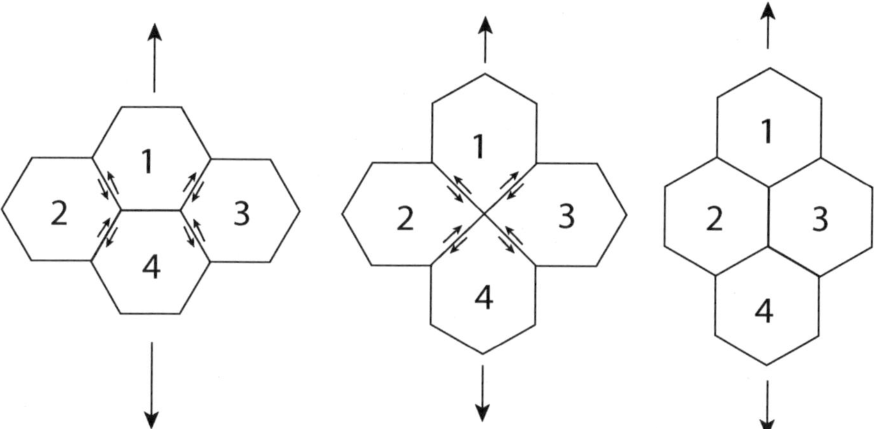

Fig. 16. Elongation of a polycrystal by intergranular glide (or grain boundary gliding) accommodated by local deformation mainly achieved by diffusion.

3.3.2. *Twinning*

Twinning is an intragranular plastic deformation mechanism alternative to dislocation glide. It generally occurs when dislocation glide is difficult, for example at low temperature (bcc), when the stacking fault energy is low (fcc) or when the number of available slip systems is not sufficient (hcp). Two crystals are in twin position if:

    i)   they have a crystallographic plane in common, called twin plane,

    ii) there exists a super lattice common to the two crystals, called coincidence super lattice

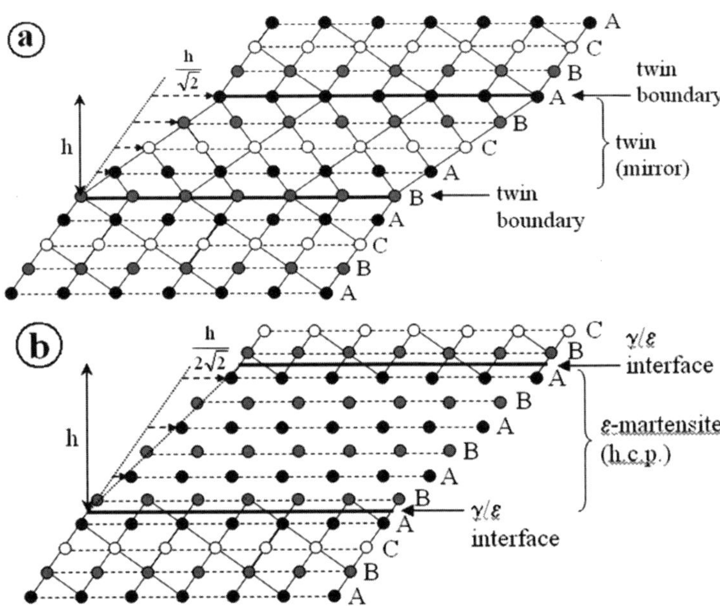

Fig. 17. Cut of the fcc structure along a {110} plane, ABCABC sequence of {111} planes. a) mirror fcc structure in a twin: BACBA sequence, b) hcp ε-martensite platelet: ABAB sequence. Glides produced by each transformation are indicated.

In high symmetry structures, such as fcc, bcc and hcp, it implies that the two crystals are in symmetrical positions from side to side of the twin plane (hence the name) and that a twin plane is a particular grain boundary delimiting two misoriented crystals. A mechanical twin produces a glide γ_t by shifting the lattice in the mirror position over the thickness of the twin. A twinning system is given by a twinning plane and direction in this plane, as for slip systems. The twinning systems are {111}<112> in the fcc structures, achieved by Shockley partial dislocations gliding in successive parallel {111} planes (Fig. 17a), in the <111> direction along {112}, {441} and {332} in the bcc structures and mainly along {10$\bar{1}$2}<10$\bar{1}$1> in the hcp structures. In the cubic structures, $\gamma_t = 1/\sqrt{2}$ and is between 0.1 and 0.2 in the hcp structures. When a twinning system is activated, the total glide produced is equal to $f\,\gamma_t$, where f is the volume fraction of twins.

Twins propagate rapidly through the crystal. When thick twins are created, especially in single crystals, a large part of the elastic strain is instantaneously relaxed and a drop of the applied stress can be observed. Figures 18 and 19 show the result of interactions between micro twinning and dislocation glide on the mechanical properties of the same Fe-22Mn-0.6C steel as presented earlier. This fully austenitic steel has a stacking fault energy equal to 20 mJ/m^2 at room temperature and thin microtwins (10 to 50 nm thick) arranged in stacks (0.1 to 1 µm thick) form along with plastic strain. Figure 18 shows optical micrographs after interrupted tensile test at different strains and then etched to reveal the grain and twin boundaries. Stacks of microtwins appear as black lines extending from a boundary to another. Secant twinning systems are progressively activated in each grain and the net result is a refinement of the microstructure along with strain. Stacks of microtwins are as strong obstacles as grain boundaries for dislocations. The mean free path of the dislocations between two strong interfaces decreases, resulting in a very high strain hardening by a dynamical Hall and Petch effect due to the refinement of the effective grain size (Fig. 19a). This phenomenon is known as TWIP effect, for twinning induced plasticity.

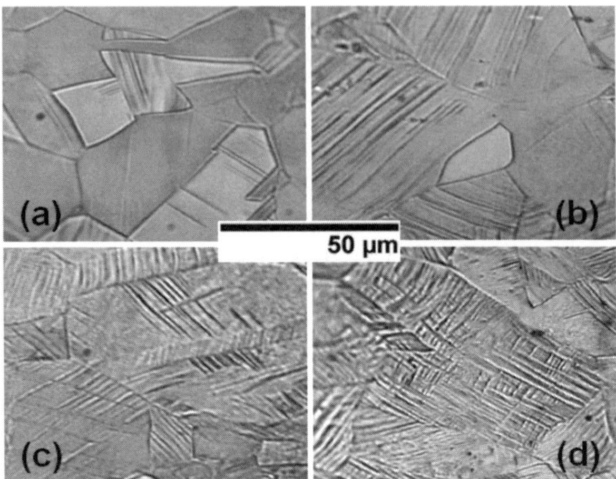

Fig. 18. Optical micrographs after etching of the Fe-22Mn-0.6C (grain size 20 μm) strained at room temperature and increasing levels: a) 0.10, b) 0.18, c) 0.26, d) 0.34. Increase of the volume fraction of stacks of microtwins along with plastic strain.

According to the Considère criterion, fracture by localised necking occurs after a large deformation, associated with a high ultimate tensile stress (Fig 19b). In comparison, the stacking fault energy of the same steel is equal to 90 mJ/m^2 at 400°C and no twinning occurs; both maximal strain and stress are much lower.

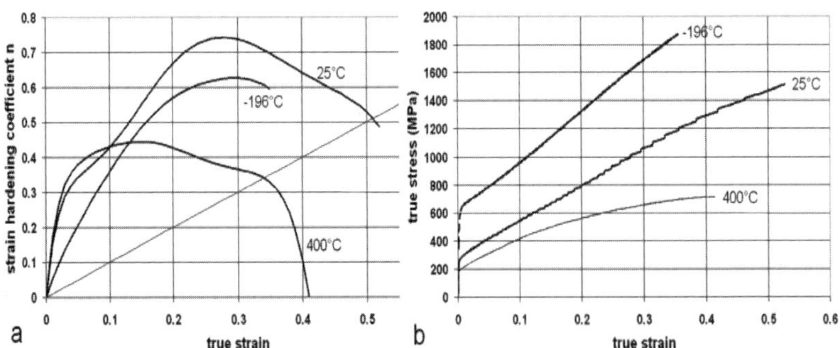

Fig. 19. Tensile test on the steel of Fig. 18 at 400°C, 25°C and -196°C. a) strain hardening coefficient n = dlnσ/dlnε, determined from b) true stress-strain curves.

3.3.3. *Martensitic transformation*

In Fig. 17.a, when the same glide between two {111} planes is achieved in the fcc structure, but every 2 planes instead of every plane, the structure is transformed into hcp (called ε-martensite in steels), instead of constructing the mirror fcc structure (Fig. 17b). This phase transformation achieved by small displacements of the atoms is a particular case of displacive transformations, as opposed to diffusive transformations, called martensitic transformations. The most well known is the γ (fcc) $\rightarrow \alpha'$ (bct) transformation in steels obtained by quenching from high temperature below a given temperature M_S (martensite start). As the transformation is displacive, it can also be triggered by applying stresses and can act as a deformation mode. The interaction between α'-martensite and dislocation glide is known as TRIP effect, for transformation induced plasticity. It is observed for instance in metastable austenitic stainless steels. Hardening results in the decrease of the dislocation mean free path and of the formation of the hard α' phase.

TRIP effect is also observed in Fig. 19. The stacking fault energy of the Fe-22Mn-0.6C steel is very low at -196°C (7 mJ/m^2), corresponding to a negative $\gamma \rightarrow \varepsilon$ transformation free enthalpy. At this temperature ε-martensite platelets replace microtwins with a glide over their thickness divided by 2 ($\gamma_t = 1/2\sqrt{2}$). The effect is similar as TWIP but with lower strain hardening and elongation due to a lower mobility of dislocations at low temperature.

Martensitic transformation is also responsible of the so-called shape memory effect observed in Ni-Ti, Cu-Al-Ni, Cu-Al-Zn or Fe-Mn-Si-C alloys. When a sample is cooled down from the initial phase called austenite, the martensitic transformation starts at M_S, the volume fraction increases when the temperature decreases and the transformation is complete at M_F (martensite finish). When the sample is heated afterwards, the reverse transformation begins at A_S (austenite start) and it is complete at A_F (austenite finish). Due to the formation of heterophase interfaces when transformations occur, there is a strong hysteresis, so that:

$$M_F < M_S < A_S < A_F$$

and the transformation free enthalpy is zero at $(M_S+A_S)/2$. When a stress is applied, the mechanical work produced by the formation of a displacive phase helps the transformation. To simplify, the transformation temperatures can be seen as increasing functions of the applied stress, which is schemed by a linear dependence on Fig. 20.

A) Shape memory effect (Fig. 20.a)

The alloy is so that room temperature is between M_S and A_S. By applying an increasing stress on a fully austenitic sample (1), the martensitic transformation begins when the M_S line is intersected and produces a deformation (2). When the stress is relaxed, the A_S line is not intersected and the deformation remains (3). In a second time, if the sample is heated, the reverse transformation begins when the A_S line is intersected, progressively vanishing the deformation produce by martensite. The sample is again fully austenitic over A_F (4) and regains its initial shape after cooling down to room temperature (5)

B) Pseudo elasticity effect (Fig. 20.b)

The alloy is so that room temperature is over A_F. When a stress is applied, the deformation is elastic until the M_S line is intersected where a large deformation begins to be achieved by martensitic transformation. If the M_F line is not intersected, all the deformation is due to martensite. When the stress is relaxed, the A_S and A_F lines are intersected successively. The entire deformation vanishes by reverse transformation and the sample regains its initial shape at zero stress. Reversible deformations up to 10% can be achieved by this mechanism, which is 2 orders of magnitude higher than the purely elastic strains available in metallic alloys.

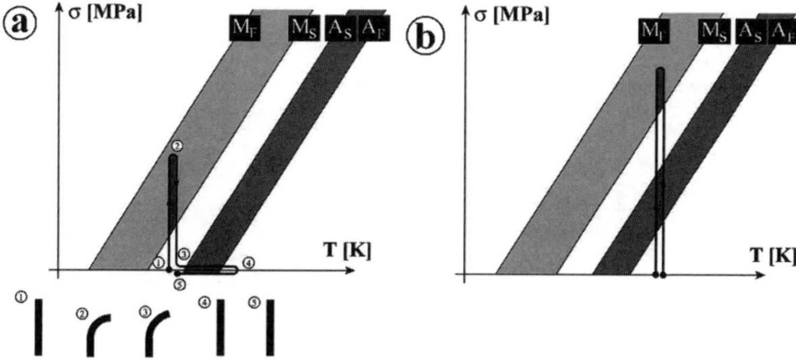

Fig. 20. Schemes on a stress-transformation temperature diagram of a) the shape memory effect, b) the pseudo elasticity effect.

Many mechanisms involved in plastic deformation are thermally activated. As a consequence, the mechanical properties of metals depend on the testing conditions.

4. Influence of Temperature and Strain Rate

4.1. *Internal and effective stresses*

Consider a single crystal oriented for single slip and tested in compression at a given imposed strain rate up to a given applied resolved shear stress (Fig. 21). At this point, the test is stopped and the stress is rapidly decreased. While the applied stress remains higher than a threshold value τ_i (1), the sample continues to deform at a constant strain rate. If the applied stress is below τ_i the sample simply undergoes an elastic unloading (3). This shows that the applied stress τ during compression had two different components:

i)   a thermally activated component $\tau^* = \tau - \tau_i$ called effective stress,
ii)   an athermal component τ_i called internal stress. This is the minimum stress required to deform the material.

                           J. P. Chateau

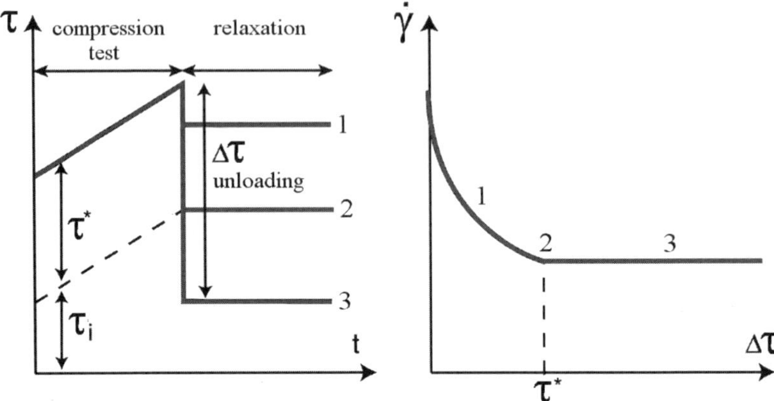

Fig. 21. Relaxation test after a compression at an imposed strain rate, measured strain rate during relaxation and definition of the macroscopic intern and effective stresses.

At the microscopic scale, this means that different types of obstacles to the motion of dislocations are present. The force required to move a dislocation through a crystal is schemed in Fig. 22. Long range interactions induce fluctuations on large distances and represent the local internal stress. In addition, short range obstacles require a local increase of the stress on short distances to be crossed over. To move a dislocation through the crystal, a stress at least equal to the maximum of the local internal stress is required. This is the internal stress τ_i. To cross a local obstacle, a free enthalpy ΔG_0 is required, called the activation energy of the obstacle. The Peach and Köhler force on the dislocation due to the effective stress τ^* furnishes a part of this energy through mechanical work:

$$W = \tau^* b\ell \cdot d = \tau^* V^*$$ (26)

where ℓ is the length of the dislocation segment and d the width of the obstacle. V^* is called the activation volume of the obstacle. The rest is furnished by thermal fluctuation:

$$\Delta G = \Delta G_0 - \tau^* V^*$$ (27)

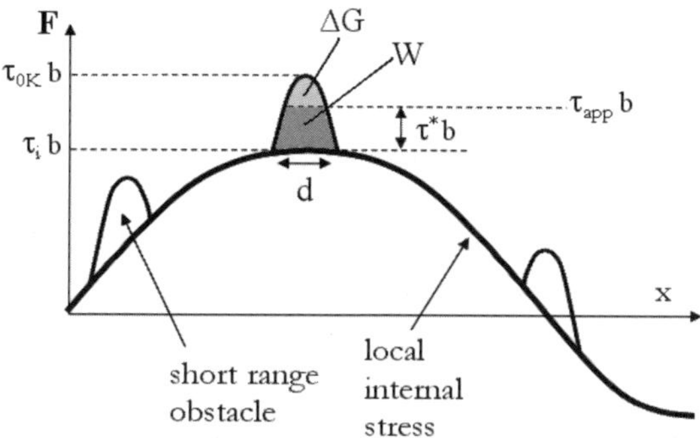

Fig. 22. Scheme of the force required to move a dislocation through a crystal vs. its position. The local force is due to long range interactions (local internal stress) and localised obstacles (short range obstacle).

A high activation energy is characteristic of a strong obstacle, while a high activation volume is characteristic of obstacles for which the major part of the energy must be furnished by mechanical work, hence with little influence of temperature. The contribution of thermal activation to cross over the internal stress on large distances is negligible.

The probability for a dislocation to cross such an obstacle depends on ΔG and the average speed of dislocations crossing identical obstacles is given by:

$$\bar{v} = v_0 \cdot \exp\left(-\frac{\Delta G}{k_B T}\right) \tag{28}$$

where k_B is the Boltzmann constant. The Orowan relation links the glide rate due to a given slip system to the average speed of the dislocations:

$$\dot{\gamma} = \rho_m b \bar{v}$$

where ρ_m is the density of mobile dislocations of the activate slip system. Thus:

$$\dot{\gamma} = \rho_m b v_0 \cdot \exp\left(-\frac{\Delta G}{k_B T}\right) \tag{29}$$

 J. P. Chateau

In Eq. 10, the internal stress is due to the interactions between dislocations:

$$\tau_i = \alpha\mu b\sqrt{\rho} \tag{30}$$

Combining Eqs. 27, 29 and 30 leads to the critical resolved shear stress:

$$\tau = \alpha\mu b\sqrt{\rho} + \frac{\Delta G_0}{V^*} - \frac{k_B T}{V^*}\ln\frac{\rho_m b v_0}{\dot{\gamma}} \tag{31}$$

The sum of the first two terms is the stress required at 0°K to move dislocations through the crystal. In Eq. 10, the contributions to the effective stress are the lattice friction and the solid solution hardening, with different activation energies and volumes. Measures of the internal stress in pure Cu, Cu-Al and pure Ge single crystals are presented in Fig. 23[8]:

- pure Cu: the only obstacles are dislocations and the flow stress is close to the internal stress. The little difference shows that dislocations crossing is slightly thermally activated, which can be attributed to dragging of jogs formed along a mobile dislocation when it crosses a secant one.

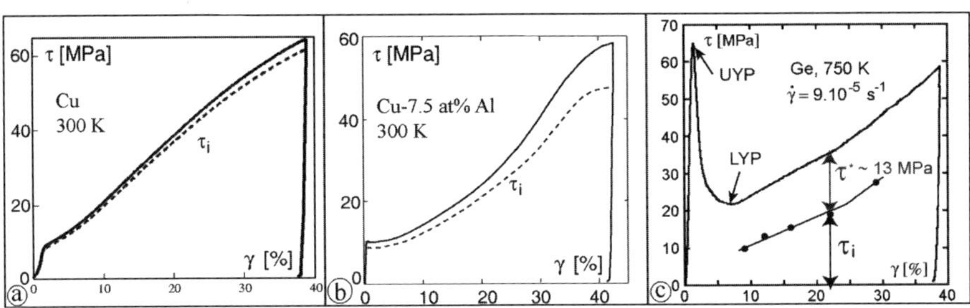

Fig. 23. Stress relaxation tests (or dip test) performed in compression on : a) Cu, b) Cu-Al at room temperature and c) Ge at 480°C.

- Cu-Al: the larger effective stress in comparison with pure Cu is due to Al solute atoms.

- pure Ge: Ge is a semiconductor with a high lattice friction. The total stress increases, due to strain hardening, while the effective stress is constant and equal to the lattice friction which contributes in a large part to the total stress. Note that the onset of plasticity is marked by a yield point. This is due to the fact that there is no dislocation in the initial crystal and dislocations sources must be created. The beginning of plastic deformation is achieved by a very low density of mobile dislocations and $\rho_m b v_0$ is lower than the imposed strain rate, which increases the critical resolved shear stress (UYS). Mobile dislocations are created and the stress decreases, until strain hardening begins (LYS).

4.2. *Thermally activated plastic deformation*

When mobile dislocations are numerous, like in metals, $\rho_m b v_0 > \dot{\gamma}$ and the last term in Eq. 31 decreases the yield stress. Figure 24 shows the dependence on temperature of the elastic limit of bcc[9] and fcc pure single crystals. According to Eq. 31, a linear decrease on temperature is observed below a given temperature, starting from the maximum elastic limit at 0°K. A plateau corresponding to the internal stress is observed at higher temperature. In this domain, the mere thermal fluctuations are sufficient to overcome the lattice friction. Due to lattice friction, the low temperature regime extends to higher temperatures in Fe than in fcc metals and the plateau stress is negligible with regards to the stress at 0°K, contrary to fcc. For Cu and Al, the plateau is reduced and a decrease of the yield stress is observed at high temperature. This is due to recovery and will be discussed further.

In the case of of polycrystals is illustrated by Fe-Mn-C steels with different compositions in Fig. 24c. The yield stress at 300K was taken as the reference to eliminate discrepancies due to the difference in composition and initial microstructure. The formation of twins or martensite platelets occurs only after a few percent of strain, so that the yield stress is characteristic of dislocation glide. Two regimes are observed. Below room temperature, the linear dependence is due to solute carbon in insertion and the slope increases with the carbon content. In the absence of carbon, a unique regime with lower temperature dependence is observed over the entire temperature range.

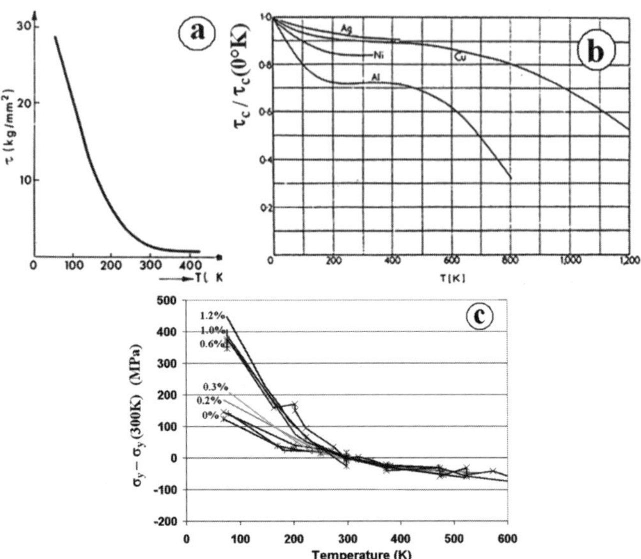

Fig. 24. Temperature dependence of the yield stress of a) pure Fe single crystal (bcc), b) fcc single crystals and c) Fe-Mn-C (fcc) polycrystals with various compositions.

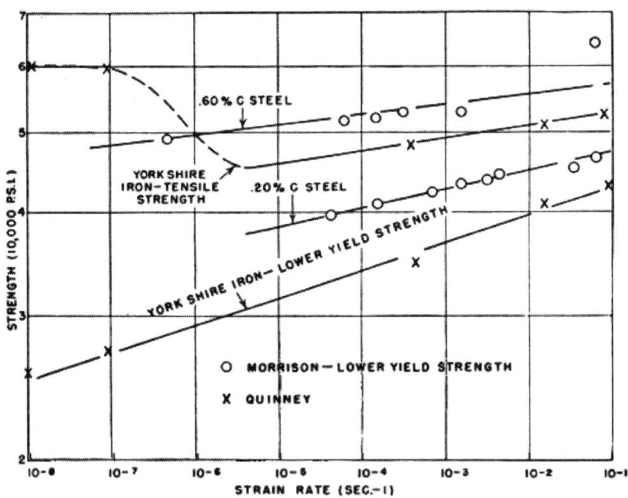

Fig. 25. Logarithmic dependence of the yield stress with regard to the strain rate in steels.

Figure 25 illustrates the increase of the yield stress with strain rate, with a logarithmic dependence, in the case of carbon steels.[10] To take account of the strain rate dependence of the true stress-strain curve during a tensile test, a local incremental behaviour law can be written as (at a given temperature):

$$\delta\sigma(\varepsilon,\dot{\varepsilon},T) = \theta(\varepsilon)\delta\varepsilon + S(\varepsilon,\dot{\varepsilon},T)\delta\ln\dot{\varepsilon}$$

where θ is the local hardening and:

$$S = \left(\frac{\partial\sigma}{\partial\ln\dot{\varepsilon}}\right)_{\varepsilon,T} \tag{32}$$

is called the strain rate sensitivity. The dependence on strain rate has the same form as in Eq. 31. An empirical description based on the Hollomon law is also usually used in simple cases:

$$\sigma = K\varepsilon^n\dot{\varepsilon}^m \tag{33}$$

at a given temperature. Similarly to the strain hardening coefficient n, the strain rate sensitivity coefficient is defined as:

$$m = \left(\frac{\partial\ln\sigma}{\partial\ln\dot{\varepsilon}}\right)_{\varepsilon} \tag{34}$$

4.3. Influence on the stability of plastic deformation

4.3.1. *High strain rate sensitivity*

During a tensile test, the decrease rate of the cross section is:

$$-\dot{A} = A\frac{\dot{L}}{L}$$

due to volume conservation. By introducing the strain rate $\dot{\varepsilon} = \dot{L}/L$ and using Eq. 31, it comes:

$$-\dot{A} = A\left(\frac{\sigma}{K\varepsilon^n}\right)^{1/m} = \left(\frac{F}{K\varepsilon^n}\right)^{1/m}A^{1-\frac{1}{m}} \tag{35}$$

where F is the tensile force. Since $0 < m < 1$, the exponent of S is negative. This means that the section decrease rate increases when the section decreases and that if a strain localisation occurs, the section reduction tends to be enhanced in the localisation zone. High values of m reduce this effect and in Newtonian fluids with $m = 1$, the section decrease rate is independent of the section. The Rossard criterion states that final necking during a tensile test occurs at:

$$\varepsilon_u = \frac{n}{1 - 2m} \tag{36}$$

$m \approx 0$ leads to the Considère criterion and high values of m increase the homogeneous strain of the sample. For $m > 0.5$, the criterion cannot be satisfied and necking never occurs, leading to a superplastic behaviour. Such values of m are observed when the main plasticity mechanism is grain boundary gliding, due to the diffusion dependence of the strain accommodation process at the grain boundaries (Fig. 16). It can be observed for small grains (typically < 1 µm), at high temperatures (typically $> 0.4\ T_m$, where T_m is the melting temperature) and slow strain rates (typically $< 10^{-3}\ \mathrm{s}^{-1}$).

4.3.2. *Negative strain rate sensitivity*

On the contrary, when a localisation band initiates, its strain rate becomes higher than the macroscopic applied strain rate. If the strain rate sensitivity S is negative, the flow stress in the band becomes lower than in the rest of the sample and the localisation is enhanced. A negative strain rate sensitivity can be observed when solute atoms dynamically interact with mobile dislocations, which is called dynamic strain aging. In a strain hardened material, mobile dislocations have to cross secant dislocations, which requires a waiting time t_w due to the slight thermal activation. When a mobile dislocation leaves an obstacle, it quickly moves to the next one. In the absence of other obstacles, the "flight" time can be neglected and the average speed of mobile dislocations is:

$$\bar{v} = \frac{1}{t_w \sqrt{\rho_f}} \tag{37}$$

where ρ_f is the density of secant dislocations, $1/\sqrt{\rho_f}$ is the average distance between two obstacles and the strain rate is given by:

$$\dot{\varepsilon} = \alpha \frac{\rho_m b}{t_w \sqrt{\rho_f}}$$ (38)

where ρ_m is the density of mobile dislocations and α a geometric parameter. In this case, the tensile stress is $\sigma = \sigma_i + \sigma^*$ (with the same definitions as for shear stresses before) and the strain rate sensitivity:

$$S_0 = \left(\frac{\partial \sigma^*}{\partial \ln \dot{\varepsilon}} \right)_\varepsilon$$ (39)

is low but positive.

Strain aging occurs when solute atoms diffuse during a heat treatment under their interaction force with dislocations (following Eq. 9, for instance) and segregate on them (Fig. 26a). During a following tensile test, the segregating solute atoms induce a local extra force required to move a dislocation (Fig. 26b), which vanishes once the dislocation has started to glide. This phenomenon is at the origin of the yield point observed in Fe-C steels after aging or annealing (Fig. 10c. and 27). The extra drag force on dislocations due to solute atom segregation increases with diffusion time.

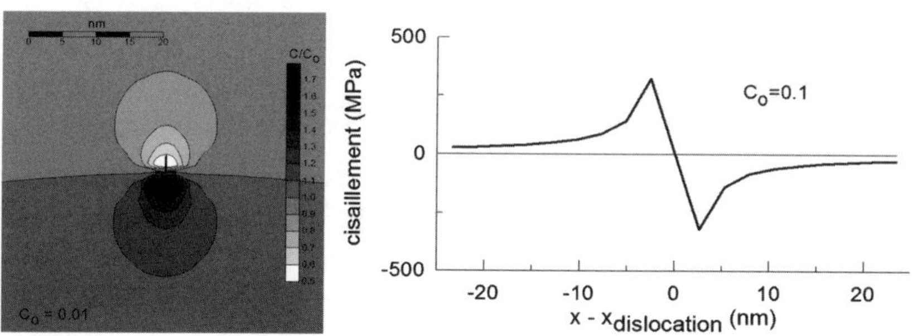

Fig. 26. a) Cottrel atmosphere of solute atoms around a dislocation. Concentration profile around an edge dislocation in the case of hydrogen in an austenitic steel at room temperature (initial uniform concentration $C_0 = 1\%$). b) Resolved shear stress in the slip plane of the dislocation due to the Cottrel atmosphere ($C_0 = 10\%$) From Ref. 11.

Dynamic strain aging (DSA) occurs when the material is deformed at high temperature which allows solute atom to segregate on dislocations during their waiting time on obstacles. An extra drag stress σ_{drag} is required and the total flow stress is:

$$\sigma = \sigma_i + \sigma^* + \sigma_{\text{drag}}$$

Solute atoms diffuse on dislocations during the waiting time on obstacles. The strain rate sensitivity is thus:

$$S = S_0 + S_{\text{dsa}}$$

where

$$S_{\text{dsa}} = \left(\frac{\partial \sigma_{\text{drag}}}{\partial \ln \dot{\varepsilon}} \right)_{\varepsilon} \propto -\left(\frac{\partial \sigma_{\text{drag}}}{\partial \ln t_w} \right)_{\varepsilon} \tag{40}$$

using Eq. 38. As σ_{drag} increases with t_w, S_{drag} is negative. In a given $T(\dot{\varepsilon})$ or $\dot{\varepsilon}(T)$ domain the DSA effect can be dominant and $S < 0$. In this case, the entire stress-strain curve is a succession of yield points due to unstable plastic deformation, called Portevin-Le Châtelier effect (Fig. 27a).[12] A localisation band forms and the applied stress drops. Then the material hardens in the band and the stress increases until another band forms. The bands formed can be either static at high temperature or slow strain rate, or propagating through the sample at low temperature or high strain rate. Serrations on the tensile curve of the Fe-22MN-0.6C are observed at 25°C on Fig. 19b and are associated with the formation and propagation of moving bands of strain localisation (Fig. 27b).[13] In this material, the DSA mechanism is more complex than dynamic pinning of dislocations by carbon and is believed to include twinning.

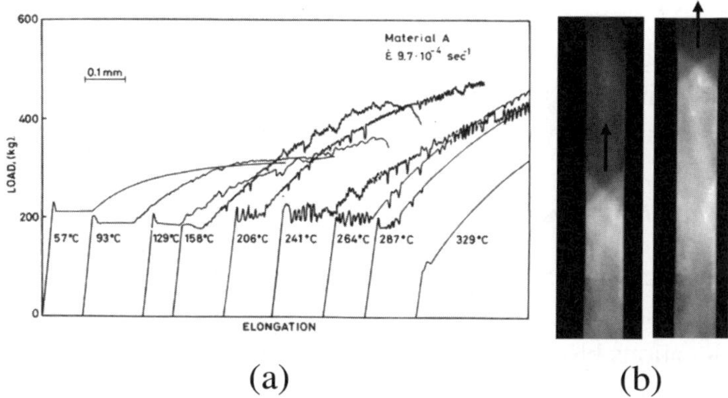

(a)   (b)

Fig. 27. a) Engineering stress-strain curves for a Fe-C steel at different temperatures. Serrations are observed from 93°C up to 287°C. b) Propagation of a moving deformation band observed by infrared camera during tensile test at room temperature in the Fe-22Mn-0.6C austenitic steel.

4.4. *Dynamic recovery and recrystallisation*

Due to the presence of high densities of stored dislocations, the crystal is not in equilibrium after plastic deformation. The excess in free enthalpy is at the origin of thermodynamical forces which tend to make the crystal evaluate towards a more stable state. The time needed to reach the more stable state will depend on the rate at which defects can be eliminated, in which temperature is attended to play a role.

The equilibrium state is the state minimising the free enthalpy $G = H - TS$. In solids, the enthalpy H is close to the elastic intern energy U. H and the configuration entropy S are increasing functions of the defect density, so that the equilibrium state corresponds to a given concentration of defects. This is the case for vacancies, for instance, for which the equilibrium proportion at a temperature T is:

$$c \approx \exp\left(-\frac{H_f}{k_B T}\right) \tag{41}$$

where H_f is the formation enthalpy of one vacancy, of the order of 1 eV. The vacancy concentration is negligible at room temperature but of the order of 10^{-4} close to T_M. In the case of dislocations, the configuration entropy in negligible with regards to the elastic energy and even a small strain hardening leads to an out of equilibrium state. As a consequence, the dislocation density is expected to evolve rapidly at high temperature.

4.4.1. *Dynamic recovery*

Different mechanisms are able to decrease the elastic energy stored in a hardened crystal (Fig. 28):

- dislocations blocked by obstacles in their slip plane can bypass them by moving out of their plane,

- dislocations of opposite signs and gliding in parallel planes can meet to annihilate themselves,

- dislocations can rearrange themselves under their mutual interactions.

The net effect is a decrease of the total dislocation density and the formation of stable structures where low dislocation density cells are separated by high dislocation density walls called sub-grain boundaries. The formation of this stable microstructure is called polygonisation. The internal stress in such microstructure is lower than in the hardened material. A deformed metal can recover its previous properties by such mechanisms, hence the term recovery.

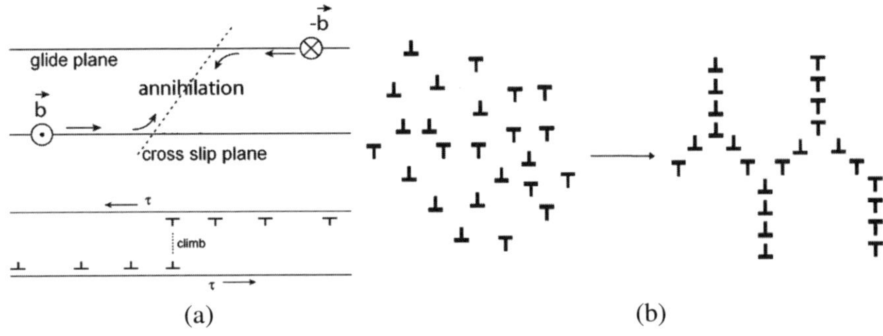

Fig. 28. a) Annihilation of screw (resp. edge) dislocations of opposite signs gliding in parallel planes by cross-slip (resp. climb). b) Arrangement of randomly distributed dislocations to form an heterogeneous structure of less total elastic energy.

Each mechanism requires the displacement of dislocations out of their slip plane. This is done by cross slip for screw dislocations and by climb for edge dislocations, both enhanced at high temperature. The stress τ_{III} defined in Fig. 8.a and corresponding to the onset of cross slip decreases when temperature increases.[14] Climb of edge dislocations requires the diffusion of vacancies which both concentration and diffusion coefficient increase with temperature.

When a metal is deformed at high temperature, dynamic recovery enters in competition with strain hardening. During an increment of strain, the increment of stored dislocations is:

$$d\rho = d\rho_+ - d\rho_-$$

where $d\rho_+$ is the strain hardening term, increasing with strain rate and $d\rho_-$ is the recovery term, increasing with temperature. Mecking and Kocks proposed the following empirical expression[7]:

$$\frac{d\rho}{d\varepsilon} = \frac{1}{\Lambda b} - f(T)\rho \tag{42}$$

where Λ is the mean free path of dislocations between obstacles. In the case of a hardened polycrystal, for instance:

$$\frac{1}{\Lambda} = \frac{1}{d} + k\sqrt{\rho} \tag{43}$$

where d is the grain size and $k < 1$ expresses the relative resistance of dislocation junctions and grain boundaries.

Figure 29 shows the typical stress-strain curve obtained by torsion test in a metal undergoing dynamic recovery.[15] The torsion test is preferred to the tensile test in this case because early necking in tension would not allow observing the behaviour at large strains. Three stages are observed:

I) the dislocation density increases and the material hardens. The average distance between dislocations is large and annihilations or rearrangements occur with a low probability,

II) the dislocation density is high enough, recovery becomes dominant and the material softens,

III) hardening and recovery compensate themselves dynamically and a steady state is reached.

When the strain rate is increased or when temperature is decreased, the hardening term is favoured and the curve is shifted to high stresses and dilated along the strain axis.

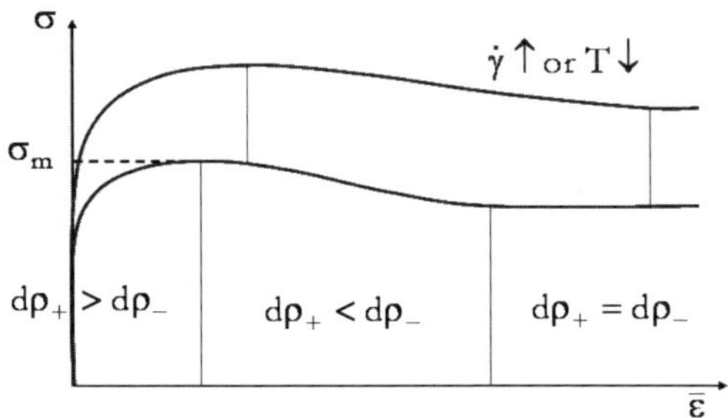

Fig. 29. Scheme of the stress-strain curve on a sample undergoing dynamic recovery during a torsion test at high temperature.

4.4.2. *Dynamic recrystallisation*

Another way to eliminate dislocations created by hardening is to grow new unstrained grains in place of the hardened ones. Consider a plane grain boundary delimiting a hardened grain with a dislocation density ρ and a new grain with a dislocation density close to zero. The free enthalpy stored per unit volume in the first grain is:

$$g = \rho \frac{\mu b^2}{2} \tag{44}$$

which leads to a migration force on the grain boundary per surface unit $P = g$. For $\rho = 10^{16}$ m^{-2}, $\mu = 50$ GPa and $b = 0.2$ nm, $P = 10$ MPa. Under this force, the grain can migrate and consummate the hardened grain. The migration occurs locally by jumps of atoms from the hardened grain to the new grain on a distance of the order of the atomic distance, with a potential barrier to overcome. The migration velocity is proportional to g

and, as it depends on the jump probability of atoms, it increases with temperature.

In the case where a spherical volume of the hardened metal has been replaced by an unstrained grain, the free enthalpy of the crystal decreases in the volume but a new grain boundary is created. The evolution of the free enthalpy is:

$$\Delta G = -\frac{4}{3}\pi r^3 \times \rho \frac{\mu b^2}{2} + 4\pi r^2 \times \Gamma \tag{45}$$

where r is the radius of the sphere and Γ is the energy per surface unit of the grain boundary. The force acting on the boundary to extend the new grain is:

$$F = -\frac{\mathrm{d}\Delta G}{\mathrm{d}r} = 2\pi \rho \mu b^2 r(r - r_c) \qquad \text{where} \qquad r_c = \frac{4\Gamma}{\rho \mu b^2}$$

Below a critical radius r_c, the new grain cannot develop. Typically, for $\Gamma = 500$ mJ/m^2 and the previous data, $r_c = 1$ μm, which is far too large to be reached by thermal fluctuations and a homogeneous nucleation is highly improbable.

Nucleation occurs by mechanisms similar to recovery: a dislocation cell created by rearrangement of dislocations can become a nucleus when it reaches the critical radius. The sub-grains formed by recovery are very stable, their migration is low and the nucleation stage is long. When they grow, they absorb new dislocations; their misorientation and their mobility increase. Once the critical size is reached, the nuclei transform into new grains which growth until the hardened microstructure has disappeared. Both nucleation and growth are thermally activated. Besides, the critical radius increases when the dislocation density stored decreases: below a critical strain hardening, cells cannot reach the nucleation size and recrystallisation does not occur.

As for recovery, when a metal is deformed at high temperature, dynamic recrystallisation can enter in competition with strain hardening. Fig. 30 shows torsion test curves of metals undergoing dynamic recrystallisation.[15,16] The curves are similar to the case of dynamic recovery. The main difference resides in oscillations on the curves when the recrystallisation kinetic is favoured, due to successive recrystallisations of the material. After the initial hardening, recrystallisation occurs and the stress decreases. When the sample is entirely recrystallised, strain hardening increases the stress, until a new recrystallisation takes place. Oscillations disappear when a recrystallisation begins while the previous one has not been complete. At steady state, hardening and recrystallisation dynamically compensate themselves. Again, strain rate and temperature have opposite influences.

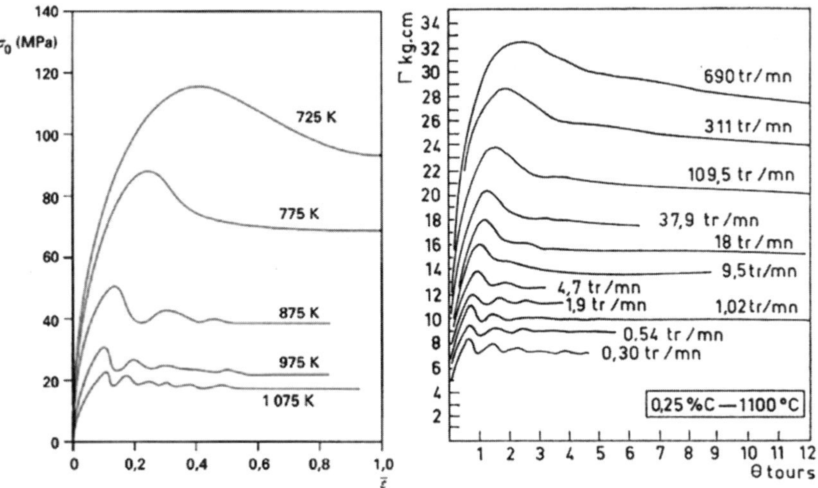

Fig. 30. Dynamic recrystallisation during torsion tests at different temperatures for : left panel) Cu at $\dot{\bar{\varepsilon}} = 2.10^{-3}\,s^{-1}$ and various temperatures[15] and right panel) Fe-0.25C steel at 1100°C and various strain rates.[16]

In metals with high stacking fault energy, cross slip and climb occur with a high probability. The polygonised structures formed by recovery are very stable and the dislocation density remains low. No recrystallisation is observed. This is the case for Al, Zn, Mg, Sn and for α-Fe when $\dot{\varepsilon}\exp(Q/RT) > 5.10^{12}$ s^{-1},[16] where Q is the activation energy of the recovery process. In metals with low stacking fault energy, the recovery stage is less efficient and dynamic recrystallisation is observed in Ni, Cu, Pb, γ-Fe and in α-Fe when $\dot{\varepsilon}\exp(Q/RT) < 5.10^{12}$ s^{-1}.

4.5. *Creep*

Creep is a slow plastic flow along with time even at very low stress. As plasticity is thermally activated, it can occur over large time scales at stress levels well below the yield stress, which has a physical meaning only at the short time scale of a tensile test (typically several minutes). The kinetics of obstacle crossing, recovery and recrystallisation induce plastic strain at a rate increasing with temperature. For industrial applications, creep has to be taken into account in the estimation of the lifetime of a mechanical part typically when the temperature in service is higher than $T_m/2$. Furthermore, at high temperature, self diffusion can contribute to a permanent plastic deformation.

Creep also designates the uniaxial tensile test performed at constant applied load used to characterise such behaviour. A constant stress or force is applied and the tensile deformation is measured along with time t, as shown on a typical creep curve in Fig. 31. At intermediate and high temperatures, three stages are observed on the curves:

- stage I or primary creep: the strain rate decreases due to hardening,

- stage II or secondary or stationary creep: the strain rate remains constant, due to competing hardening and softening mechanisms,

- stage III or tertiary creep: the strain rate increases due to ductile damage by cavitation leading to final fracture. In this process, the stress triaxiality is of great importance. The main part of the curve corresponds to stage II. The competing hardening and softening mechanisms depend on the applied stress and temperature.

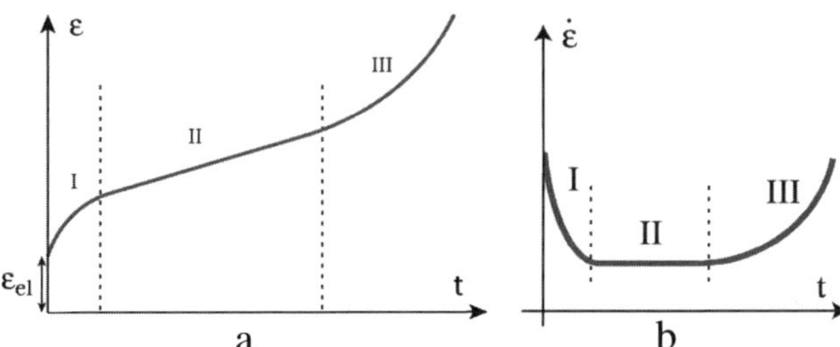

Fig. 31. a) Scheme of the strain-time curve of a metallic alloy during creep test; b) Corresponding strain-rate dependence with time. Three stages are observed.

The main part of the curve corresponds to stage II. The competing hardening and softening mechanisms depend on the applied stress and temperature. The general strain rate dependence can be represented on maps, as illustrated in the case of nickel in Fig. 32.[17]

A) Power-law creep

The main behaviour at intermediate and high temperature and stress. Hardening by dislocation multiplication is compensated by recovery. Recovery is controlled by dislocation climb and strain rate depends on the diffusion coefficient of vacancies D. In stage II:

$$\dot{\varepsilon} = \alpha \frac{D\mu b}{k_B T} \left(\frac{\sigma}{\mu} \right)^n \tag{46}$$

with α a material's constant. The exponent n in the power law varies from 3 to 5. At high temperature (H.T. creep), vacancies diffuse in volume (bulk diffusion). At intermediate temperature (L.T. creep), the kinetic of bulk diffusion is too low and a significant diffusion of vacancies can only be obtained by shortcut in the dislocations core (pipe diffusion); in this case n must be replaced by $n+2$.

At high temperature and stress recovery is replaced by dynamic recrystallisation.

B) Diffusional flow

At low stress, intermediate and high temperature. The dislocations kinetics is very slow and strain is mainly achieved by direct diffusion of matter. Vacancies diffuse from the grains towards the grain boundaries. At high temperature, they diffuse in the volume (Nabarro-Herring mechanism) and the strain rate is:

$$\dot{\varepsilon} = \alpha \frac{D\sigma b^3}{d^2 k_B T} \tag{47}$$

where d is the mean grain size and $\alpha \approx 10$ for spherical grains is a constant. At intermediate temperature, vacancies diffuse in the grain boundaries (Cobble mechanism) and the strain rate is:

$$\dot{\varepsilon} = \alpha \frac{D_{GB}\sigma b^3 \delta}{d^3 k_B T} \tag{48}$$

where D_{GB} is the vacancy diffusion coefficient in the grain boundaries of thickness δ and $\alpha \approx 150$.

C) Logarithmic creep

At low temperature, intermediate and high stress. Plasticity is due to dislocation glide with low recovery rate. No stationary stage is observed and the strain is of the form:

$$\varepsilon = \alpha \ln\left(1 + \beta t\right) \tag{49}$$

with α and β material, temperature and stress dependent constants.

At high stresses, close or slightly above the yield stress, the material deforms plastically at a strain following Eq. 29.

In most mechanisms, the creep rate increases when the grain size decreases. It can be of great interest to use single crystals for applications at high temperature. For instance, monocrystalline nickel base superalloys are used in turbine blades for aircraft reactors. But as a nickel-aluminium solid solution would have a poor mechanical resistance at high temperature it is reinforced by precipitates of Ni_3Al intermetallic phase. This phase has the particularity of showing a yield stress anomaly[18]: its yield stress increases with temperature below 800°C to reach about 700 MPa at 800°C. This anomaly is due to a complex structure of dislocations core in the $L1_2$ structure (fcc ordered lattice), composed of two dissociated a/2<110> dislocations separated by an antiphase boundary. At low temperature, such superdislocations are planar and glissile and become non-planar by cross slip and sessile when the temperature increases. Up to 70% volume fraction of Ni_3Al is precipitated as cuboids in the Ni solid solution matrix for the best mechanical and creep resistance.

                                    J. P. Chateau

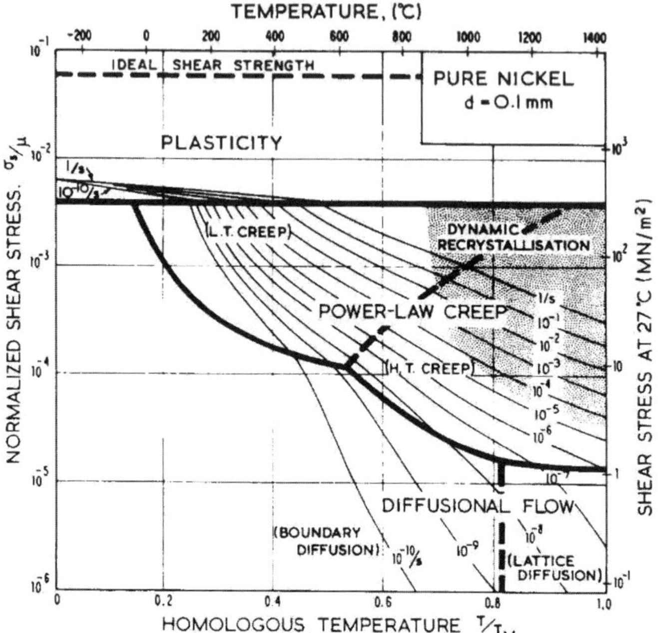

Fig. 32. Deformation mechanism map as a function of applied stress and temperature in the case of Ni. The horizontal plain line is the yield stress during tensile test.

5. Conclusion

Below a critical stress level, metals behave like linear elastic materials. Above the elastic limit or yield stress, plastic deformation takes place by glide along crystallographic planes, performed by the displacement of linear defects called dislocations. The crystallographic nature of the glide planes and directions depend on the atomic structure and on the crystal orientation with regards to the applied stress. The lattice friction, the presence of solute atoms and the strain hardening due to interactions between dislocations contribute to the total yield stress.

The mechanical properties are characterised by tensile testing: yield stress, tensile strength, elongation to fracture, strain hardening. The latter is of great importance in the onset of plastic instability or necking. When plastic deformation occurs inside the grains, the macroscopic yield stress inversely scales like the square root of the grain size. Below a critical value of the grain size, intergranular glide becomes predominant and leads to strongly strain rate dependent behaviour. Twinning and martensitic transformations are alternative intragranular plasticity mechanisms and lead to peculiar properties (high hardening rate, shape memory effect).

The crossing of localised obstacles by dislocations is thermally activated while long range obstacles induce an athermal internal stress. The effective stress decreases when the temperature increases or when the strain rate decreases. High strain rate sensitivity leads to superplastic behaviour, while negative strain rate sensitivity induces plastic instabilities characterised by serrations on the tensile curves. The internal stress in a deformed polycrystal can be reduced by reorganisation of the dislocations inside the grains (dynamic recovery) or by growth of new unstrained grains (dynamic recrystallisation). Both lead to a reduced flow stress at high temperature. Even when loaded below their yield stress, metallic alloys undergo large plastic strains over long time scales by creep mechanisms. The creep rate depends on the dominant mechanism which varies with the applied stress and temperature.

References

1. S. Allain, Ph.D. Thesis, INPL, Nancy, 2004.
2. H. Neuhauser and Ch. Schwink, in "Material Science and Technology: A comprehensive Treatment", Eds. R.W. Cahn, P. Haasen and E.J. Kramer, Vol.6 (VCH Weinheim, New York, 1993) p. 191
3. K. Lücke and H. Lange, *Z. Metalk.*, **55** (1952).
4. P. Haasen, *Phil. Mag.*, **3** 384 (1958).
5. F. D. Rosi, *Trans. AIME*, **200** 1009 (1954).
6. U. F. Kocks, *Acta. Met.*, **6** 85 (1958).
7. S. Allain, J. P. Château and O. Bouaziz, *Mat. Sci. Eng.* **A, 387-389** 143 (2004); S. Allain, J. P. Château, O. Bouaziz, S. Migot and N. Guelton, *Mat. Sci. Eng.* **A, 387-389** 158 (2004) ; S. Allain, J. P. Chateau, D. Dahmoun and O. Bouaziz, *Mat. Sci. Eng.* **A, 387-389** 272 (2004).
8. T. Kruml, O. Coddet and J. L. Martin, *Mat. Sci. Eng.* **A, 387-389** 72 (2004).
9. J. J. Cox, R. F. Mehl and G. T. Horne, *Trans. ASM,* **49** 123 (1957).

10. J. H. Hollomon and L. D. Jaffe, in *"Ferrous metallurgical design"* (John Wiley and Sons, 1947) p. 112.
11. J. P. Chateau, D. Delafosse, T. Magnin, *Acta Mater.*, **50** 1507 (2002).
12. Y. Bergström and W. Roberts, *Acta Met.*, **19** 815 (1971).
13. S. Allain, P. Cugy, C. Scott, J. P. Chateau, A. Rusinek and A. Deschamps, *Int. J. Mat. Res.* **99** 734 (2008).
14. T.E. Mitchell, *Prog. Applied Materials Research*, **6** 117 (1964).
15. F. Montheillet, *Métallurgie en mise en forme*, Techniques de l'ingénieur, ref M600.
16. C. Rossard and P. Blain , *Rev. Met.*, **55** 573 (1958).
17. H. J. Frost and M. F. Ashby, *Deformation-Mechanism Maps*, (Pergamon Press, 1982).
18. S. M. Copley and B. H. Kear, *Trans. Met. Soc. AIME*, **23** 977 (1967).

General books

Engineering materials, M.F. Ashby, D.R.H. Jones, Int. series on Materials Science and Technology, vol 34, Pergamon press (1980).

Dislocations et plasticité des cristaux, J-L. Martin, Presses Polytechniques et Universitaires Romandes (2000).

Déformation plastique des métaux et alliages, G. Champier et G. Saada, Masson et Cie éd. Paris (1968).

Mechanical metallurgy, Metallurgy and metallurgical engineering series, G.E. Dieter, Jr., McGraw-Hill (1961).

Introduction to the mechanics of solids, M. A. Eisenberg, Addison – Wesley (1980).

Métallurgie générale, J. Bénard, A. Michel, J. Philibert et J. Talbot, 2nd Ed., Masson, Paris (1984).

The inhomogeneity of plastic deformation, Seminar of the Am. Soc. for Metals, October 1971, ASM (1973).

Plasticité à haute température des solides cristallins, J-P. Poirier, Editions Eyrolles, Paris (1976).

Corrosion sous contrainte, ed. by D. Desjardins et R. Oltra, Bombannes 1990, les Editions de Physique (1992).

CHAPTER 3

MICROSTRUCTURE - PROPERTIES RELATIONSHIPS IN METAL-BASED ALLOYS

Alexis Deschamps

SIMAP, Grenoble INP, Université de Grenoble
BP 75, 38 402 Saint Martin d'Hères Cedex, France
Email : alexis.deschamps@grenoble-inp.fr

This course aims at summarising the relationship that exists between the microstructure of a metallic alloy and its properties related to plastic strain. In a first part, a few techniques that can give access quantitatively to the parameters of the microstructure (phase proportion & size, dislocation density, etc.) will be presented. A second part will be devoted to the relationship between microstructure and yield strength. Next, the material's behaviour during plastic strain will be presented. The final part will briefly introduce the relationship between microstructure and fracture.

1. Introduction

Metal-based alloys are designed to response to given properties, by the way of a controlled microstructure. Of course, the range of properties that metallic materials have to withstand is extremely vast, ranging from standard mechanical properties (such as strength) to chemical properties (such as corrosion resistance), electrical (e.g. resistivity), optical, ... The restriction of this chapter will be to consider only mechanical properties, and in this topic only those concerned with monotonic loading (this excluding fatigue properties) and at relatively low homologous temperature (defined as T/T_m, where T is the absolute temperature and T_m the melting point, lower than 0.5).

73

These mechanical properties: the ability of a material to resist some level of stress without plasticity (strength), its ability to show increasing flow stress during plastic strain (strain hardening), its ability to resist the presence of defects and fracture (ductility, toughness), are vastly different, depending on the arrangement of the different atomic species of the alloy, so-called the alloy microstructure. A microstructure will be defined by several important parameters:

- The alloy composition; the mechanical properties of pure metals are almost always very low, and useful alloys may contain 2, 3 or more species.

- The number and nature of crystalline phases; their respective volume fractions, which can be found at equilibrium on a phase diagram, but which are generally out-of-equilibrium.

- The chemical composition of the phases that may be determined by their crystal structure (stoechiometric phases) or may show some degree of freedom (substitution of chemical species on some of the crystal sites).

- The scale and arrangement of the phases: size, morphology, percolation.

- The presence of defects in the different phases, and especially crystal defects such as dislocations.

All of these features of the microstructure will have strong consequences on the mechanical properties. In this chapter, we aim at presenting some laws that relate the mechanical properties to the main parameters of the microstructure. Since most of these laws relate the properties to some quantitative parameters of the microstructure (such as precipitate size, density of defects, etc.), we will first present some experimental techniques that are able to characterize these parameters in a quantitative way. In a second section we will investigate the relationship between microstructure and strength. In a third section we will investigate the influence of microstructure on strain hardening, and in the fourth section the effect of microstructure on fracture will be shortly discussed.

2. Quantitative Characterisation of Microstructures

As written above, the main parameters needed for the characterization of microstructures are the proportion of phases, the presence of crystalline defects and the scale of phases. We will concentrate on these three topics in this section.

2.1. *Characterization of phase proportions*

Most metallic materials are multi-phased, that is are made of several types of crystalline phases. For instance Ni-based superalloys are made of an FCC phase (Ni) and an $L1_2$ ordered structure (Ni_3Al), both structures showing extensive substitution of the other atomic species present in the material. Steels are made of several phases, that may be austenite (FCC Fe), ferrite (BCC Fe), martensite (BCT Fe-C), cementite (Fe_3C) or many other types of carbides (M_6C, $M_{23}C_6$, ... M being a combination of metallic species present in the Fe-based alloy). Aluminium alloys are made of FCC Al, and many other phases depending on the alloy considered (Al_2Cu, $MgZn_2$, Mg_2Si, Al_3Zr, etc.).

A number of experimental techniques are available to measure quantitatively the relative proportion of these phases. First, microscopy is of course a useful tool. Depending on the scale of the microstructure, optical microscopy may be used (scale > 1 µm), or Scanning Electron Microscopy (scale roughly above 10 to 50 nm), or Transmission Electron Microscopy (no limits of scale). However these techniques are faced with difficulties for the determination of the phases, and it is always difficult and tedious to convert the images in numbers (volume fraction of phases for instance). Moreover, these techniques are in essence local techniques, namely offer a view of the microstructure that is limited in space.

Similarly, indirect techniques are widely used for monitoring the phase proportions of alloys. Dilatometry exploits the fact that when a material changes its crystal structure (e.g. Fe transforming from austenite to ferrite), it changes volume. Differential Scanning Calorimetry exploits the heat release associated with phase transformations from one crystal structure to another (latent heat associated with changes in entropy, or changes in heat capacity for second order phase transformations).

One method of choice for characterizing the phase proportions is powder X-ray diffraction. It consists in illuminating a sample with an X-ray beam and recording the diffraction images by a detector. It can be both carried out in reflexion or in transmission, although the latter is preferable since it is then easier to extract quantitative data from it. If the material's texture is sufficiently homogeneous, and if a sufficient number of grains are probed by the X-ray beam, then the diffraction spots appear as diffraction rings (so-called Debye-Scherrer rings), at an angle characteristic of the reticular planes of the crystal. The analysis of the position and relative intensity of the different rings can give access to the crystallographic nature of phases and to their proportions, using for instance the classical Rietveld analysis of the powder diagram.[a,1] Figure 1 gives an example of such rings and the extracted powder diagram in the case of a Fe-Mn-C alloy showing some martensitic phase after deformation at low temperature.

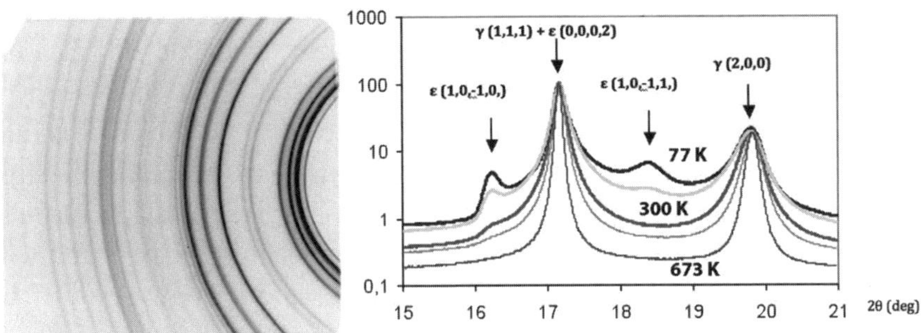

Fig. 1. Part of Debye-Scherrer rings recorded in transmission with a CCD camera on a deformed Fe-Mn-C steel, and associated part of powder diffraction pattern, as a function of deformation temperature. At low deformation temperature, the presence of the ε-phase (hexagonal martensite) can be detected and quantified, whereas at higher temperatures only the γ-phase (austenitic steel, FCC) is present.[2]

---

[a] *Software Fullprof.2k : J. Rodriguez-Carvajal.*
http://www.llb.cea.fr/fullweb/fp2k/fp2k.html

Such experiments need to be carried out with suitable conditions of X-ray wavelength, beam size and measurement resolution and angles. The use of synchrotron sources make it possible to tune precisely the X-ray wavelength, to use high energy X-rays (in the range 20-100 keV), and thus to use relatively thick samples in transmission and obtain for a given angle a large number of diffraction peaks. It makes it also possible to increase the measurement resolution, both in reciprocal space, and in time for time-resolved experiments (e.g. in situ heating or cooling). Figure 2 shows the evolution of austenite and ferrite during heating of a steel, measured in-situ by high energy X-ray diffraction on a synchrotron source.

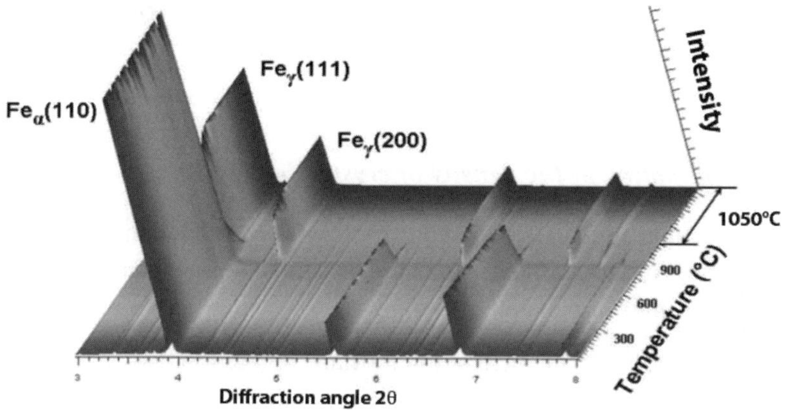

Fig. 2. Series of powder diffraction diagrams recorded during cooling of a steel from 1050°C to room temperature. It is possible to visualise and quantify the austenite (γ) to ferrite (α) transition around 900°C. In addition, smaller peaks (visible as faint lines) provide information on the formation of various precipitates during this heat treatment.[3] (Courtesy of A. Benneteau and E. Gautier, Ecole des Mines de Nancy, France).

It is however important to be aware of the limitations of this technique. In fact, several problems can arise when translating the diffraction peak heights into volume fractions of phases. First, diffraction peaks may overlap in materials where crystal structures are close or where a large number of phases exist. The measurement of volume fractions will then be made difficult, because it will depend somewhat on the deconvolution procedure used. As a general rule, a successful measurement can be made with highly convoluted peaks only if the shape of the peaks in reciprocal space is well known a priori. Second, if the material is textured (which means that the crystal grain orientations are not randomly distributed), some diffraction peaks may have a lower or higher intensity, depending on the sample orientation. Part of this problem can be overcome by sample rotation in order to average the diffraction intensity. Lastly, the intensity of diffraction peaks depends not only on the volume fraction of the phases, but also on their chemical composition, that is on the scattering factors of the different chemical species on the sites of the crystal lattice. Thus, if the crystal phase studied is non stoechiometric and the composition is not known, some uncertainty in the determination of the respective fraction of the phases will remain.

2.2. *Determination of the density of crystal defects*

The mechanical properties of metal based materials depend crucially on the presence of crystal defects. Crystal defects are usually classified according to their dimension:

- 0: point defects; vacancies, interstitials, solute atoms
- 1: linear defects; dislocations
- 2: surface defects; stacking faults, grain boundaries
- 3: volume defects; voids, precipitates

We will here concentrate on the crystal defects responsible for plastic deformation at low temperature, namely dislocations & stacking faults (including twins). The first information generally needed is the density of these crystalline defects. For dislocations the density ρ is defined as the length of linear defect per unit volume and thus is numbered in m^{-2}. For planar defects such as stacking faults it is generally defined as the probability of a given crystal plane to be a fault and is thus without unit. Alternatively it can be defined as the average distance between two planar defects. Some more refined defect properties can also be sought.

For instance, one may be interested in the nature of the dislocations (respective proportions of edges and screws), in their arrangements (random or in cells), in the nature of the stacking faults (respective proportions of intrinsic, extrinsic faults and twin boundaries).

Crystalline defects are generally observed directly in the transmission electron microscope due to their diffraction contrast. However, determining quantitatively their density is generally extremely difficult and tedious. Few other techniques are available. In heavily deformed materials, dislocation densities can be evaluated by a precise measurement of the material's density due to the volume change that they induce in the material, provided that no other microstructure phenomena (such as phase transformations) induce density changes of larger magnitude.

Again, X-ray diffraction can be a useful tool for studying the density of crystal defects. Crystal defects induce perturbations of diffraction conditions as compared to a perfect crystal, and thus change the diffraction pattern.[4] Their presence increase the width of the diffraction peaks (peak broadening), change their shape (Gaussian or Lorentzian character, asymptotic behaviour), their symmetry, and induce peak displacement. Moreover, all these effects are peak-dependent and thus the study of several diffraction peaks of a given phase can lead to an evaluation of many parameters of the crystal defect microstructure.

The analysis of the crystal defects[5] can be done alternatively in the reciprocal space (that is, directly on the diffracted peaks) or in direct space (after Fourier transform of the diffracted peaks). In the literature, integrated models (so-called full pattern modelling techniques) exist that extract from a powder diagram the density and distributions of the crystalline defects. More simply, one can give some general rules for relatively simple cases. The peak broadening has several components. One is due to the crystal size: the smaller the crystals, the larger the induced broadening. The magnitude of this effect is peak-independent. The second component is due to lattice distortions due to the presence of dislocations. This one is proportional to the Bragg angle of each diffraction peak, and thus can be separated from the first one. Its magnitude depends on the contrast factor of the dislocations, which depends on the crystal structure investigated and on the nature of the dislocations (e.g. screw / edge proportion). Classical models exist also on the Fourier transforms of the peaks to extract characteristics of the crystal defects, such as the Wilkens model or the Williamson-Hall model. In some cases the situation is complicated by the fact that many defects are present at the same time. For instance in Fe-Mn-C steels where the stacking fault energy is low, plastic deformation occurs by the planar glide of dislocations, and by twinning. The final microstructure consists of dislocation pile-ups, of twin planes and stacking faults. Moreover, dislocation pile-ups show stress fields that are fundamentally different from that of randomly distributed dislocations, and result in the presence of large internal stresses, that influence the peak shape and broadening in a specific manner. A full analysis of this problem has lead to the quantitative determination of the main parameters of the microstructure, as shown in Fig. 3. Details of the procedure can be found in Ref. 2.

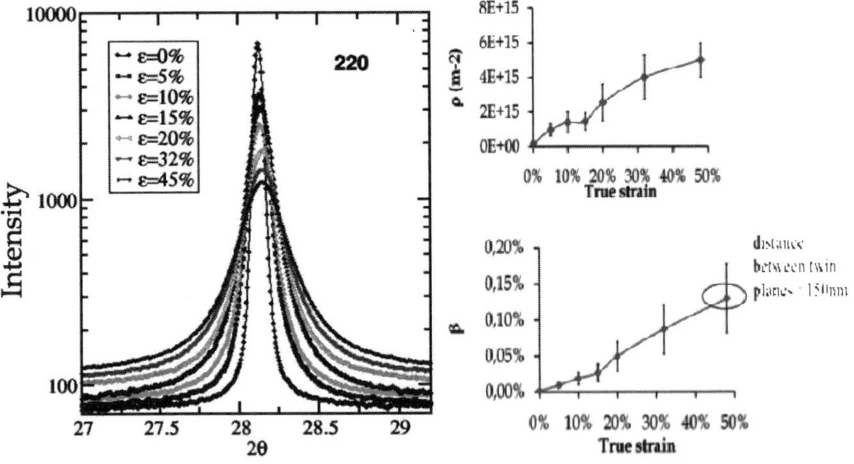

Fig. 3. Evolution of the shape of a diffraction peak (here the (220) peak of austenite) with strain in an Fe-Mn-C steel as a function of deformation. It is apparent that the peak broadens, becomes asymmetrical, and shows some displacement. A full analysis of many diffraction peaks provides the density of dislocations (ρ) and the probability of presence of stacking faults (β) as a function of strain (from reference 2).

2.3. *Characterisation of precipitates*

Among hardening mechanisms, precipitation at the nanoscale is one of the most widely used and most efficient. It is a particular case of a material constituted of several phases, where the second phase is distributed as dense, small objects, which intersect the glide planes of dislocations. In order to understand the mechanical properties, the main information that needs to be obtained is morphological, that is the average size of the precipitates (and the dispersion of this size around the mean value), their morphology, their volume fraction, and number density. Many other parameters may be also valuable: their chemical composition, the crystallographic relationship between the matrix and precipitates (orientation relationship, habit plane), the elastic misfit between matrix and precipitates, etc.

Many experimental techniques are available to access this information. Among direct techniques that image the precipitates, the two techniques of choice are transmission electron microscopy (morphology, crystal structure, spatial distribution, relation with crystal defects, etc.) and Atom Probe Tomography (chemical composition, morphology). However, accessing average, quantitative data on precipitate size and number density with these techniques is again very difficult and tedious, since the analysed volumes are extremely small.

Average, indirect techniques are also widely used for obtaining information on precipitates; one can cite resistivity, differential scanning calorimetry, thermo-electric power. However these techniques are generally restricted to an estimation of the volume fraction of the precipitates and do not provide a measurement of the precipitate scale.

The most versatile technique to obtain rapidly statistical information on precipitate size and volume fraction is Small Angle Scattering (SAS), which can be alternatively carried out using X-rays (SAXS) or neutrons (SANS).[6] A small angle scattering experiment consists in illuminating a sample in transmission with a collimated beam, which is then scattered (at small angles) by the objects present in the material. A detector is placed behind the sample and records the scattering spectrum. The scattering spectrum depends on the size distribution of the objects (smaller objects scatter at larger angles), on the contrast between the objects and the matrix (electronic density contrast for X-rays, nuclear or magnetic scattering lengths for neutrons) and on their respective volume fractions. The scattering functions are known exactly for simple objects such as spheres or ellipsoids, and thus one can fit models of precipitate size distributions on experimental data when such objects are known to be present. Complications arise however when the concentration of precipitates becomes such as the distance between precipitates is of the same order as their size. Contrast variation can be used for obtaining information on the chemical composition of precipitates. For X-rays, it consists in recording the scattered intensity for different wavelengths close to an absorption edge (such as K_α) of the element of interest. The scattering efficiency, and thus the contrast with the matrix, of this element will change with the wavelength, and thus the presence of this element in the precipitates can be proven if the recorded intensity is wavelength-dependent.

Figure 4 gives an example of the analysis of a scattering curve in order to obtain precipitate characteristics in a Fe-Cu alloy containing Cu spherical precipitates.

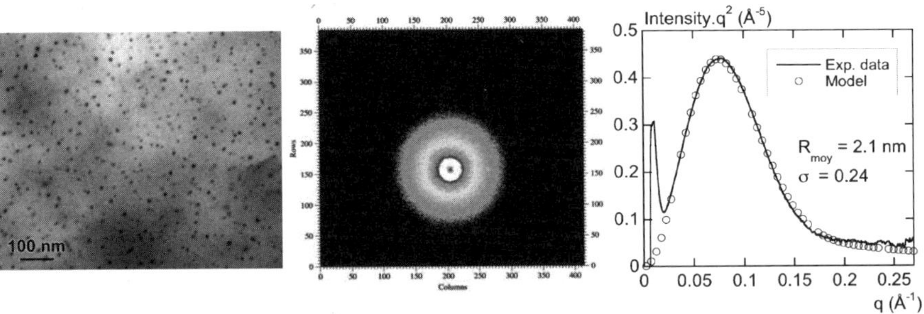

Fig. 4. a) Transmission Electron Micrograph of Cu precipitates in Fe ; b) corresponding Small Angle X-ray Scattering Image ; c) Krattky plot (I.q² vs q, where I is the scattered intensity and q is the scattering vector), along with a model of scattering by a log-normal distribution of spheres of average size R_m and distribution width σ.

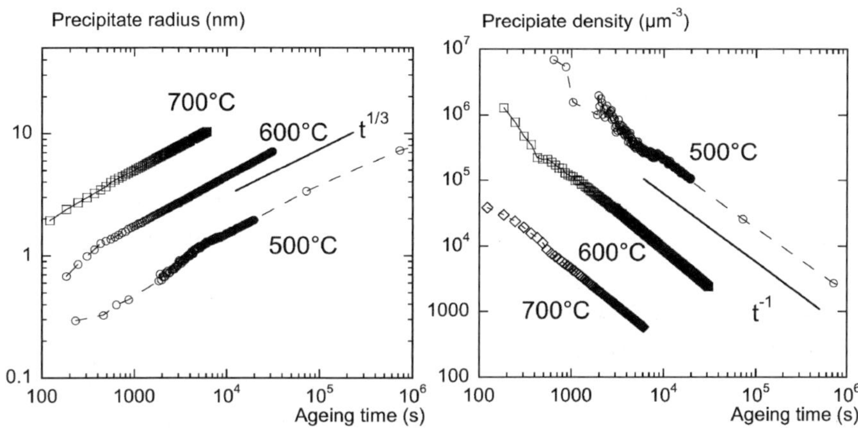

Fig. 5. Evolution of precipitate radius and precipitate density in a Fe-Cu alloy measured in-situ at various temperatures by Small-Angle Scattering, showing that the LSW theory of coarsening describes well the experimental behaviour.

Figure 5 shows how in-situ heating of the sample, while measuring the scattering behaviour, enables to obtain precipitation kinetics, namely here the evolution with time at three temperatures of the precipitate size and number density.

3. Relationship between Microstructure and Yield Strength

Now we will assume that the main parameters of the microstructure are accessible, and we will discuss the main scaling laws that relate these microstructure characteristics to the yield strength of the material. We will discuss the scaling laws depending on the dimension and morphology of the obstacles that the microstructure offers to dislocation glide. The material's yield strength will be written as:

$$\sigma_y = \sigma_o + \Delta\sigma \tag{1}$$

where σ_o is the Peierls (or lattice friction) stress, that describes the stress necessary to move a dislocation in the perfect matrix crystal, and $\Delta\sigma$ is the contribution of the microstructure components that will be reviewed below.

3.1. *Grain boundaries and multi-phase materials*

The yield strength is first influenced by the presence of impenetrable components of the microstructure. These can be the grain boundaries of polycrystals, or the phase boundaries of multi-phase materials. Dislocations cannot cross these boundaries, and thus the extension of plastic deformation needs the activation of plastic flow in each phase or each grain. When the scale of the microstructure decreases, it becomes more difficult to activate plasticity in all the grains or phases and the yield strength increases. This increase is generally well described by the Hall-Petch law[7, 8]:

$$\Delta\sigma = K_{HP}d^{-1/2} \tag{2}$$

where K_{HP} is a material dependent constant and d is the characteristic size of the microstructure (grain size, interlamellar distance, …). Actually this law is extremely well followed in a very wide range of situations, but is not yet very well understood on theoretical basis. Three main theories exist to explain this hardening. The first describes the hardening effect as the stress necessary to activate dislocations sources in neighbouring grains, due to dislocation pile-ups forming close to the grain boundaries.[9] One of the main shortcomings of this most classical interpretation of the Hall Petch law is that it s not applicable to the materials where dislocation glide is not planar (for which the Hall Petch relationship still holds). The second theory relies on the geometrical incompatibilities between neighbouring grains during plastic deformation.[10] These incompatibilities need to be accommodated by polarized dislocations that are called geometrically necessary dislocations. The density of such dislocations is modelled to be inversely proportional to the grain size, which causes the yield stress to increase according to the Hall Petch law. The third theory considers that the dislocations need to be emitted by the grain boundaries and that the Hal Petch law derives from an increasing difficulty of such emission when the grain size decreases.[11]

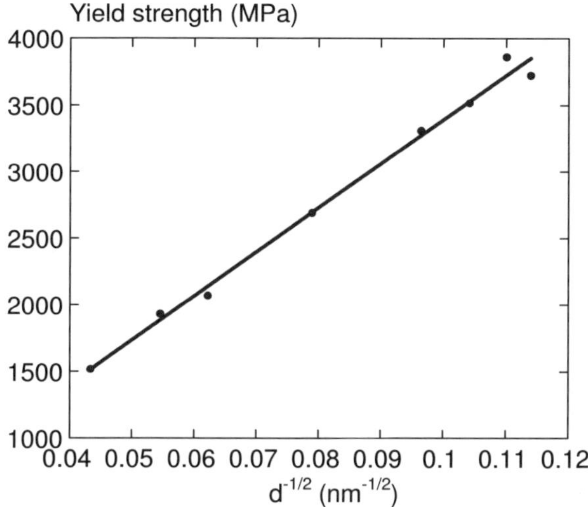

Fig. 6. Evolution of wire strength of wire drawn pearlite with $d^{-1/2}$, d being the inter-lamellae spacing, showing that the Hall-Petch law describes precisely the influence of microstructure scale on the strength.

Figure 6 shows a plot of the strength of extruded pearlite (Fe/Fe$_3$C composite) as a function of the interlamellar spacing. The strengthening effect of the phase boundaries is extremely large, leading to strength of 3.5 GPa, and follows well the Hall Petch law.

Figure 7 shows the effect of interphase spacing on the strength of multilayer metals, as measured by their hardness (nano-indentation measurement). Below an interlamellar spacing of about 20 nm, the Hall Petch law describes very well the evolution of strength. However, for very small scales, this relationship is not valid anymore, and the strength saturates. The same behaviour is found on metals of nanocrystalline structure, and is attributed to changes in the plasticity mechanisms, like the activation of deformation at the grain boundaries.

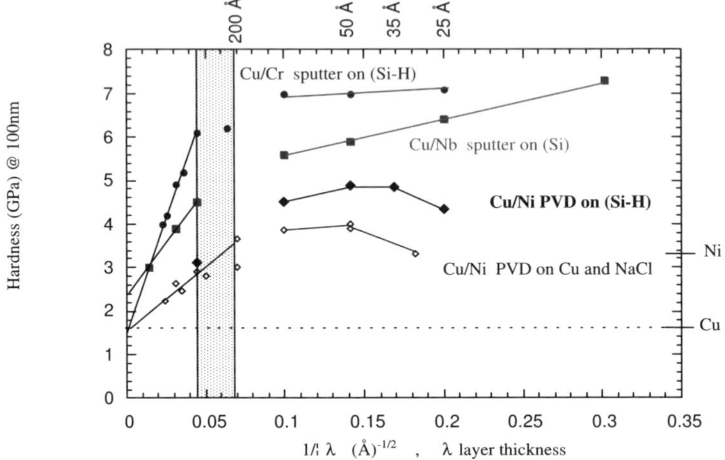

Fig. 7. Measurement (by nanoindentation) of the hardness of multilayers made of different pairs of materials, as a function of $\lambda^{-1/2}$ where λ is the layer thickness. Above a thickness of 20 nm (left of the graph) The Hall Petch law is shown to be valid, whereas hardness saturates for layer thickness below this value (courtesy of M. Verdier, SIMAP laboratory, Grenoble, France).

3.2. *Localized obstacles*

In this section we will discuss the effect on yield strength of the presence of obstacles to the dislocation glide that can be considered as almost point-like in terms of their interaction with dislocations. Figure 8 shows schematically the sequence of events that leads to the overcoming of such obstacles by a dislocation submitted to a shear stress in the glide plane, that provokes its bowing between the obstacles. The action of a stress tensor applied on the material on a small segment dL of the dislocation line (of Burgers vector b) is the so-called Peach-Koehler force, always normal to the dislocation line:

$$\overrightarrow{dF}_{PK} = \left(\overline{\overline{\sigma}} \cdot \vec{b} \right) \times \overrightarrow{dL} \; here \; dF_{PK} = \tau b \left(dL \right) \tag{3}$$

and the dislocation line tension pulling on the obstacles (always parallel to the dislocation line):

$$\Gamma \cong \frac{1}{2}\mu b^2 \qquad (4)$$

where μ is the material's shear modulus. If the obstacles are impenetrable by the dislocation (e.g. incoherent precipitate), the dislocation bows between them until configuration (d) where the bowing segments annihilate each other and leave a dislocation loop around the unsheared obstacle.

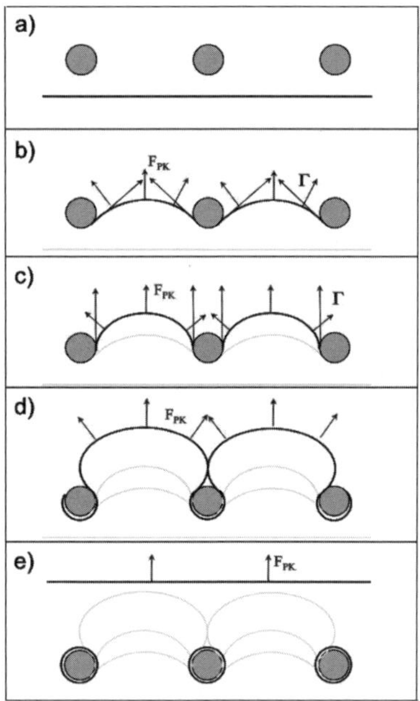

Fig. 8. a) to e) mechanism of overcoming of an impenetrable obstacle by a dislocation. The line tension Γ is the vector parallel to the dislocation line, and the Peach Koehler force is the vector normal to the dislocation line, which results from the effect of the applied stress. In a) the dislocation is under zero stress and is thus straight. From b) to c) the stress increases and so the radius of curvature of the dislocation decreases, inversely proportional to the applied stress. In d) the two segments of the dislocation on each part of the obstacle attract each other and annihilate mutually, leaving in e) a loop around the obstacle and the dislocation going on towards new obstacles. When obstacles can be sheared by the dislocation the process stops in b), and the mechanical balance defines the obstacle strength F_c as a function of the line tension and the bowing angle of the dislocation.

                    A. Deschamps

On the other hand, if the obstacle can be sheared after it has been submitted to a critical force F_c by the line tension of the dislocation, the obstacle is sheared in a configuration like (b). Then, it is possible to trace a scheme of the critical configuration at the stress necessary for obstacle shearing, as shown in Fig. 9.

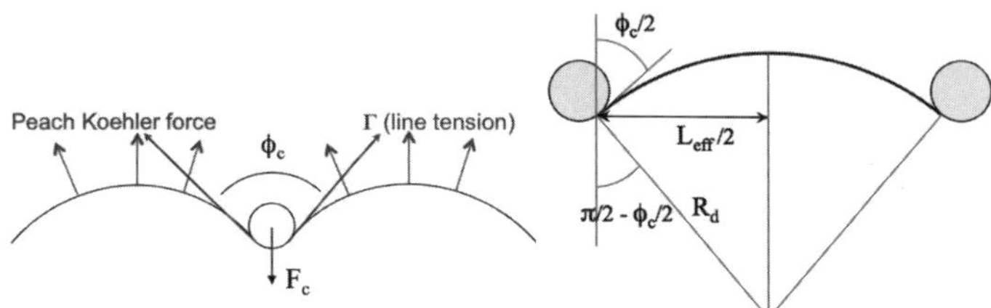

Fig. 9. (left) Mechanical equilibrium between the Peach Koehler Force acting on a dislocation and the obstacle strength, defining the critical breaking angle ϕ_c. (right) Geometrical relationship between the dislocation radius of curvature R_d, breaking angle and effective distance between obstacles along the dislocation line L_{eff}.

The balance between the line tension and the critical force F_c, characteristic of the obstacle strength, dictates that:

$$F_c = 2\Gamma\cos\left(\frac{\phi_c}{2}\right) = \mu b^2 \cos\left(\frac{\phi_c}{2}\right) \qquad (5)$$

where ϕ_c (comprised between 0 and 180°) is the critical breaking angle of the dislocation in its interaction with the obstacle.

Depending on the value of F_c, obstacles can be classified in weak or strong character. As a rule of thumb, weak obstacles have a breaking angle larger than 100°, and strong obstacles lower than that. Depending on the character of the obstacles, the propagation of dislocations through a forest of obstacles may show different characteristics. Weak obstacles tend to favour a homogeneous movement of the dislocations through all obstacles, whereas for strong obstacles the movement of dislocations is more localised in the regions that show a lower obstacle density.

The balance between the dislocation line tension and the Peach-Koehler force imposes the radius of curvature of the dislocation R_d as follows:

$$R_d = \frac{\Gamma}{b\tau} \cong \frac{\mu b}{2\tau} \tag{6}$$

Therefore, it is possible to calculate the shear stress necessary to overcome the obstacles from simple geometrical considerations as shown in Fig. 9, right panel:

$$\frac{L_{eff}}{2R_d} = \sin\left(\frac{\pi}{2} - \frac{\phi_c}{2}\right) = \cos\left(\frac{\phi_c}{2}\right) = \frac{F_c}{2\Gamma} \tag{7}$$

where L_{eff} is the average *effective* distance between two obstacles along the dislocation line. This equation yields:

$$\frac{L_{eff}\,\tau}{\mu b} = \frac{F_c}{\mu b^2} \tag{8}$$

and finally:

$$\tau = \frac{F_c}{b L_{eff}} \tag{9}$$

The calculation of this distance L_{eff} is a rather complicated task, since it depends on the details of the flexibility of the dislocation, as seen in Fig. 10. A perfectly rigid, straight dislocation will only see a low number of obstacles along its line, whereas a very flexible one will need to overcome a much larger number of obstacles. We have seen on the other hand that the radius of curvature of the dislocation depends on the applied shear stress.

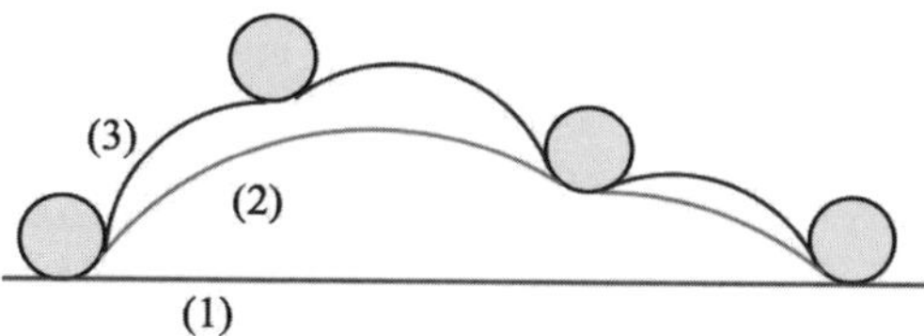

Fig. 10. Influence of dislocation flexibility on the distance between obstacles along the dislocation line. (1) straight, rigid dislocation, (2) dislocation of low flexibility, (3) dislocation of large flexibility (and low radius of curvature). A higher flexibility of the dislocation results in a higher density of obstacles along the dislocation line, and thus an increased hardening.

A convenient distance to consider is the average distance between obstacles in the glide plane L, which is related to the density of obstacles in the glide plane N by $N=L^{-2}$. For a given obstacle strength, L_{eff} must be related to L and if we write

$$L_{eff} = \alpha_{eff} L \qquad (10)$$

we obtain the general equation for strengthening:

$$\Delta \tau_c = \frac{\mu b}{\alpha_{eff}} \cos\left(\frac{\phi_c}{2}\right) \sqrt{N} \qquad (11)$$

Then the increase in yield strength related to this increase in resolved shear stress is calculated by taking into account the polycrystalline nature of the material through the Taylor factor M, that describes the inverse of the average Schimd factor necessary for generalised plasticity. This factor is about 3 for non textured materials, and can become closer to 2 in highly textured materials where crystals are oriented favourably for plastic deformation.

$$\Delta \sigma = \frac{M \mu b}{\alpha_{eff}} \cos\left(\frac{\phi_c}{2}\right) \sqrt{N} \qquad (12)$$

Now we will make an estimate of the effective distance L_{eff} for two limiting cases: small obstacle strength (weak obstacles) and impenetrable obstacles.

Let us first concentrate on weak obstacles. A sequence of two successive dislocation positions during obstacle overcoming is shown in Fig. 11.

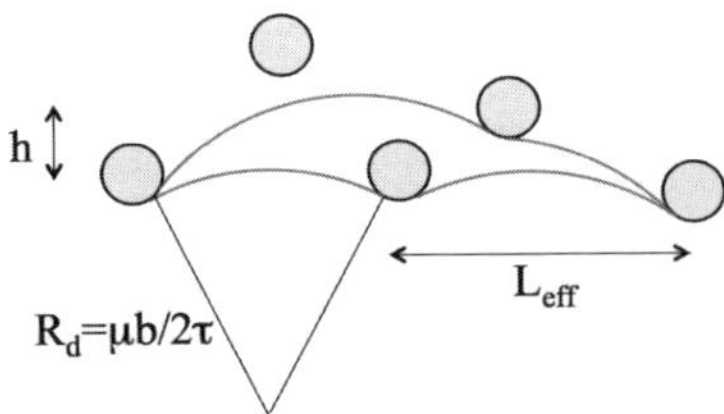

Fig. 11. Geometrical parameters of obstacle overcoming in the case of weak obstacles. L_{eff} is the average distance between obstacles along the dislocation line, h is the average distance travelled by the dislocation when an obstacle has been overcome, and R_d is the radius of curvature of the dislocation under the applied shear stress τ.

The dislocation is bowed with a radius of curvature proportional to the inverse of the applied stress, and will travel in average a distance h between the overcoming of two obstacles. Friedel's so-called "stationary" assumption[12] states that the area swept during this process must be equal to L^2, L being the average distance between obstacles in the glide plane. Simple geometrical considerations give then:

$$hL_{eff} = L^2 \qquad (13)$$

Assuming that the dislocation's curvature radius is much larger than the inter-obstacle distance (and therefore that the obstacles are very weak), it is possible to calculate a relationship between h, L_{eff} and the dislocation radius of curvature R_d:

$$L_{eff}^2 \cong 2hR_d \qquad (14)$$

which leads by a combination with the preceding equation to:

$$L_{eff}^3 \cong L^2 \frac{\mu b}{\tau} \qquad (15)$$

and from that the strengthening effect of weak point-like obstacles can be evaluated to:

$$\Delta\sigma_c^{weak} = \chi^{3/2} M \frac{\mu b}{L} = \chi^{3/2} M\mu b \sqrt{N} \; where \; \chi = \cos\left(\frac{\phi_c}{2}\right) \qquad (16)$$

For strong, impenetrable obstacles, it could be assumed that the effective distance L_{eff} would simply be equal to the average distance between obstacles in the glide plane L. Actually, statistical calculations by Kocks for instance[13, 14] have shown that the dislocation flexibility is not large enough to attain this value of L, and that one needs to increase it by about 15%, leading to the following strengthening for impenetrable obstacles:

$$\Delta\sigma_c^{strong} \cong \frac{1}{1.15} M \frac{\mu b}{L} \cong 0.87 M \frac{\mu b}{L} \qquad (17)$$

It is possible to apply now these equations to different obstacles. The first one that we will consider is forest dislocations, namely dislocations

92   *A. Deschamps*

lying in a different glide plane than the moving dislocation. In this case, the density of obstacles in the glide plane N_d is simply:

$$N_d = \rho \tag{18}$$

where ρ is the dislocation density. If we assume that forest dislocations are weak obstacles, the increase of yield stress due to the presence of dislocations writes:

$$\Delta\sigma = M\mu b\chi^{3/2}\sqrt{\rho} = M\alpha\mu b\sqrt{\rho} \quad where\, \alpha = \left[\cos\left(\frac{\phi_c}{2}\right)\right]^{3/2} \tag{19}$$

α is a material-dependent constant that describes the strength of the dislocation-dislocation interactions. Its value is generally considered to be close to 0.3, which translates into a breaking angle of about 130°. This confirms the assumption that dislocations are weak obstacles. Figure 12 shows that this relationship is well followed in an Al-Mg alloy, where the dislocation density was measured by resistivity. It must be stressed that recent X-ray diffraction experiments[15] have indicated that the α factor may change during plastic deformation, non monotonically and in a range as large as 0.1-0.3.

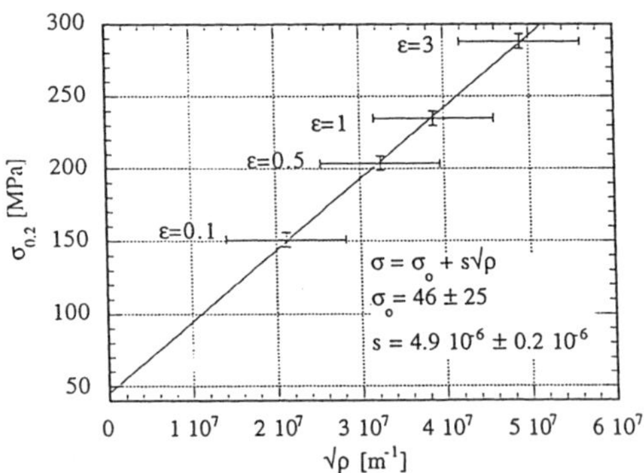

Fig. 12. Relationship between yield strength (measured at 0.2% offset strain) and the square root of the dislocation density in a cold rolled Al-Mg solid solution, showing that equation (7) describes well the effect of dislocations on strengthening. Dislocation density was measured by resistivity (Courtesy of M. Verdier, SIMAP Laboratory, Grenoble, France).

The second case that we can consider is that of precipitates. Precipitates can have two types of interactions with dislocations. Either they can be sheared by moving dislocations (they will undergo the same plastic deformation as the matrix) or they can be by-passed (in which case the dislocation will leave a so-called Orowan loop around the precipitate, which still has its undeformed shape but is now elastically loaded) (see Fig. 13).

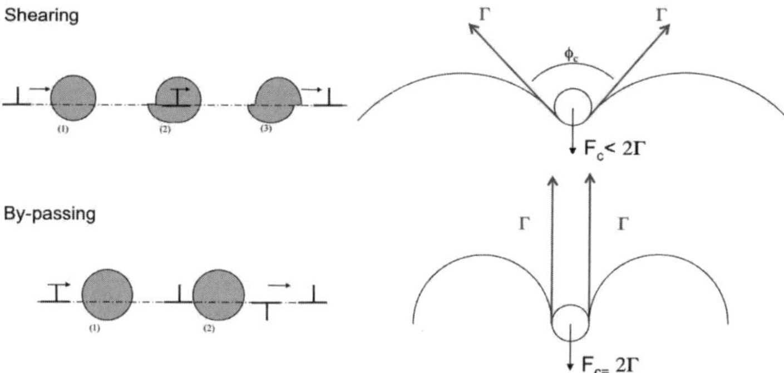

Fig. 13. Scheme of the two possible mechanisms of overcoming of precipitates by dislocations, and of the configuration of highest stress when the obstacle is passed.

The choice of mechanism depends on several factors. The Orowan mechanism of precipitate by-passing requires the dislocation to completely bow around the precipitate. The critical configuration to obtain this by-passing requires the angle ϕ_c to be equal to 0 and thus the obstacle strength to be equal to the maximum value of 2Γ. Thus Orowan by-passing occurs only if shearing is not possible: either if the transmission of slip is impossible between the matrix and precipitate (because of incompatible crystal structures) or if the force required to pass the dislocation through the precipitate is larger than 2Γ. Generally, small precipitates, of metastable character are coherent with the matrix and are sheared, and larger ones are by-passed.

In order to establish the contribution of precipitates to the flow stress, it is first necessary to derive the density of obstacles on the glide plane, as a function of the accessible parameters of the precipitate microstructure that are the average precipitate size and volume fraction. For spherical precipitates of radius R, the obstacles that cross a given glide plane are those with their centre at a distance smaller or equal to R from the glide plane. A box of size X*X in the glide plane and 2R in the perpendicular direction contains a number n of precipitates defined by:

$$f_v = \frac{\frac{4}{3}\pi R^3 n}{2RX^2} \tag{20}$$

and thus the density of obstacles in the glide plane can be written as:

$$N = \frac{n}{X^2} = \frac{3f_v}{2\pi R^2} \tag{21}$$

If we add that $F_c = 2\Gamma$ then the increase in yield strength for non-shearable precipitates can be written:

$$\Delta\sigma \cong 0.87 M\mu b \sqrt{\frac{3}{2\pi}} \frac{\sqrt{f_v}}{R} \approx 0.6 M\mu b \frac{\sqrt{f_v}}{R} \tag{22}$$

For shearable precipitates the situation is much more complicated, since the obstacle strength is not known a priori. Generally speaking it depends on the precipitate size, but also on many other parameters such as the nature of the precipitate, the coherency strains, the orientation relationship, etc. It is useful to first review the main mechanisms that can explain the hardening related to the presence of precipitates in a material:

- A first mechanism is the interaction of the dislocation with the strain field due to the elastic misfit between precipitate and matrix. If this interaction is repulsive dislocations will need an extra stress to approach the precipitate, if it is attractive the dislocations will need an extra stress to detach from the precipitate.

- Similarly, the difference in elastic modulus between the precipitate and matrix changes the value of the line tension of the dislocation as it travels through the precipitate. If the modulus of the precipitate is lower that that of the matrix the energy of the dislocation will be lower and the interaction will be attractive. If it is higher the interaction will be repulsive. In both cases, similarly to the case above, an extra stress will be required for precipitate shearing.

- When the dislocation has finished shearing the precipitate, an extra interface between the precipitate and matrix of $2\pi Rb$ has been created, which has required to spend an energy of $2\pi Rb\gamma$, where γ is the interfacial energy between precipitate and matrix. This energy results from the work of a force F along the distance 2R, and thus an estimation of the obstacle strength in this case is $F_c=\pi b\gamma$.

- If the precipitate has an ordered structure, its shear by a single matrix dislocation leaves an anti-phase boundary inside the precipitate, that has also an energy cost associated to it.

These different mechanisms are not mutually exclusive and together result in the global obstacle strength. It is usual to approximate the obstacle strength by a linear dependence with obstacle size, which simply translates in the simplest form that small precipitates are easier to shear than larger ones:

$$F_c = k\mu bR \Rightarrow \cos\left(\frac{\phi_c}{2}\right) = \frac{kR}{b} \tag{23}$$

where k is a numerical constant, that depends on the efficiency of a given precipitate to resist shearing by dislocations, and is generally comprised between 0.01 and 0.1. Using the same value for the average distance between obstacles in the glide plane L as for precipitate by-passing, the predicted strengthening (under the hypothesis of weak obstacles) is:

$$\Delta\sigma = M\mu\sqrt{\frac{3k^3}{2\pi b}}\sqrt{Rf_v} \tag{24}$$

When the precipitates grow in size, they are usually initially shearable at very small sizes and undergo a shearable - non shearable transition at a given precipitate size R_t. Comparing Eqs. (21) and (23), it is apparent that if one assumes a constant volume fraction (which is not the case in practice) during precipitate growth, the yield stress will first increase as $R^{1/2}$ and then decrease as R^{-1}. At the transition radius one can equal the two equations and obtain an estimate of the value of k:

$$k \cong 0.9 \frac{b}{R_t} \qquad (25)$$

Figure 14 shows the respective relationships between the precipitate size and the strength within the two regimes of shearing and by-passing. Figure 15 shows the evolution of yield strength with ageing time in an Al-Zn-Mg-Cu alloy, where precipitates grow in size and volume fraction (as measured by Small Angle X-ray Scattering), showing a well-defined peak in yield strength.

The two equations provide a very good description of the two main stages of shearing and by-passing, but fail to describe the strength at the transition radius.

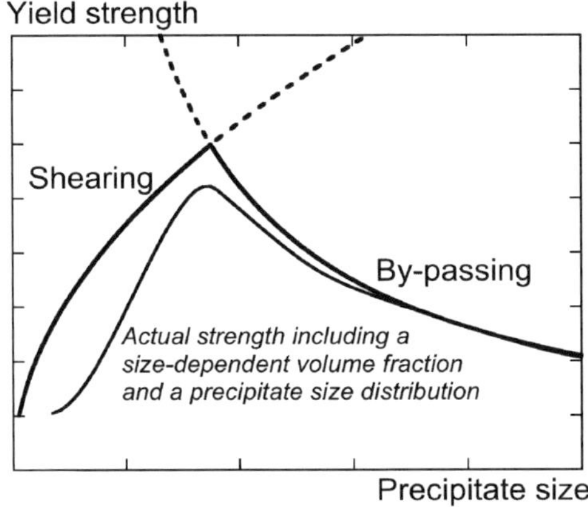

Fig. 14. Precipitate size dependency of the strength for the two overcoming mechanisms. The cross-over of the two contributions defines the peak strength. However, the fact that volume fraction is size dependent (begins at zero and finishes at thermodynamic equilibrium), and that precipitates are distributed in size for a given microstructure, decreases the actual strength and changes the sharp maximum predicted into a smoother dependency.

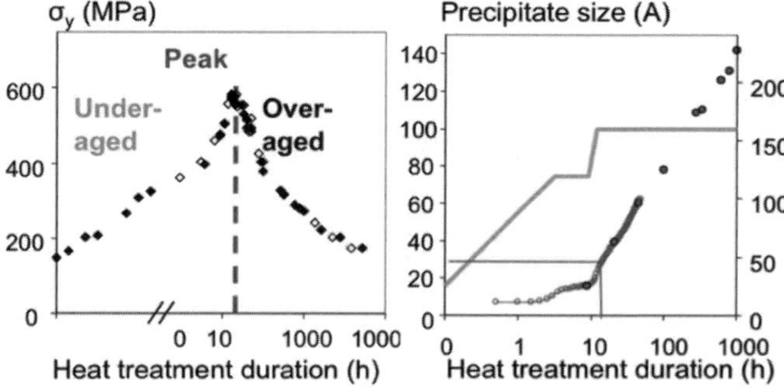

Fig. 15. Example of the evolution of yield strength along a heat treatment in an Al-Zn-Mg-Cu alloy. Peak strength is observed for a precipitate size of about 3 nm (30 Å).

The main reason is that the material actually shows a precipitate size distribution and thus that very few precipitates actually provide the maximum strength when the average radius is at the transition radius. In any case, precipitation is a very effective way of hardening materials, as Fig. 16 illustrates by showing the stress-strain curves of pure Aluminium and of a high strength, precipitation hardened, peak strength Al-Zn-Mg-Cu alloy.

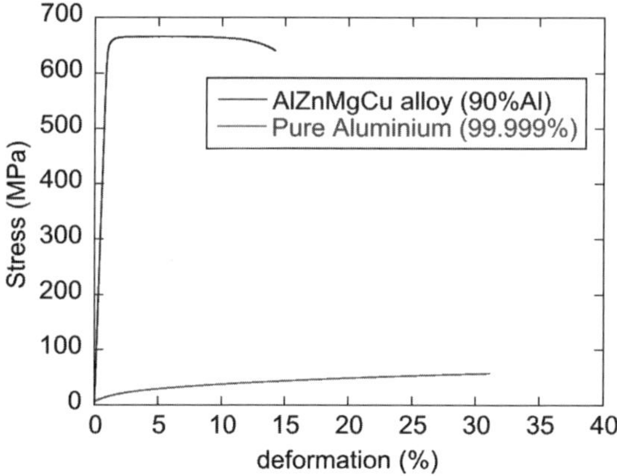

Fig. 16. Comparison between the stress-strain curves of a peak strength, precipitation hardened aluminium alloy (yield strength of about 650 MPa) and of pure aluminium (yield strength less than 7 MPa).

As concluding remarks for this section, one should point out that a maximum in strength with precipitate size (called peak-ageing state) does not require a shearing – to by-passing transition. In fact, two other mechanisms can lead to the presence of a maximum in strength:

- An evolution of volume fraction with aging time, starting with zero, and finishing at thermodynamic equilibrium;

- A saturation of obstacle strength in the shearing mode, lower than the value of 2Γ that triggers Orowan by-passing.

3.3. *Diffuse obstacles*

Diffuse obstacles are obstacles of size sufficiently small and density sufficiently high so that they cannot be considered as point obstacles on the dislocation line. Thus their contribution to strengthening has to be calculated in an average way, and not considering individual obstacle-dislocation interactions. Similarly to shearable precipitates, there can be several mechanisms by which solutes can harden a material. The first obvious one is the size effect. When a solute atom is on an interstitial site, it induces compressive stresses in the matrix and thus has an attractive interaction energy with the tension zone of edge dislocations. Substitutional atoms of larger size than matrix atoms show the same behaviour, whether atoms that are smaller than matrix ones are attracted to the compressive zone of edge dislocations. At first order the size effect does not give a significant contribution to hardening for screw dislocations, which do not induce hydrostatic stress in the material. However, other effects are operative for any dislocation type, and especially the local change of bonding caused by the presence of the solute atom, which acts on the local line energy of the dislocation.

Generally, it is observed that the yield strength increases non-linearly with solute content, and one writes[16]:

$$\Delta\sigma = k_{ss} C^n \tag{26}$$

where C is the concentration in solute and n is an exponent, generally comprised between ½ and 2/3. For instance in the case of a pure size effect the following contribution has been proposed:

$$\Delta\sigma = M\, 2^{1/3}\, \mu(\eta/10)^{4/3}\, C^{2/3} \tag{27}$$

where $\eta=\Delta V/V$ is the relative volume change between the solute and matrix atoms.

Figure 17 shows the solute strengthening effect in the case of fully miscible solid solutions, namely here Au-Ag and Cu-Ni. In the low concentrations regions an equation of type (25) is appropriate to describe the hardening effect.

An additional complexity actually arises when considering solute atoms that are not immobile in the material, but can segregate to the dislocations cores to which they are attracted. Two possibilities can be encountered. The first is that dislocations are mobile before the material is deformed plastically, and immobile afterwards. The second situation is that solute atoms have a comparable mobility to that of the dislocations during plastic deformation.

In both cases, the concentration of solute atoms at the dislocations is initially larger than the average value, and thus the initial yield strength is larger than expected.

An extra stress is required to free the dislocations from the solute cloud (so called Cottrell atmosphere). If the solute atoms are later immobile, the dislocations can move at a lower stress once they are freed from their initial solute atmosphere.

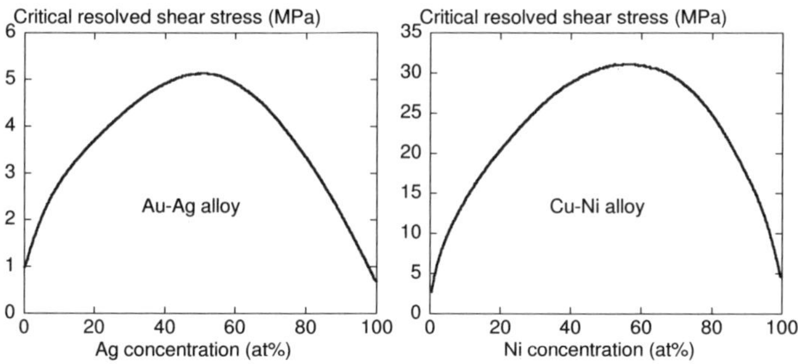

Fig. 17. Influence of solute content on yield strength (critical resolved shear stress) of Au-Ag and Cu-Ni alloys that show complete miscibility.

Thus, after a so-called upper yield stress (UYS) (sharp point in the stress-strain curve, see Fig. 18), deformation goes on at a lower stress (lower yield stress, LYS) by the propagation of a Piobert-Lüders band (localised deformation) until strain hardening has taken place in the whole material and that the flow stress becomes everywhere larger than the upper yield stress, after which the plastic deformation proceeds normally.

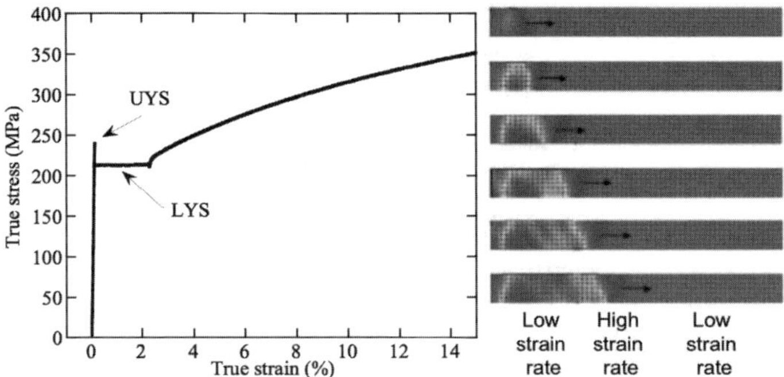

Fig. 18. a) Stress-strain curve of a low carbon steel showing the Lüders plateau corresponding to the propagation, at constant stress, of a deformation band through the tensile sample, followed by homogeneous deformation at larger strains; b) strain maps on the tensile sample at different stages of the propagation of the deformation band.

In the second case, the material shows continuous plastic instabilities, called Portevin-le-Châtelier (PLC) effect, related to the negative strain rate sensitivity (SRS) of the material in a given range of strain rates around the applied strain rate. Schematically, the dependency of the materials' flow stress with strain rate shows an N-curve (Fig. 19). At very low strain rates the flow stress is quite high because the solute clouds follow continuously the dislocation motion. At high strain rates the flow stress follows a lower, parallel line because the dislocations interact only with immobile solute atoms (solutes cannot move to segregate to dislocations). In both cases a normal behaviour is followed, which means that increasing the strain rate increases the flow stress. However, in an intermediate range of strain rate, the reverse is observed.

In this range, increasing the strain rate results in increasing the dislocation velocity, and thus decreasing the amount of segregation of solute to the dislocations, and finally decreasing the flow stress.

If one imposes an average strain rate in this range of negative SRS, plastic flow will separate in two alternating regimes, one of very low strain rate, and one of very high strain rate. Bands carrying plastic deformation travel continuously through the material, causing the apparition of surface lines, and the stress oscillates between the flow stresses at the different strain rates occurring in the material. Such a stress-strain curve is shown in Fig. 19.

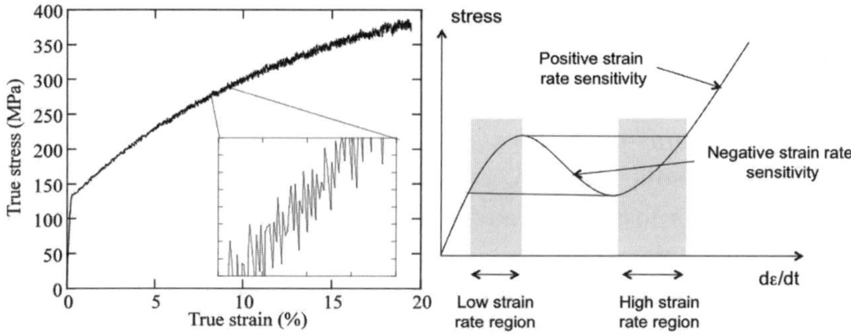

Fig. 19. a) Stress-strain curve of an Al-Zn-Mg solid solution showing continuous Portevin-Le Châtelier serrations; b) schematic of the influence of strain rate on flow stress in such a solid solution, due to solute-dislocations interactions, and resulting in plastic instabilities that favour the separation of the material in regions with high strain rate and regions of low strain rate.

3.4. *Addition laws*

When several obstacles to dislocation motions are present in a given material, the question arises to add their contributions to the yield strength. The first case to consider is that of obstacles operating at vastly different length scales. For instance the Peierls lattice friction, precipitates and the grain size. In this case the contribution at the smaller scale acts as a friction stress, reducing the effective stress acting on the next type of obstacles, and thus the different contributions to the flow stress can be added linearly. When obstacles are of comparable strength and scales, one needs to consider their contributions in a common scheme. For localised obstacles, we have seen that the stress contribution is related to the obstacle density: $\Delta\sigma \propto N^{1/2}$. Thus if two comparable families of obstacles are present of densities N_1 and N_2, the total density will be $N=N_1+N_2$ and thus the addition law between the contributions to the yield strength must be quadratic:

$$\Delta\sigma^2 = \Delta\sigma_1^2 + \Delta\sigma_2^2 \rightarrow \Delta\sigma = \sqrt{\Delta\sigma_1^2 + \Delta\sigma_2^2} \qquad (28)$$

If we consider the series of obstacles considered above, we can propose a general addition law:

$$\sigma_y = \sigma_o + \Delta\sigma_{HP} + \Delta\sigma_{ss} + \sigma_i + \sqrt{\Delta\sigma_d^2 + \Delta\sigma_p^2} \qquad (29)$$

where σ_o is the lattice friction stress, $\Delta\sigma_{HP}$ is the contribution of the Hall Petch law (grain boundaries and phase boundaries), $\Delta\sigma_{ss}$ is the contribution of solid solution, σ_i is the internal stress if present (which can be related for instance of a composite effect or to former deformation with impenetrable obstacles), $\Delta\sigma_d$ is the contribution of forest dislocations and $\Delta\sigma_p$ is the contribution of precipitates. One must be careful however about the validity of this last addition between dislocations and precipitates, since it depends on the strength of precipitates and cannot take in principle such a general form.

4. Relationship Between Microstructure and Strain Hardening

4.1. *General framework*

We will now consider what happens after the yield point, namely when a generalized movement of dislocations happens. We will place ourselves in the framework of polycrystalline plasticity, where several, non coplanar, slip systems are activated simultaneously in each grains. The major effect that can be observed in the stress-strain curve is strain hardening (see Fig. 20), namely that the flow stress (defined as the stress required to move dislocations at a given level of strain) increases monotonically with strain until fracture. Figure 20 shows that strain hardening is very much dependent on the conditions of the plastic deformation, namely here temperature.

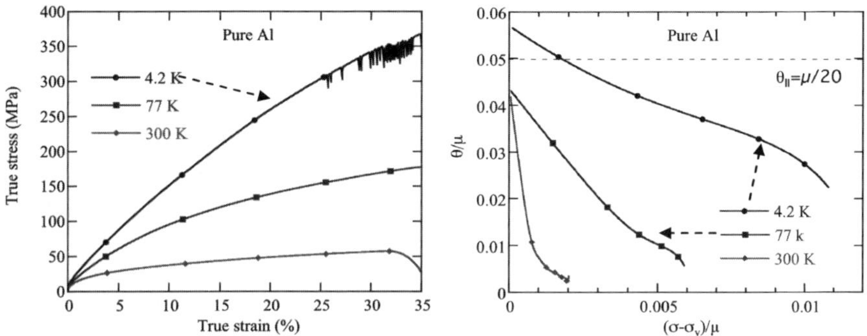

Fig. 20. (left) Stress-strain curves of a pure Aluminium sample at three temperatures; (right) strain hardening rate ($\theta=d\sigma/d\varepsilon$) curves, normalised by the temperature-dependent shear modulus.

Following the yield point, in a material that does not initially contain a significant density of dislocations, and making the approximation of a linear addition law between the contribution of dislocations to the flow stress and the contributions responsible for the yield strength, the increase in flow stress can be written as a function of the dislocation density:

$$\sigma - \sigma_y = M\,\alpha\mu b\sqrt{\rho} \tag{30}$$

A common way to study strain hardening, based on the Kocks, Mecking and Estrin approach[17-19], is to follow the evolution of an internal variable with strain, the value of which is sufficient to determine the flow stress. In simple microstructures a single internal variable is sufficient and a natural choice is the dislocation density ρ. Following the evolution of the dislocation density is equivalent as following the evolution of flow stress, and the evolution rate $d\rho/d\varepsilon$ is of course related to the strain hardening rate $d\sigma/d\varepsilon$. Thus the most common way to study strain hardening is to represent the strain hardening rate $d\sigma/d\varepsilon$, representative of the evolution rate of dislocation density, as a function of the reduced flow stress $\sigma-\sigma_y$, representative of the dislocation density. Figure 20(right) shows such a plot (called Kocks-Mecking plots) for the three deformation curves presented in Fig. 20(left). We can see that in pure Aluminium, the initial strain hardening rate is more or less constant with temperature (in values reduced by the temperature-dependent shear modulus). As strain goes on and the flow stress increases, the strain hardening rate decreases towards zero. It is observed to decrease much faster as temperature gets higher, which is equivalent to say that the strain hardening capability becomes much smaller.

Before describing the mechanisms responsible for this evolution of strain hardening rate along the deformation process, it is necessary to define the general framework of the constitutive model that will be used. Its general form is the following:

$$\sigma = f(\dot{\varepsilon},T)\,\hat{\sigma}(microstructure) \tag{31}$$

In this equation, we separate the effect of microstructure evolution during straining from the effect of the deformation conditions (temperature and strain rate) at constant microstructure. Influence of deformation conditions at constant microstructure can be investigated by changing abruptly the deformation conditions, using strain rate or temperature jumps. A stress strain curve is shown in Fig. 21, where such jumps are regularly made between two values of strain rate. Each time the strain rate is varied, so does the stress, and the magnitude of the stress jump can be used to obtain the first part of equation (30). Activation theory of plastic deformation (presented elsewhere in this book) enables us to formalise this equation:

$$\frac{\dot{\varepsilon}}{\dot{\varepsilon}_0} = \left(\frac{\sigma}{\hat{\sigma}}\right)^m \Leftrightarrow \sigma = \hat{\sigma}\left(\frac{\dot{\varepsilon}}{\dot{\varepsilon}_0}\right)^{\frac{1}{m}} \tag{32}$$

is a reference strain rate and m is the strain rate sensitivity that can be written as:

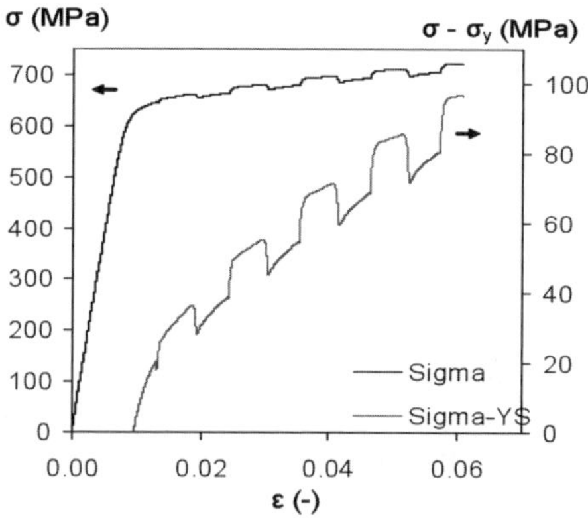

Fig. 21. Stress strain curve of a precipitation hardened aluminium alloy with strain rate jumps between 10^{-2} and 10^{-4} s^{-1} regularly performed during the tensile test.

$$m = \frac{V^* \sigma}{kT} \tag{33}$$

where V^* is defined as the activation volume for the overcoming of obstacles controlling plastic deformation. V^* is itself defined as:

$$V^* = bA^* \tag{34}$$

A^* is called the activation area for obstacle overcoming by dislocations and is defined as the area swept by the dislocation during the overcoming process, as shown in Fig. 22. This means that the larger is this area, the larger is the activation volume, and the lower is the effect of strain rate (or temperature) on flow stress. Thus large, strong obstacles will lead to a less thermally activated flow stress, less sensitive to temperature and strain rate changes. Actually the activation volume is accessible experimentally by strain rate jumps using the following equation:

$$\frac{d\sigma}{d\ln\dot{\varepsilon}} = \frac{kT}{V^*} \tag{35}$$

Now in the following we will assume that plastic deformation takes place at a constant strain rate equal to the reference strain rate so that the flow stress equals the microstructure-dependent stress.

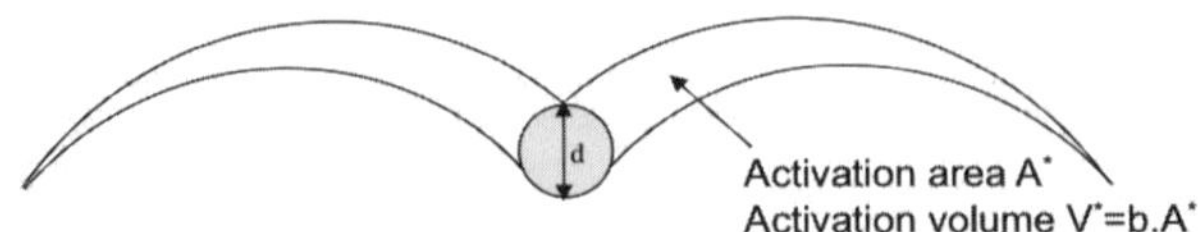

Fig. 22. Definition of the activation area and of the activation volume for the overcoming of an obstacle of diameter d by a dislocation.

4.2. *Strain hardening in a pure metal*

We have seen in Fig. 20 that the strain hardening rate is initially high at the onset of plastic deformation, and then decreases with strain. This means that initially dislocation storage (i.e. the increase of dislocation density with strain) is high, and that it becomes less efficient as plastic deformation goes on. This behaviour will now be interpreted in terms of dislocation storage (which increases the dislocation density) and dislocation annihilation (which decreases the dislocation density), the latter being called dynamic recovery.

Dislocation storage occurs in a pure metallic polycrystal mainly by interactions between dislocations lying in different slip planes. Let us consider a fixed dislocation. Moving dislocations can interact during a time dt with this dislocation at two conditions: they should be sufficiently close to this fixed dislocation (at a distance less than $v.dt$ where v is their velocity) in the direction of their movement, and also in the perpendicular direction (or else they will interact with other dislocations). For this last criterion, the dislocations should be closer than the inter-dislocation distance $\rho^{-1/2}$ from the fixed dislocation. The length of mobile dislocations crossing the area defined by the distance $(v.dt.\rho^{-1/2})$ is $(v.dt.\rho^{-1/2}\rho_m)$.

If one considers that a proportion K of the moving dislocations will be stored by their interaction with the fixed dislocation, then the storage rate of dislocation length writes:

$$d\rho_{disl}^{+} = \frac{K\rho_m v dt}{\rho^{1/2}} \tag{36}$$

Now, if we consider that all dislocations can lead to dislocation storage, and not only the one fixed dislocation that we have considered, we have to multiply Eq. (35) by the total dislocation density, which yields the dislocation storage term:

$$d\rho^+ = K\rho_m v dt \rho^{1/2} \tag{37}$$

Finally, one needs to use the equation relating the density of mobile dislocations, the dislocation velocity and the Burgers vector to the strain rate, so-called Orowan equation:

$$\frac{d\varepsilon}{dt} = \rho_m b v \tag{38}$$

and we can obtain the final equation for the dislocation storage rate:

$$\frac{d\rho^+}{d\varepsilon} = \frac{K}{b}\rho^{1/2} \tag{39}$$

This equation shows that for a constant storage efficiency K, the dislocation storage rate evolves as the square root of the dislocation density. Translating this in terms of strain hardening rate using Eq. (29), one obtains:

$$\frac{d\sigma_{storage}}{d\varepsilon} = \frac{M\alpha\mu K}{2} \tag{40}$$

Thus dislocation storage by mutual interaction results (for a constant storage efficiency) in a constant strain hardening rate, that corresponds to the stage II hardening rate of single crystals. Considering that this maximum strain hardening rate is of the order of $\mu/20$, this gives an estimate of the efficiency factor $K \sim 1/10$.

Now it is necessary to model in a similar way the dynamic recovery, which is the mutual annihilation of dislocations with strain. We will consider the same fixed dislocation, which interacts during time dt with mobile dislocations situated at a distance smaller than $v.dt$ in the direction of dislocation motion. Annihilation will be considered to happen if dislocations are situated at a distance smaller than a critical distance y_o in the direction normal to the direction of slip. For screw dislocations capable of cross-slip, this distance would be the critical distance for the triggering of this mechanism. Then the length of dislocation annihilating in time dt with the fixed dislocation is :

$$dp_{disl}^- = -\rho_m y_o v dt \qquad (41)$$

We have now, like above, to multiply this term by the total dislocation density and use the Orowan equation. The annihilation dislocation rate then writes:

$$\frac{d\rho^-}{d\varepsilon} = -\frac{y_o}{b}\rho \qquad (42)$$

Finally, this dislocation annihilation rate can be translated into a strain hardening rate:

$$\frac{d\sigma^-_{annihilation}}{d\varepsilon} = -\frac{y_o}{2b}\sigma \qquad (43)$$

This equation shows that annihilation becomes more and more efficient as the dislocation density increases, or equivalently when the flow stress increases. Finally the total storage rage of dislocations can now be obtained from Eqs. (38) and (41):

$$\frac{d\rho}{d\varepsilon} = \frac{d\rho^+}{d\varepsilon} + \frac{d\rho^-}{d\varepsilon} = \frac{K}{b}\rho^{1/2} - \frac{y_o}{b}\rho \qquad (44)$$

and correspondingly the total strain hardening rate:

$$\frac{d\sigma}{d\varepsilon} = \frac{d\sigma_{storage}}{d\varepsilon} + \frac{d\sigma_{annihilation}}{d\varepsilon} = \frac{M\alpha\mu K}{2} - \frac{y_o}{2b}\sigma \qquad (45)$$

More precisely this equation should be applied to the flow stress increase after the yield point and thus :

$$\frac{d\sigma}{d\varepsilon} = \frac{M\alpha\mu K}{2} - \frac{y_o}{2b}\left(\sigma - \sigma_y\right) \qquad (46)$$

Figure 20 shows that this equation, which predicts a linear dependence of strain hardening rate with the flow stress increase, describes very well the behaviour of a pure polycrystal like aluminium. The slope of these curves gives access to the critical annihilation distances, which can be expressed in units of b: 23.b for room temperature, 4.b at 77K, and 1.3.b at 4.2K.

Equation (45) can be integrated to obtain the evolution of flow stress with strain

$$\sigma = \sigma_y + \frac{A}{B}(1 - \exp(-B\varepsilon)) \tag{47}$$

where $A=M\alpha\mu K/2$ and $B=y_c/2b$. This equation is equivalent to write:

$$\sigma = \sigma_y + (\sigma_s - \sigma_y)(1 - \exp(-B\varepsilon)) \tag{48}$$

where $\sigma_s=\sigma_y+A/B$ is the saturation stress that would be reached by the material at large strains if necking and fracture did not occur, and corresponds to the dislocation density for which dislocation storage is exactly balanced by dislocation annihilation by dynamic recovery. This exponential-based constitutive law is also known as the Palm-Voce equation[20], and was proposed as a phenomenological morel a long time before these dislocation-based models were developed by Kocks, Mecking and Estrin.[17-19]

4.3. *Influence of impenetrable obstacles on strain hardening rate*

When impenetrable obstacles (grain boundaries, phase boundaries, non shearable precipitates) are added to a pure metal, they affect not only the yield strength as seen in section 3. During plastic strain they can be considered as introducing two contributions to the flow stress. One arises because the introduction of geometrically necessary dislocations (which need to be introduced to compensate for the plastic incompatibilities of the different regions of the material) provide new forest obstacles. To describe it, one needs to add a new term of storage on the dislocation density evolution, which is constant with strain since the impenetrable obstacles are not modified by strain:

$$\frac{d\rho}{d\varepsilon} = K' + \frac{K}{b}\rho^{1/2} - \frac{y_o}{b}\rho \tag{49}$$

which translates in an additional term in the strain hardening rate equation:

$$\frac{d\sigma}{d\varepsilon} = \frac{K''}{d\sigma} + \frac{M\alpha\mu K}{2} - \frac{y_o}{2b}(\sigma - \sigma_y) \tag{50}$$

where d is the inter-obstacle distance. Such a term can be experimentally determined by comparing two materials showing a different density of impenetrable obstacles, and drawing a so-called Haasen plot of $(\theta\sigma)$ as a function of σ, where θ is the strain hardening rate. The additional factor K''/d is then simply measured as a translation of the plot on the y axis.

One other contribution needs also to be considered. The geometrically necessary dislocations that are introduced (for instance as Orowan loops around non shearable precipitates, see Fig. 23) are polarized, and induce an internal stress in the material that the applied stress needs to overcome so that plasticity proceeds.

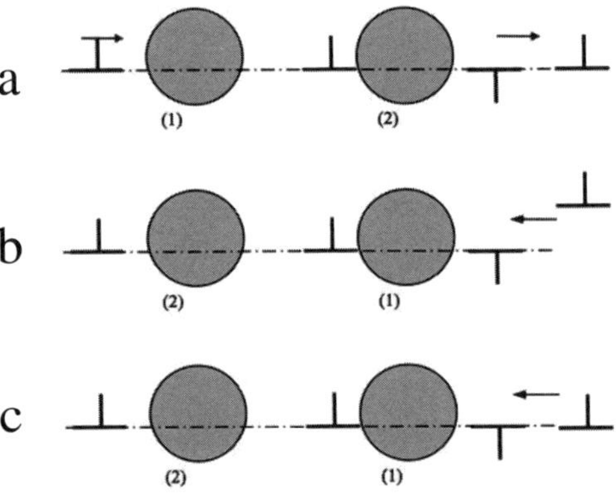

Fig. 23. a) Storage of an Orowan loop around a precipitate, generating an internal stress in the material; b) upon strain reversal, a dislocation is attracted by the internal stress towards the precipitate, but still has to overcome the particle if it is not in the same glide plane as the original dislocation; c) if the strain is reversible upon strain reversal (the dislocation travels back the same path), the dislocation annihilates with the Orowan loop and thus needs no additional stress to by-pass the precipitate.

This internal stress results in an extra contribution to the flow stress during monotonic loading. If however strain is reversed, the internal stress now helps plastic deformation and will reduce the flow stress for plastic deformation. Thus this internal stress contribution can be evaluated by performing so-called Bauschinger tests, consisting in a first stage of tensile testing, followed by a stage of compression testing. The half difference between the flow stress in tension and compression is the internal stress σ_i (see Fig. 24). As shown schematically in this figure, upon strain reversal the yield stress is often initially very small. This is caused by a partial reversibility of strain (scheme (c) of Fig. 23), which prevents the precipitates or other impenetrable obstacles to play their strengthening role. When the geometrically necessary dislocations around these obstacles are removed, the obstacles recover their effectiveness and the difference between the forward and reverse stresses become much smaller, simply related to the average value of the internal stress in the material. Experimental data shows that a high value of internal stress is obtained only when a large amount of plastic incompatibility is present in the material. This is especially the case in precipitation-hardened materials with unshearable particles, in multi-phase materials that do not have compatible slip systems, in materials with a low number of active slip systems like HCP materials, or in materials showing extensive twinning.

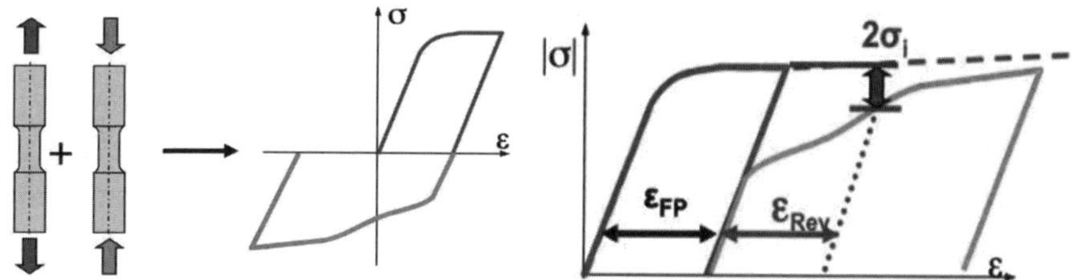

Fig. 24. Scheme of the Bauschinger test, and of the way the internal stress is calculated from the half difference, at a given reversal strain, of the forward and reverse stress.

4.4. *Influence of shearable precipitates and solute atoms on strain hardening*

As we have seen, just above, shearable precipitates cannot induce additional dislocation storage in the material and thus additional strain hardening. Actually, when precipitates are very small, plastic deformation can lead to their complete shear (as seen for a 1 atomic layer thick GP zone in Aluminium in Fig. 25). In this case, a dislocation arriving on the same glide plane will not see the obstacle anymore and can move freely through the matrix. In some cases precipitates can even be dissolved by the movement of dislocations that split them into sizes where they become subcritical.

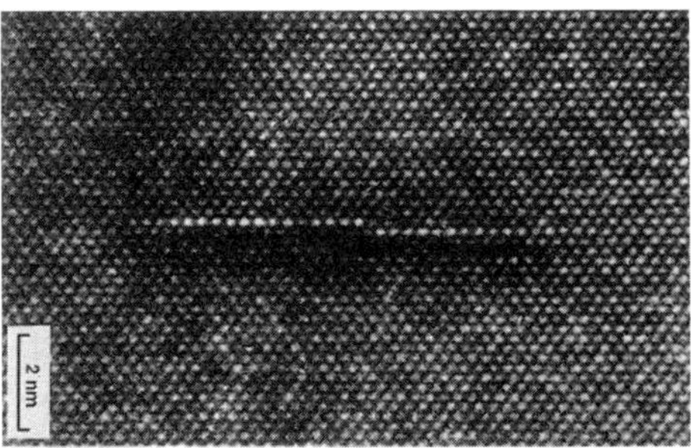

Fig. 25. High Resolution Electron micrograph of an Al-Cu GP zone sheared by a dislocation (from Ref. 21).

If the obstacles to dislocation motion are destroyed by plastic strain one can expect to observe strain softening, as the contribution of precipitates to the flow stress will decrease with strain. It is actually observed that the strain hardening rate in some materials with shearable precipitates (particularly those close to peak hardening where the precipitates have grown to relatively large sizes) is very low, leading to early necking and ductile fracture. However, some other materials with shearable precipitates, and particularly those with very small precipitates and high levels of residual solid solution, show in contrast very high strain hardening rates, as shown in Fig. 26.

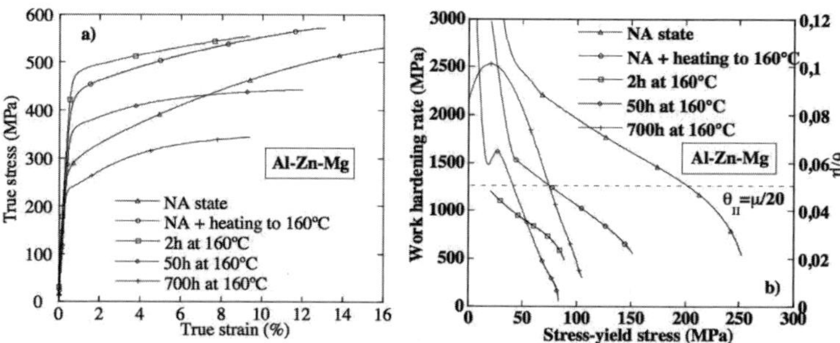

Fig. 26. Stress-strain curves and strain (or work) hardening rate curves as a function of precipitation state in an Al-Zn-Mg alloy. In the first curve the material contains shearable GP zones with a high level of residual solid solution. The strain hardening rate is very high; the third curve (2h at 160°C) corresponds to peak strength, namely shearable precipitates, and shows the lowest strain hardening rate. The two last curves correspond to unshearable particles (over-aged state) and the strain hardening rate raises again.

Now if only solute atoms are present in the material, and no precipitates (at least initially), the strain hardening rate is strongly modified as well. Solute atoms generally increase the strain hardening rate, as shown in Fig. 27. In the Kocks-Mecking-Estrin framework, it is assumed that their effect relies in the modification of the stress required for cross slip, that controls the dynamic recovery. This can happen for instance if solutes change the stacking fault energy of the material, and therefore the splitting distance of partial dislocations.

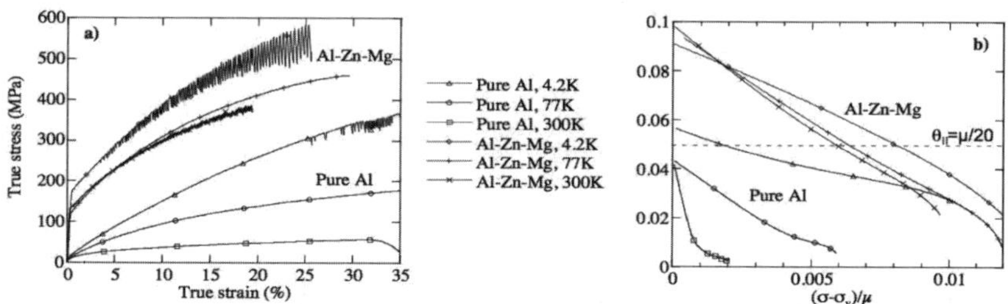

Fig. 27. Stress-strain curves and strain (or work) hardening rate curves of pure aluminium and a supersaturated Al-Zn-Mg solid solution as a function of temperature, showing that the strain hardening capability of the solid solution is not only much larger, but also shows a very different temperature dependence.

However, there is now good evidence that this description is not sufficient to account for the strain hardening of many solid solutions. Notably, in concentrated, supersaturated solid solutions, initial hardening rates are found to be much higher than the value of $\mu/20$ that would be expected in a material that does not contain any impenetrable obstacles to store additional dislocations as compared to the forest interactions. This effect can be actually understood by the dynamic change in microstructure induced by the straining. The solute atoms, the movement of which is helped by the movement of dislocations, can group to make clusters or small precipitates during the plastic strain. Two situations may arise that can explain the additional strain hardening observed: the precipitates can contribute themselves to the flow stress, and if they form on dislocation junctions they can increase the efficiency of the dislocation storage and thus increase the contribution of dislocations to the flow stress. Formation of precipitates during straining is called dynamic precipitation and has been directly observed by in-situ small angle X-ray scattering in aluminium alloys.[22] Without going into the details of this complex coupling between microstructure evolution and flow stress, the flow stress must be expressed in this case in the general form:

$$\sigma = \sigma_o + \sigma_p(\varepsilon) + \sigma_d(\varepsilon) \tag{51}$$

Many other kinds of couplings can happen between microstructural evolution and flow stress. One very good example is when straining induces a phase transformation. This is the case in the so-called TRIP (transformation induced by plasticity) steels, where metastable austenite is transformed upon straining into martensite, and where this martensite considerably hardens the material, resulting in an extremely high strain hardening rate.

5. Relationship between Microstructure and Fracture

In this last section the relationship between microstructure and fracture of metals will be briefly presented. This is a very complicated subject, and would need many developments to be fully treated. Thus this presentation will deliberately stay qualitative and aims only at giving simple, general ideas. Also, fracture will only be discussed in terms of monotonic tension.

5.1. *Relationship between strain hardening rate, flow stress and ductile fracture*

Along the lines of the two former sections, it is worth to start with the relationship that exist between the values of flow stress, strain hardening rate and the stability of plastic flow in tension, which is related to the uniform elongation of a material. The uniform elongation is a property related to ductility. It is normally a lower bound for elongation at fracture, the latter including also post-necking deformation. This relationship can be evaluated very simply using the Considère's criterion, which is derived as follows. We consider now a tensile sample of section S and length L. We will assume that locally the section is reduced to S-dS, and will investigate the stability of this reduced area.

This reduced area induces locally a higher stress exerted on the material, since the load applied to the sample is still the same $F = \sigma\,S$, where σ is the value of flow stress. The increase of flow stress exerted on the material is simply:

$$d\sigma_{geometrical} = \frac{F}{S-dS} - \frac{F}{S} = \frac{\sigma}{1-dS/S} - \sigma \cong \sigma\frac{dS}{S} \qquad (52)$$

In the reduced section, the volume conservation of matter during plastic deformation imposes that the material is subjected to a small increment of deformation $d\varepsilon$:

$$LS = (L+dL)(S-dS) \Leftrightarrow \frac{dS}{S} = \frac{dL}{L} = d\varepsilon \qquad (53)$$

and thus the increment in flow stress imposed to the material in the reduced section is:

$$d\sigma_{geometrical} = \sigma d\varepsilon \tag{54}$$

On the other hand, the material is capable of resisting to this extra stress in the limits of its strain hardening caused by the extra deformation, that expresses in an additional flow stress:

$$d\sigma_{strain\,hardening} = \frac{d\sigma}{d\varepsilon}d\varepsilon = \theta d\varepsilon \tag{55}$$

Plastic deformation becomes unstable if the increase of flow stress caused by the extra strain hardening becomes smaller than the geometrical increase in flow stress, which defines Considère's criterion:

$$d\sigma_{strain\,hardening} < d\sigma_{geometrical} \Leftrightarrow \theta < \sigma \tag{56}$$

Thus an instable, localized deformation occurs when the strain hardening rate becomes lower than the value of flow stress. This localized deformation, called necking, eventually causes ductile fracture. Figure 28 shows that this criterion works well for predicting the ductility of pure aluminium at two temperatures.

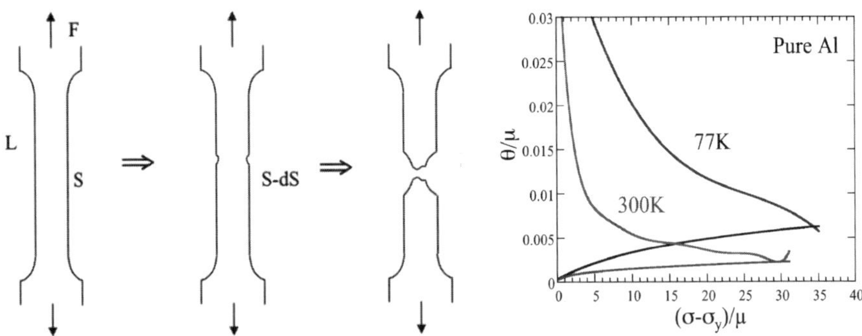

Fig. 28. Definition of necking as an unstable localization of plastic deformation in tension, and plot of the stress and strain hardening rates as a function of strain (light curves for strain hardening, dark ones for flow stress). When the two curves meet, Considère's criterion is verified and necking starts.

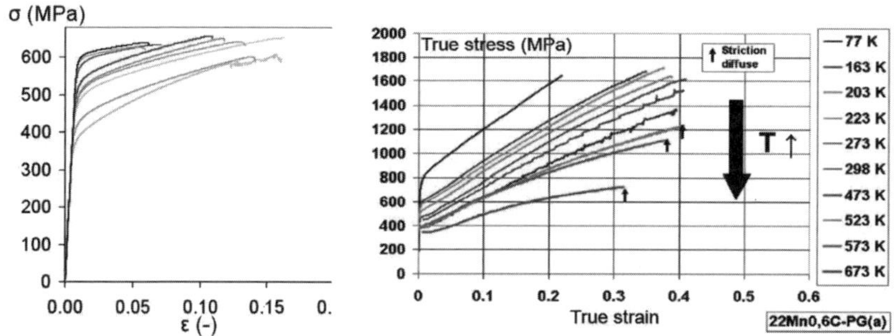

Fig. 29. (a) Stress-strain curves of an Aluminium alloy, for different states of precipitation. The increase of yield stress is accompanied by a decrease of strain hardening rate, resulting in a strong reduction of ductility. (b) Stress strain curves of a TWIP Fe-Mn-C steel as a function of deformation temperature. When temperature is reduced, the extension of twinning is increased, resulting in a very high level of strain hardening rate. This compensates for the increase in flow stress, and ductility is actually observed to increase along with flow stress (except at the lowest temperatures).

A few comments can be made on the consequences and limits of this criterion:

- Fracture can occur before the occurrence of Considères' criterion if an alternative fracture mode is triggered (like intergranular fracture).

- Considère's criterion defines the occurrence of necking (called uniform elongation) and not of fracture (called ductility or fracture strain). Although the two are related, materials can show very different post-necking behaviours. The most important parameter controlling this aspect is the material's strain rate sensitivity. If it is high this will tend to stabilize deformation (by increasing the flow stress of zones deforming more rapidly, like the necking zone), whereas if it is close to zero or, even worse, negative, deformation will lead catastrophically to fracture as soon as the Considère criterion will be met.

- In the framework of Considère's criterion, a high ductility will be favoured by low values of flow stress and by high values of strain hardening rate. This is typically met for solid solutions. Thus, increasing the yield strength of the material generally results in a loss of ductility, except if it is compensated by an increase in strain hardening rate. This is illustrated in Fig. 29 in two cases. The classical one (different states of precipitation in an aluminium alloy), where an increase in flow stress, associated to a decrease of strain hardening rate, results in a strong diminution of ductility, and a less classical case (TWIP steel) where extensive twinning during deformation (which induces a strain hardening rate) enables to simultaneously increase the flow stress and the ductility when temperature becomes lower.

5.2. *Ductile transgranular fracture*

Ductile transgranular fracture can be identified on a fracture surface by the presence of dimples, which are craters left by the material's failure around voids that have nucleated and grown in the material during plastic deformation. Such dimples can be seen in Fig. 30.

Voids are generally nucleated around second phase particles (see Fig. 30, right panel) that can then be observed in the centres of the dimples. Second phase particles can nucleate voids via several mechanisms, the most important being fracture of the particle or decohesion of its interface with the matrix (see Fig. 31). Thus the nucleation of voids will be favoured by the following characteristics of the microstructure:

- high volume fraction of second phase particles (increasing the density of voids)

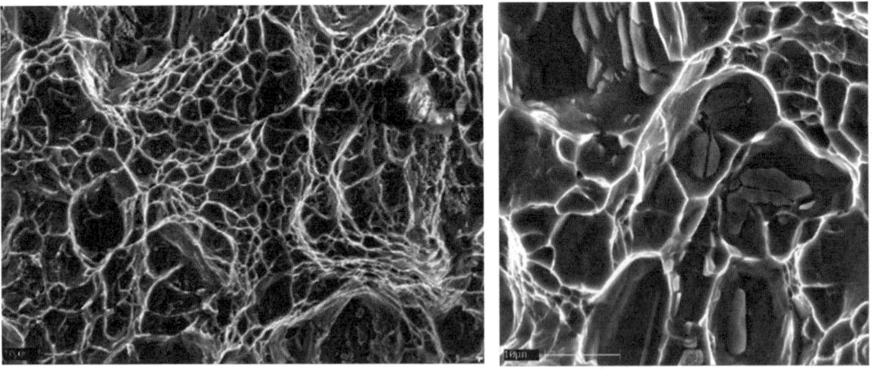

Fig. 30. Ductile fracture surfaces of an aluminium alloy observed in the scanning electron microscope, and broken particles lying in the centre of dimples.

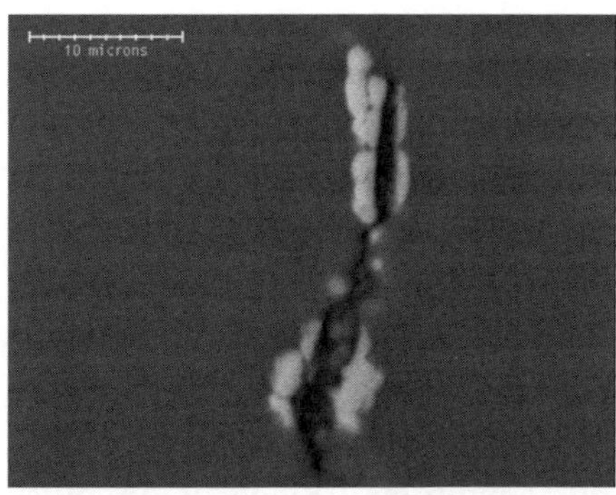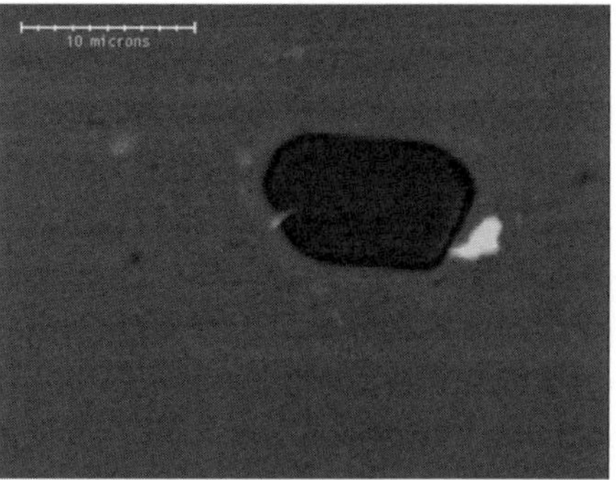

Fig. 31. Nucleation of voids during plastic deformation on second phase particles, by particle fracture or interfacial decohesion.

- large size and aspect ratio (this increases both the brittleness of the phases and the load transfer from the matrix, and results in a fracture at smaller values of strain in the matrix); any brittle phase above a □m in size will have a detrimental effect on fracture

- weak interface with the matrix (resulting in early void nucleation by interfacial decohesion)

- clustering and alignment (that can result in favoured crack propagation paths)

- high yield stress of the matrix (which increases the load transfer from the matrix to the particles)

Once voids have nucleated in the material, they will grow until void coalescence occurs leading to fracture. Void growth is mainly controlled by the stress triaxiality T which is defined as the ratio between the hydrostatic stress (pressure in the material) and the Von Mises equivalent stress (component of stress controlling plastic deformation):

$$T = \frac{\sigma_H}{\sigma_{VM}} = \frac{1}{3} tr\left(\overline{\overline{\sigma}}\right)\Big/\sigma_{VM} \tag{57}$$

The stress triaxiality in uniform tension is 1/3. Close to a notch or a crack, it can reach more than 1. This can result in accelerated void growth, and thus reduced ductility, in the cases where samples have abrupt changes in geometry.

                    A. Deschamps

5.3. *Cleavage fracture*

Cleavage fracture is a brittle fracture mode, occurring via the propagation of a sharp crack in crystallographic planes with (almost) no plastic deformation in the surrounding material. This mode results in planar fracture surfaces, except for steps that link different zones of the grains in the case where the fracture surface is not planar. These steps result in river-like lines converging to the nucleation site of fracture (see Fig. 32).

Cleavage fracture requires several conditions. Since it involves reaching the cleavage stress (theoretical fracture stress) locally, which can be extremely high, it requires high stress concentrations in the material. One good indication of this necessity is that cleavage fracture often happens in cascade in neighbouring grains, one cleavage crack serving as the initiation site for the crack in the next grain and so forth (Fig. 32b). Such high stress concentrations cannot be obtained in materials where the lattice friction stress (Peierls stress) is low, and thus cleavage generally does not appear (apart from very special circumstances) in FCC materials. In BCC materials, especially Fe and steels, it is observed at low temperature where the lattice friction becomes sufficiently high. It is also quite frequent in HCP materials, like Zn.

Actually, increasing the lattice friction stress is generally not sufficient to cause cleavage as this stress will generally stay below the theoretical stress for planar separation of the material along crystallographic planes. Cleavage is therefore highly dependent on other microstructural features such as:

- Microstructural elements favouring brittle intergranular failure may help trigger cleavage by the associated stress concentration at the tip of an intergranular fracture zone.

- Grain to grain plastic incompatibility (especially in materials like HCP crystals that have a low number of slip systems) may lead to high local stresses.

- Microstructural elements such as solute atoms that favour unstable plastic deformation, and especially avalanche movements of dislocations, may help the triggering of cleavage. It is particularly striking that extremely pure Fe can be ductile down to very low temperatures whereas a few ppm carbon make cleavage occur at much higher temperatures.

- A smaller grain size will generally make cleavage more difficult.

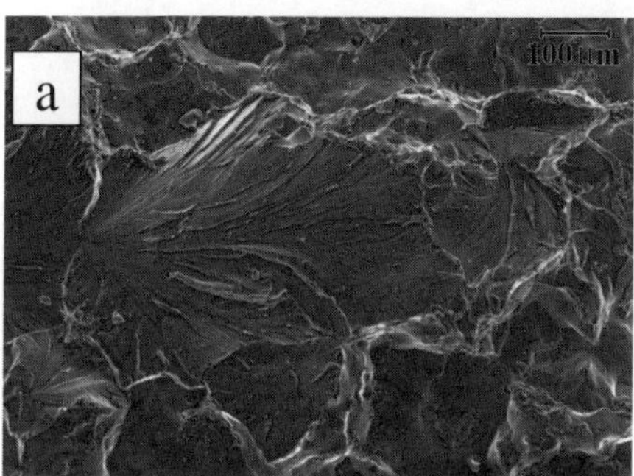
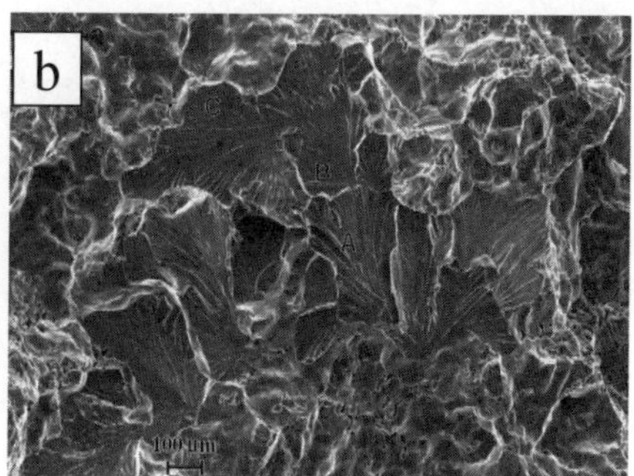

Fig. 32. a) Example of cleavage fracture in the Scanning Electron Microscope, showing the planar nature of this fracture mode, and the steps (river lines) diverging from the fracture initiation site. b) Cascade of cleavage fractures in neighbouring grains.

5.4. *Intergranular fracture*

Intergranular fracture is defined as fracture occurring along the grain boundaries. It can either be a ductile fracture mode (if accompanied by ductility in the material surrounding the grain boundary) or a brittle fracture mode (if the separation of the grains occurs without measureable plasticity). In either case, it can happen when the material is globally elastic, or after some generalized plasticity has occurred.

Figure 33 shows an example of brittle intergranular fracture. The grain boundaries are extremely well defined on the fracture surface, and no trace of plasticity can be observed. This type of fracture is particularly favoured when chemical elements that decrease the cohesion of the grain boundaries are present, either due to the processing of the material or to the chemical environment (like the penetration of hydrogen or liquid metals into the grain boundaries).

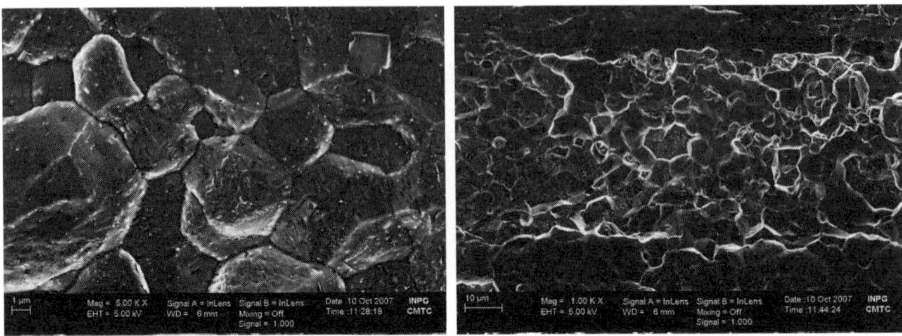

Fig. 33. Examples of brittle inter-granular fracture in an Aluminium alloy, related to hydrogen embrittlement.

Figure 34 shows an example of ductile intergranular fracture. In this case, fracture occurs at grain boundary only because damage accumulation (nucleation and growth of voids) occurs more rapidly there as compared to the grain interiors. This is caused by microstructural features such as:

- The presence of a high density of large particles at the grain boundary. These can be due either to the nucleating role of these particles during recrystallization, to the pinning role of particles during grain growth or to precipitation during heat treatments (imperfect quench, ageing).

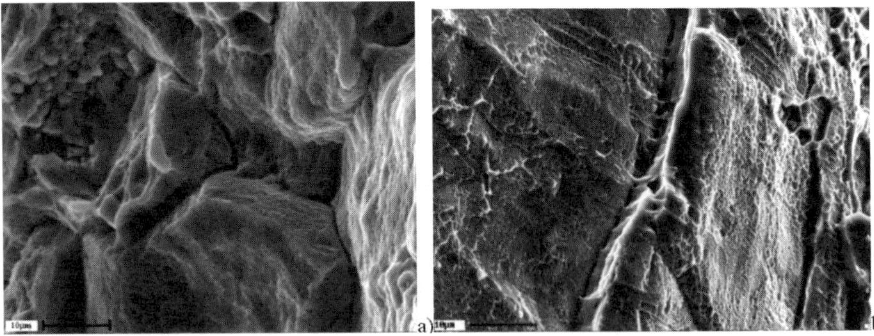

Fig. 34. Examples of ductile inter-granular fracture in an Aluminium alloy. Very small dimples can be observed at high magnification on the fractured grain boundaries.

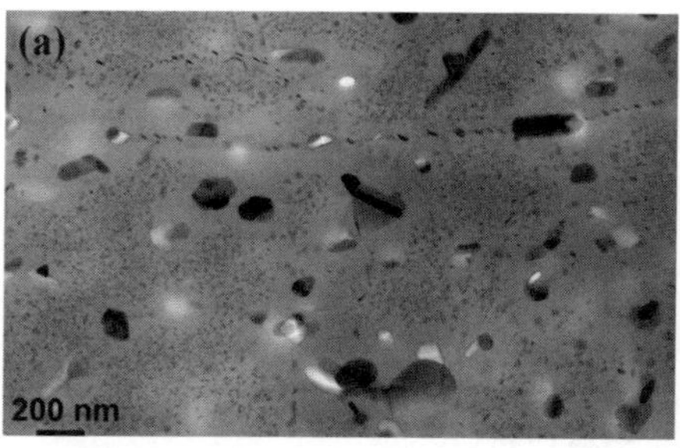
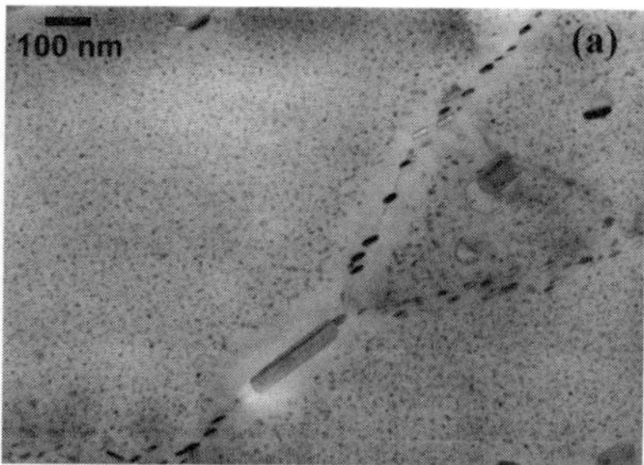

Fig. 35. Precipitate-free zones in an Al-Zn-Mg alloy after a relatively slow quench from the solutionizing temperature, followed by a precipitation ageing treatment. The grain boundaries and the coarse quench-induced precipitates are surrounded by a Precipitate-Free-Zone.

- Precipitation hardening materials often present a softer zone around the grain boundaries, where no particles are present because of lower solute or vacancy concentrations (Fig. 35). These regions, called Precipitate-Free Zones (PFZ), enter into plasticity before the bulk of the grains, which induces a high triaxiality ratio in the grain boundaries and thus an easy void nucleation and a fast growth of them.

References

1. H. Rietveld, Journal of applied crystallography, **2** 65 (1969).
2. J. Collet, PhD,. Institut Polytechnique de Grenoble, France (2009).
3. A. Bénéteau , PhD. Institute Polytechnique de Lorraine, France (2007).
4. T. Ungar, *Scripta Materialia*,**51**777 (2004).
5. B. Warren, *X-Ray Diffraction*, (Dover, New York, 1990).
6. P. Fratzl, J. of Applied Crystallography **36** 397 (2003).
7. E. Hall, Proceedings of the Phys. Soc. **B** 747 (1951).
8. N. Petch, The Journal of the Iron and Steel Institute, **174** 25 (1953).
9. R. Armstrong, I. Codd, R. Douthwaite and N. Petch, *Phil. Mag.* **7** 45 (1962).
10. H. Conrad,. *Acta Metallurgica* **11** 75 (1963).
11. J. Li and Y. Chou, Metallurgical and materials transactions **B** 1(1970).
12. J. Friedel, *Dislocations* (Pergamon Press, Oxford, 1964).
13. U. Kocks, *Phil. Mag.* **13** 541 (1966)
14. U. Kocks, *Canadian J. of Physics* **45** 737 (1967)

15. E. Schafler, K. Simon, S. Bernstorff, P. Hanak, G. Tichy, T. Ungarand M. Zehetbauer, *Acta Materialia* **53** 315 (2005)
16. U. Kocks, A. Argon and M. Ashby, *Progress in Materials Science* **19** 1 (1975)
17. U. Kocks, J. of Engineering Materials Technology **98** 76 (1976)
18. H. Mecking, B. Nicklas, N. Zarubova and U. Kocks,. Acta Metallurgica **26 267** (1986).
19. Y. Estrin, in Unified Constitutive Laws for Plastic Deformation, (Academic Press, 1996).
20. E. Voce, J. of the Institute of Metals **74 537** (1948).
21. M. Karlik and B. Jouffrey, *Journal de Physique* iii **6** 825 (1996).
22. A. Deschamps, F. Bley, F. Livet, D. Fabregue and L. David, *Phil. Mag.* **83** 677 (2003).

CHAPTER 4

DEFORMATION OF INTERMETALLIC ALLOYS AT HIGH TEMPERATURES

Gerhard Sauthoff

Max-Planck-Institut für Eisenforschung GmbH
40074 Düsseldorf, Germany
E-mail: g.sauthoff@mpie.de

An overview is given on the complex high-temperature deformation behaviour of intermetallic alloys which results from a tutorial for students of the 4th European School in Materials Science of the EU Network of Excellence "Complex Metallic Alloys" CMA. After a discussion of the basic effects of atomic ordering in intermetallic alloys on high temperature deformation various case studies are presented which are of interest for structural applications at high temperatures. The aim is to point out the special features of the deformation of intermetallic alloys at high temperatures which have to be considered when developing new intermetallic materials for structural applications at high temperatures. Intermetallics have been subject of a long series of excellent reviews which is not to be continued by the present overview. This overview is based on a recent compilation of the knowledge on intermetallics[1] and shorter reviews.[2,3] Data of intermetallic materials were collected in reference 4.

1. Introduction

1.1. *Definition of intermetallics*

Intermetallics is the short and summarizing designation for the intermetallic phases and compounds which result from the combination of various metals and which form a tremendously numerous and manifold class of materials. An example is given by the binary Ni-Al phase diagram (Fig. 1) which comprises the phase NiAl with compositions around 50

125

at.% among other Ni-rich and Al-rich phases. This NiAl phase has a significantly higher melting temperature than the constituent metals Ni and Al indicating a much stronger bonding between the unlike Ni and Al atoms than between the alike Al atoms and Ni atoms. Its crystal structure is known as B2 structure (Strukturbericht designation) which is a bcc structure with atomic ordering and which clearly differs from the fcc structure of the constituent metals.

Accordingly there is a simple general definition[5,6]: intermetallic phases and compounds are chemical compounds of metals the crystal structures of which are different from those of the constituent metals. The composition of an intermetallic may vary within a restricted composition range known as homogeneity range. This homogeneity range may be narrow or vanishing as is the case for a line compound and such phases are usually addressed as intermetallic compounds. Phases with a wide homogeneity range are usually addressed as intermetallic phases.

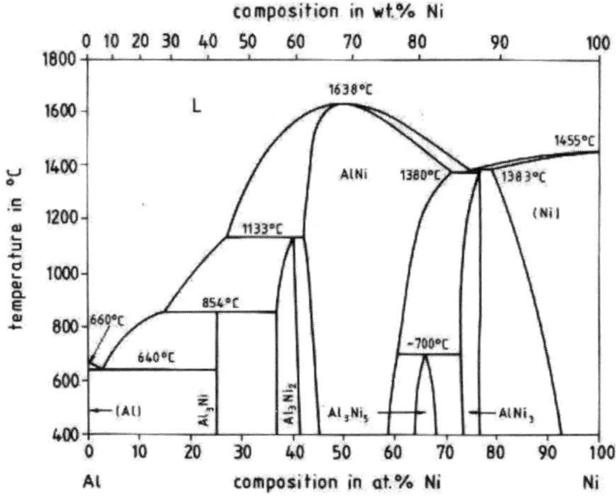

Fig. 1. Ni-Al phase diagram.[2]

Phases may exist only in a restricted temperature range. There are phases which show an order-disorder transition when heated above a critical temperature as is exemplified by the Cu₃Au phase which is fcc ordered (L12 structure) at lower temperatures and disorders above the critical disordering temperature to form the familiar fcc solid solution.[7] This case exemplifies the fact that atomic order in intermetallic compounds and intermetallic phases does <u>not necessarily</u> produce a higher stability and a melting temperature than those of the constituents.

1.2. *Applications of intermetallics*

Intermetallic materials have been used commercially since long as functional materials because of attractive physical properties, i.e. superconductor, magnetic or shape memory materials.[2,7] Atomic ordering in intermetallics is expected to provide a higher deformation resistance at high temperatures, and various intermetallics indeed show high strengths at high temperatures which induced researchers to start material developments for structural applications at high temperatures since the fifties of last century. A problem has been and still is the brittleness which usually accompanies the high strength and which makes such developments difficult. Presently there are various advanced developments as summarized in Table 1.[4]

Table 1. Intermetallic alloys of interest for structural applications at high temperatures.[4]

Material	Basic phase	Application envisaged
Advanced nickel alumi nide alloys	Ni_3Al	high-temperature wear-resistant components
alpha-2 titanium aluminide alloys	Ti3Al	jet engine components etc.
B2 nickel aluminide alloys	NiAl	gas turbine and automotive components
gamma titanium aluminide alloys	TiAl	aircraft, jet engine, automotive components etc.
iron aluminide alloys	Fe_3Al, FeAl	chemical engineering
molybdenum disilicide alloys	$MoSi_2$	aircraft, automotive components etc.

2. Basic features of Deformation of inTermetallics at High Temperature

2.1. *Dislocation effects*

The increased Burgers vector length of a perfect dislocation in an ordered structure produces dislocation splitting to form a superlattice dislocation (superdislocation for short) with planar faults – antiphase boundaries and/or stacking faults – between the partials depending on the type of crystal structure as visualized in Table 2.[8]

Table 2. Superlattice dislocations for various intermetallic alloy types (NN = disordered with respect to nearest neighbours, NNN = disordered with respect to next nearest neighbours, SF = stacking fault.[8]

Superlattice type (Strukturbericht)	Alloy types	Superlattice dislocation type	Burgers vector	Antiphase boundary type
B2	FeAl NiAl FeCo	⊥	$a_o\langle 100\rangle$	None
			$\frac{1}{2}a_o\langle 111\rangle$	//// NN
DO3	Fe3Al, Fe3Si Fe3B		$\frac{1}{4}a_o\langle 111\rangle$	//// NN　×××× NNN
L12	Cu3Au Ni3Mn Ni3Al Ni3Fe Ni3Si Ni3Ti		$\frac{1}{6}a_o\langle 112\rangle$	//// NN　×××× NN+SF

The complex structure of such superdislocations results in a migration behaviour with glide and cross-slip with mobility depending on orientation and temperature in a complex way. A most striking and important example is the Kear-Wilsdorf mechanism which produces the anomalous temperature dependence of the yield stress of Ni_3Al and other phases with L12 structure.[9] An illustrative example is given by monocrystalline Ni_3Ga

which shows a yield stress of about 200 MPa at room temperature and a relative yield stress maximum of about 700 MPa at 600 °C and which has been analysed in much detai.[10] Such flow-stress anomalies are also shown by other intermetallic phases with other crystal structures, and various mechanisms have been proposed for these phases.[11]

2.2. *Point defect effects*

Perfect atomic ordering in an intermetallic phase is possible only at the stoichiometric composition neglecting thermal vacancies. Deviations from stoichiometry produce constitutional point defects, i.e. either constitutional vacancies or antisite atoms or both.[12] The respective partial defect concentrations including the thermal defects can be determined as a function of phase composition through first-principles calculations as has been done e.g. for PdIn with B2 structure containing Pd vacancies, In vacancies, antisite Pd atoms and antisite In atoms.[13]

Point defects act as obstacles to dislocation movement at low temperatures and thus contribute to hardening as was studied in detail for the ternary B2 phase (Ni,Fe)Al with experimental determination of the concentrations of vacancies, antisite Ni atoms and antisite Fe atoms as a function of the Al content characterizing the deviation from stoichiometry.[14] The effect on hardening was indeed minimum only for the stoichimetric composition of the binary NiAl phase with 50 at.% Al where the partial or complete substitution of Fe for Ni resulted in a complex less simple variation of hardening with composition. In addition, the more Fe substituted for Ni, the more hardening increased with the temperature of prior quenching which is caused by the reduced vacancy mobility in FeAl. In FeAl the activation energy for vacancy migration is higher than that for vacancy creation which is exceptional and in contrast to NiAl and most other intermetallic phases. Consequently, quenching of FeAl from high temperatures produces excess vacancies which heal out only during very long times of annealing.

At high temperatures, point defects are mobile and enhance diffusion as was studied for NiAl (with In as tracer for Al).[15] Thus deviations from stoichiometry at high temperatures produce softening through enhanced diffusivity as is exemplified by Fig. 2. The data for three (Ni,Fe)Al alloys clearly indicate that the off-stoichiometric (Ni1.0Fe0.2)Al0.8 shows the highest yield stress at low temperatures and the lowest yield stress at high temperatures with a transition from low temperatures to high temperatures at about 700 °C.

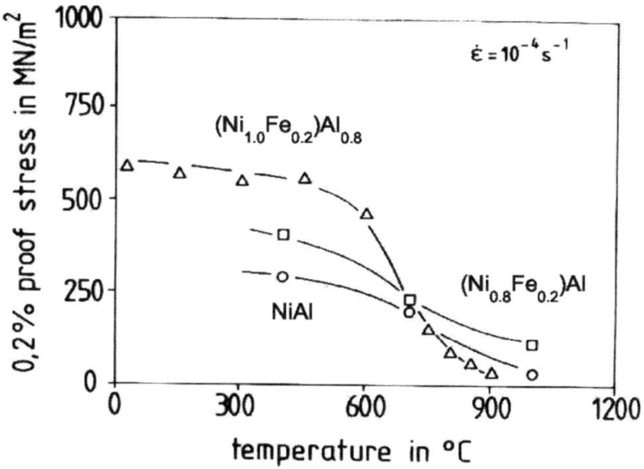

Fig. 2. Flow stress (0.2%-proof stress in compression at 10^{-4} s^{-1} strain rate) as a function of temperature for binary and ternary NiAl phases, i.e. stoichiometric NiAl (circles), stoichiometric $(Ni_{0.8}Fe_{0.2})Al$ (squares) and off-stoichiometric $(Ni_{1.0}Fe_{0.2})Al_{0.8}$ (triangles).[2]

In the transition range from low temperatures to high temperatures the mobilities of such point defects and those of moving dislocations are comparable which give rise to mutual interactions as indicated by the stress-strain curves for Fe$_3$Al alloys in Fig. 3 which show a yield stress drop at about 600 °C.

The data in Fig. 4 for FeAl alloys with varied Al content demonstrate the sensitive and complex variation of this yield stress drop with increasing Al content which corresponds to the complex variation of the point defect concentrations noted above.

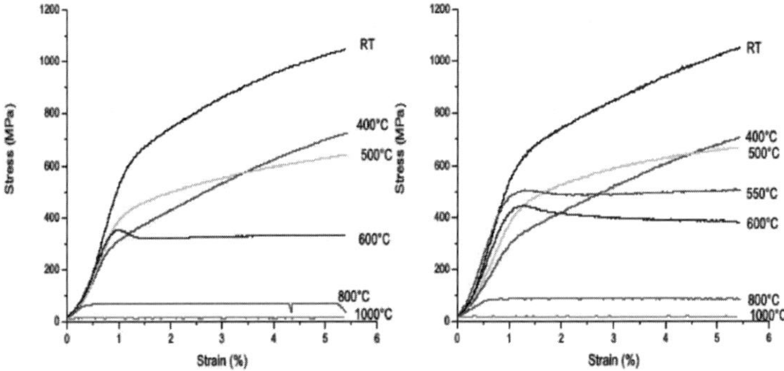

Fig. 3. Stress strain curves of Fe_3Al alloys (Fe-25Al-1Ta): as-cast (left) and heat treated at 1000 °C for 200h (right).[16]

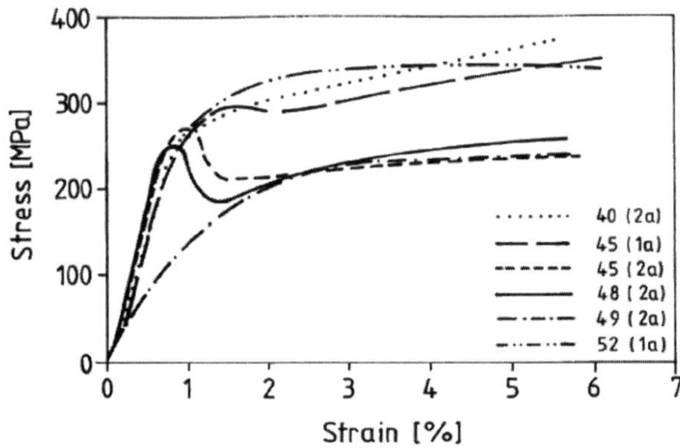

Fig. 4. Stress-strain behaviour of various polycrystalline FeAl alloys at 600 °C and 10^{-3} s-1 strain rate (the numbers indicate the nominal Al content in at.% with the alloy series number in brackets).[17]

2.3. *Creep*

Loading with constant stress at high temperatures produces continuous creep with the familiar creep stages, i.e. primary creep with decreasing creep rate, secondary creep with ideally constant creep rate corresponding to stationary creep, and tertiary creep with increasing creep rate due to internal damaging as is exemplified by Fig. 5 with creep curves for an intermetallic TiAl alloy. It is noted that alloys with complex microstructures, in particular multiphase alloys, show only a relatively short secondary creep stage corresponding to a creep rate minimum which represents a transition from primary creep to tertiary creep.

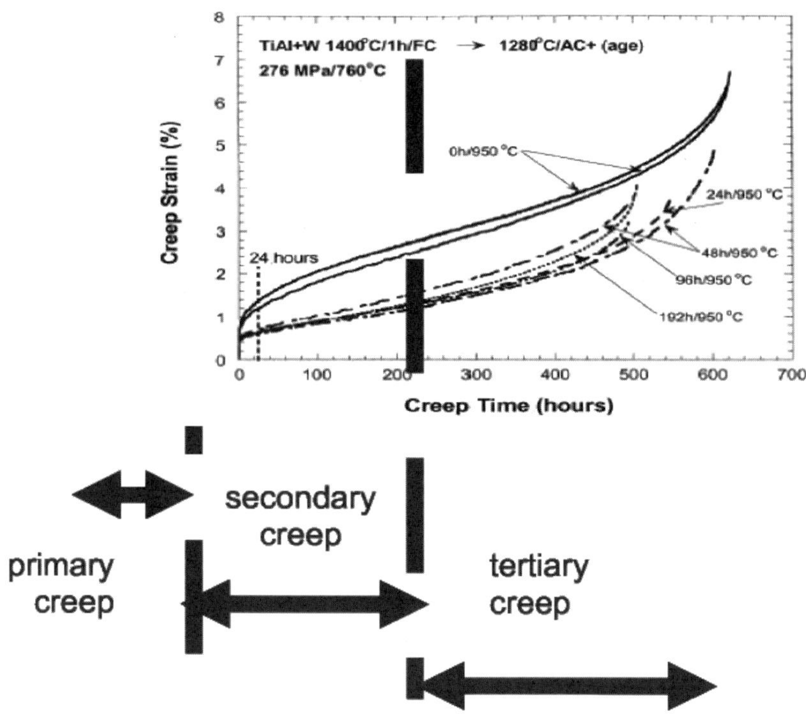

Fig. 5. Creep curves, i.e. strain as a function of time, for a TiAl alloy after various prior heat treatments.[18]

2.3.1. *Primary creep*

Primary creep is transient instationary flow and may be descsribed by the simple Mecking-Estrin Hybrid Model which considers the concurrent action of hardening and softening[19] and which describes experimental data satisfactorily as illustrated by Fig. 6.

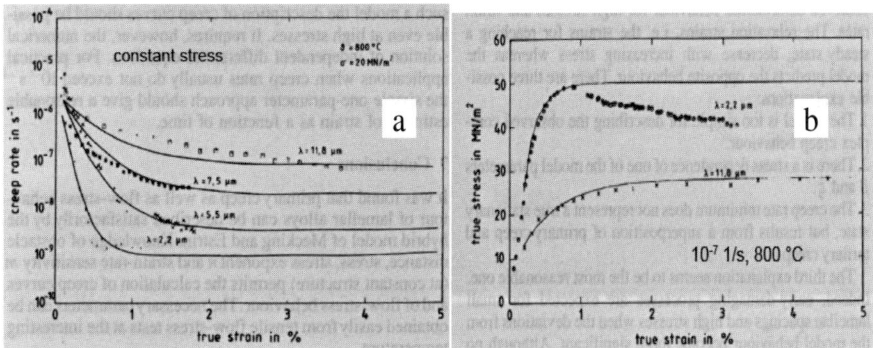

Fig. 6. Creep rate as a function of strain at constant creep stress a) and stress as a function of strain for constant creep rate b) for a two-phase Fe-42 at.% Ni-18 at.%Al alloy with a fcc-Ni and intermetallic NiAl (symbols: experimental data, drawn lines: model calculation.[20]

2.3.2. *Secondary creep*

Secondary creep is stationary plastic flow with softening balancing hardening. With mobile dislocations, there is dislocation creep with glide and climb of dislocations which is described by the familiar Dorn equation

$$\varepsilon' = (AGbD/(kT))(\sigma/G)^n \tag{1}$$

with the creep rate ε', the microstructure-dependent factor A of proportionality, Burgers vector b, diffusion coefficient D, Boltzmann constant k, absolute temperature T, creep stress σ, and stress exponent n which is in the range 3 – 5 for monophase alloys and higher for multiphase alloys – see e.g.ref. 21. Without mobile disloations, there is still diffusion creep with creep strain through diffusive flow of atoms. In the case of lattice diffusion controlling the creep rate with atoms diffusing through the grains which is known as Nabarro-Herring creep, the creep rate is described by the constitutive equation

$$\varepsilon' = A\Omega DL\sigma/(d^2 kT) \tag{2}$$

with the atomic volume Ω and the grain size d, whereas in the case of grain-boundary diffusion controlling the creep rate with atoms diffusing along the grain boundaries which is known as Coble creep, the creep rate is described by

$$\varepsilon' = A\Omega wDgb\sigma/(d^3kT) \tag{3}$$

with the grain boundary thickness w.[21] The transition from dislocation creep with stress exponents 3 or higher to diffusion creep with stress exponent 1 can only be observed at very low stresses and strain rates as is illustrated by Fig. 7. The action of the various creep mechanisms can best be visualized by Ashby's deformation mechanism maps.[21] The respective map for the case of Fig. 7 is shown in Fig. 8 for a grain size of 60 μm.[22] It is noted that this map additionally considers small apparent threshold stresses for both dislocation creep and diffusion creep.

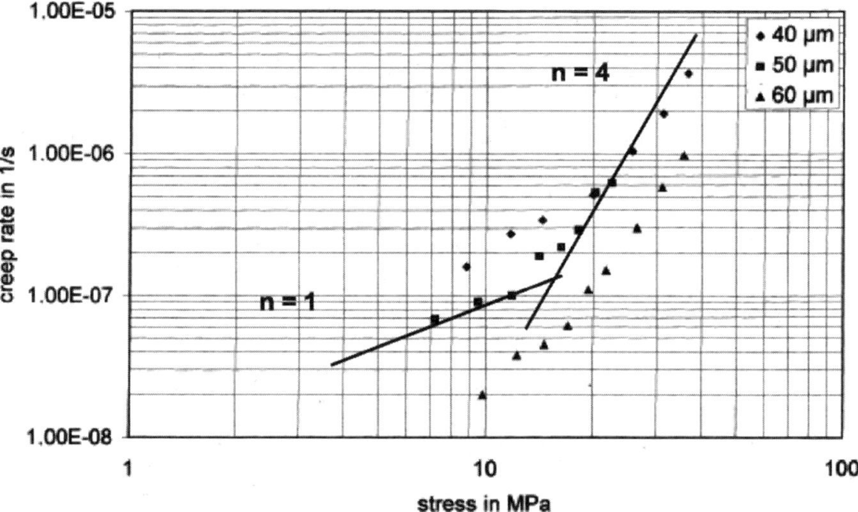

Fig. 7. Creep rate as a function of stress at 1000 °C for TiAl alloys with 51 at.% Al with precipitated Ti3Al and various grain sizes.[22]

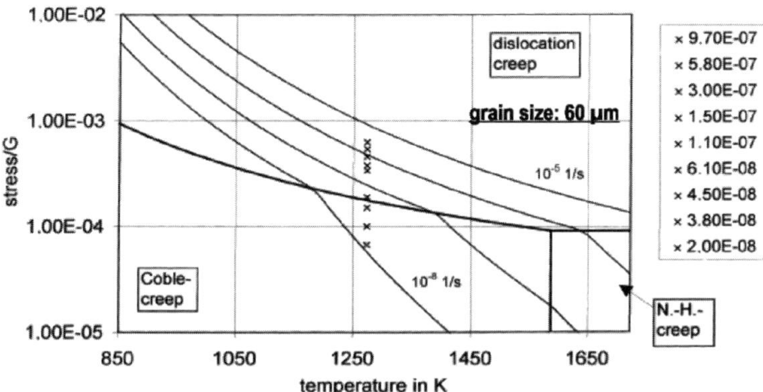

Fig. 8. Deformation mechanism map for the case of Fig. 7 with small apparent threshold stresses for both dislocation creep and diffusion creep.[22]

As already discussed for the yield stress in the preceding section, deviations from stoichiometry produce constitutional defects which are mobile at high temperatures leading to softening. Accordingly such deviations from stoichiometry enhance creep and reduce the creep resistance (stress for a given creep rate) as is exemplified by Fig. 9.

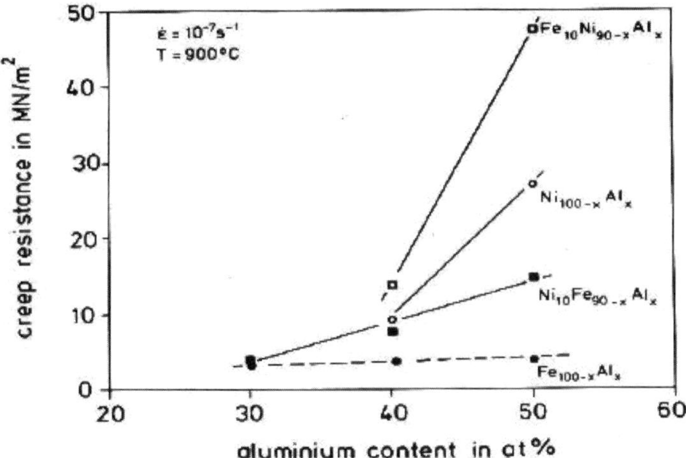

Fig. 9. Creep resistance, i.e. creep stress for 10-7 s-1 strain rate (in compression), of binary and ternary B2 aluminides at 900 °C as a function of Al content.[2]

Solute atoms affect the diffusion coefficient and thereby the creep rate according to the above constitutive equations as is demonstrated by Figs. 10 and 11.

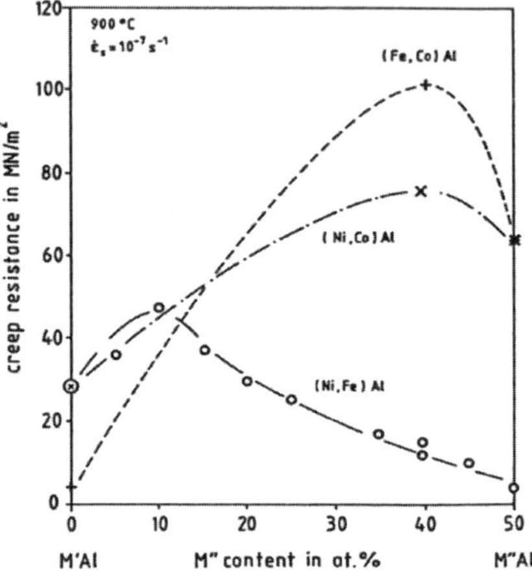

Fig. 10. Creep resistance, i.e. creep stress for 10-7 s-1 strain rate (in compression), of ternary stoichiometric B2 aluminides (M',M")Al with M', M" = Fe, Co, Ni, respectively at 900 °C as a function of M" content.[23]

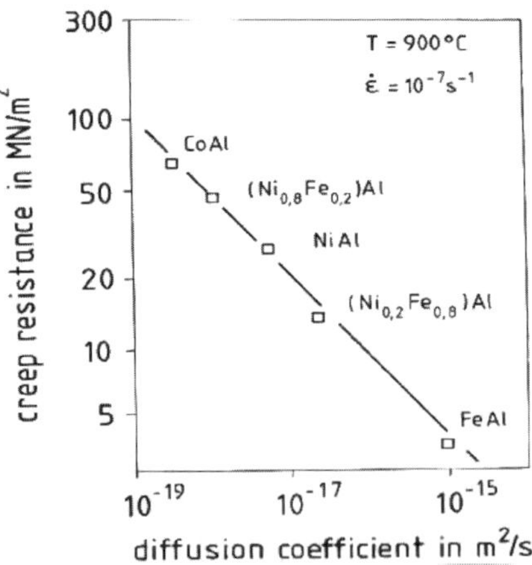

Fig. 11. Creep resistance, i.e. creep stress for 10-7 s-1 strain rate (in compression), of ternary stoichiometric B2 aluminides (M',M")Al at 900 °C as a function of temperature.[2]

2.3.3. *Tertiary creep*

Creep in the tertiary stage accelerates due to internal damaging with cavitation and microcracking leading to final rupture, i.e. creep life is finite. As a rule, the creep life time tr is correlated with the secondary creep rate εs' according to the Monkman-Grant relation:

$$\varepsilon_s' - t_r = \text{constant} \tag{4}$$

- see e.g. ref. 24. A physically based constitutive equation for tertiary creep with consideration of the microstructural processes is not yet available because of lack of reliable data on the kinetics of damage evolution. Creep damaging is thus described up to now only through using a phenomenological damage parameter as proposed e.g. by M. Kachanov.[25]

Only recently the damage evolution during creep in a brass alloy (Cu-40 wt.% Zn-2 wt.% Pb) was successfully observed in-situ using synchrotron radiation microtomography.[26] As an example, Fig. 12 clearly shows that the number of visible cavities is increasing with increasing creep time and that these cavities grow and coalesce. Such microtomographic measurements allow for a quantitative determination of the size distributions of the cavities during creep as shown in Fig. 13.

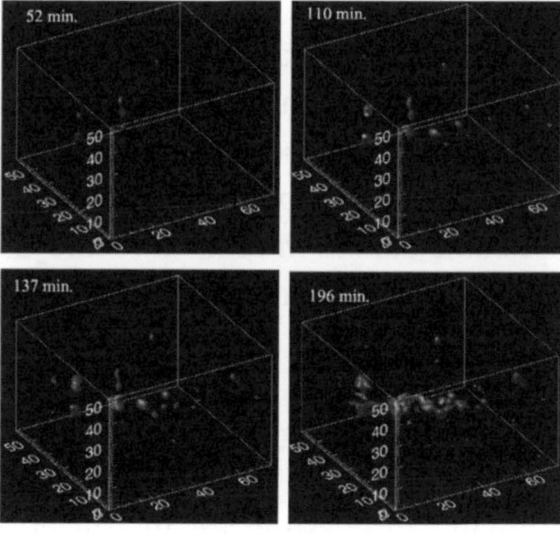

Fig. 12. Cavity evolution, i.e. cavity distribution after various creep times in a selected volume during creep of Cu-40 wt.% Zn-2 wt.% Pb with an applied stress of 25 MPa at 375°C (scale units are pixels (1 pixel = 1,6μm)).[27]

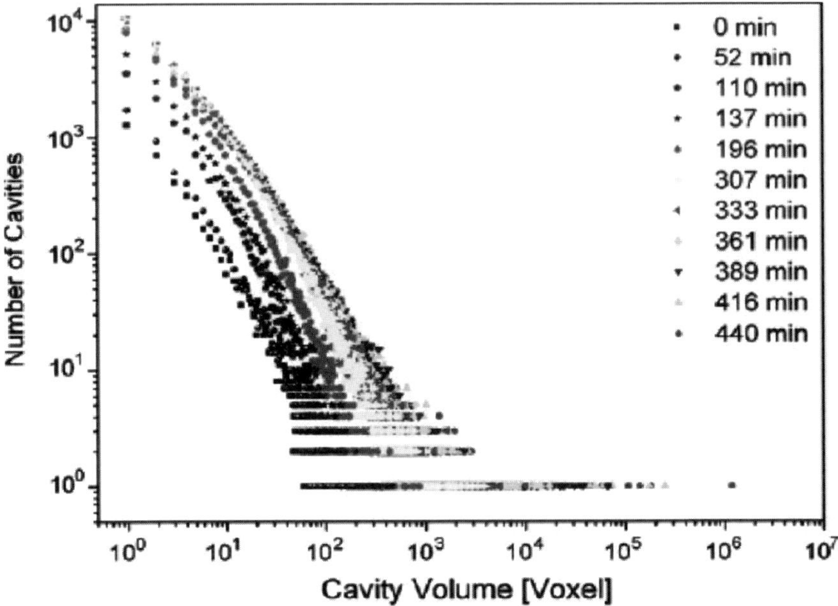

Fig. 13. Cavity size size distributions, i.e. number of cavities as a function of cavity volume (1 voxel = 1.6 μm^3) after various creep times for Cu-40 wt.% Zn-2 wt.% Pb with an applied stress of 25 MPa at 375°C.[26]

The present discussion of the basic features of plastic deformation of intermetallics at high temperatures including creep shows that the familiar theoretical models, which have been developed for describing and understanding the plastic behaviour of metallic materials at high temperature, are applicable also to intermetallic alloys if the formation of superlattice dislocations and constitutional point defects is properly considered.

3. Case Studies

Subjects of the following discussion are intermetallic phases with Al which have been considered since long as candidate phases for developing novel high-temperature materials. The reason for this is that such phases with sufficient contents of Al can form protective adherent oxide scales of

α-Al_2O_3 at temperatures above 1000 °C which provide a sufficient oxidation resistance and is a prerequisite for high-temperature application. Phases with Cr are not useful for applications above 1000 °C since they form scales of Cr oxides which evaporate above 1000 °C.[7]

3.1. *Titanium aluminides and related phases*

3.1.1. *Ti₃Al*

The Ti-rich aluminide Ti_3Al with hexagonal D019 structure was selected already in the 1950s as candidate phase for materials development because of its low density. Figure 14 shows the temperature dependence of the strength and ductility of this phase. Obviously this phase is brittle up to a brittle-to-ductile transition temperature (BDTT) of about 600 °C. Only above this temperature a yield stress can be determined. Below the BDTT only a fracture stress was determined without prior plastic deformation. The observed deformability was accompanied by microcracking which illustrates the difficulties of determining the plastic ductility in a reliable way.

A necessary condition for plastic ductility is given by the von Mises criterion , according to which 5 independent slip systems are necessary for homogeneous plastic deformation.[29]

In the D019 structure of Ti_3Al, 5 independent slip systems can be activated for the movement of dislocations[30] which would satisfy the Von Mises criterion. However, single-crystal studies on Ti_3Al have shown that the yield stresses for the different slip systems are widely different[31, 32] and thus not all 5 slip systems are activated during the deformation of polycrystalline Ti_3Al. Since there is no stress-relieving twinning as in hexagonal metals, the insufficient number of slip systems leads to strain incompatibilities and stress concentrations at grain boundaries from which cleavage fracture results.[30, 33, 34]

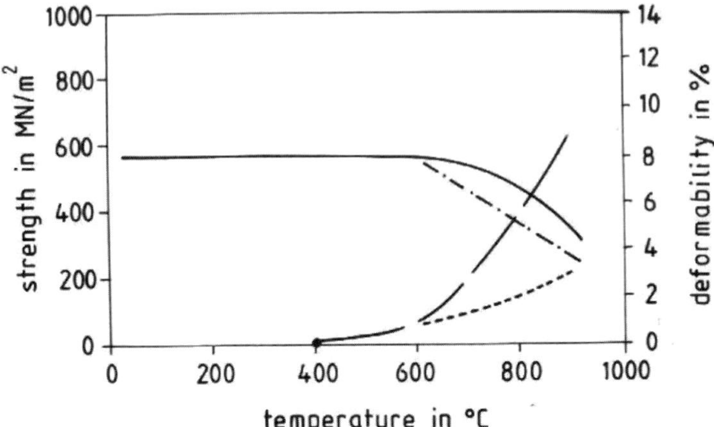

Fig. 14. Tensile strength and plastic deformability as a function of temperature for single-phase polycrystalline stoichiometric Ti3Al (—— fracture stress, — · — yield stress, — - apparent deformability including microcracking, - - - - estimated deformability without microcracking. [2, 28]

Various materials developments on the basis of Ti3Al were carried out in order to increase both the strength and plastic deformability.[34, 35] The most advanced Ti$_3$Al alloys were obtained through alloying with large amounts of Nb as is exemplified by Fig. 15. Ti$_3$Al-base alloys with engineering significance are known as α2 alloys and super- α2 alloys with 10-30 at.% Nb.[4, 35] Such large amounts of Nb lead to multiphase alloys which additionally contain β-Ti in the disordered state with fcc structure or in the ordered state with B2 structure and/or the orthorhombic O phase. The deformation behaviour of these Ti$_3$Al-Nb-base alloys and in particular the balance of strength and ductility is controlled by the nature and distribution of the various phases in the alloys, i.e. by the multiphase microstructure which is produced by the thermal and mechanical processing of the alloys.[36] Advanced developments are in progress – see e.g. refs.37 and 38.

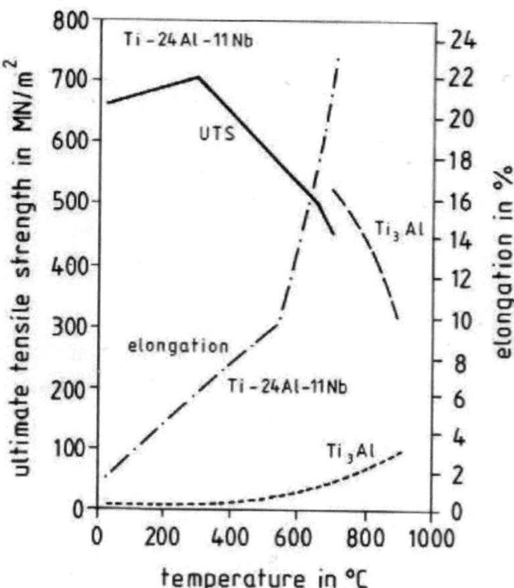

Fig. 15. Temperature dependence of the ultimate tensile strength UTS and the tensile elongation of Ti-24at.%Al-1 1at.%Nb (——_and —·—·—·-, respectively) in comparison to that of Ti3Al (———— and ----, respectively.[2, 35]

3.1.2. *TiAl*

The titanium aluminide TiAl - known as γ phase - crystallizes with the tetragonal L10 structure which is basically an fcc lattice with atomic ordering and tetragonal distortion. Figure 16 shows the temperature dependence of the strength and ductility of the TiAl phase. As in the case of the Ti3Al alloys, the phase is brittle up to a BDTT of about 700 °C, above which plastic deformation through thermally activated dislocation motion occurs. The materials developments on the base of TiAl for improving strength and ductility rely primarily on a reduction of the Al content which produces two-phase TiAl alloys with Ti₃Al as second phase.

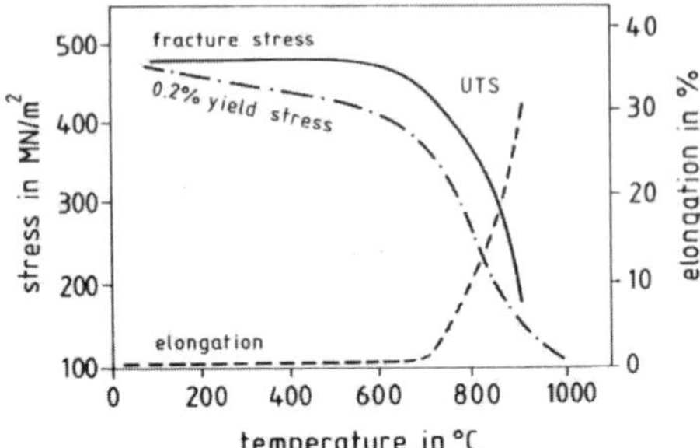

Fig. 16. Tensile strength - i.e. fracture stress, ultimate tensile strength (UTS), and 0.2 % yield stress - and plastic deformability - i.e. tensile elongation - as a function of temperature for single-phase polycrystalline TiAl with 54 at.% Al.[2, 39]

The balance of strength and ductility is then controlled by the amounts and distribution of phases which is optimized by appropriate thermal and mechanical pretreatments. Representative data are shown in Fig. 17 which illustrates the variation of properties as a function of microstructure variation.

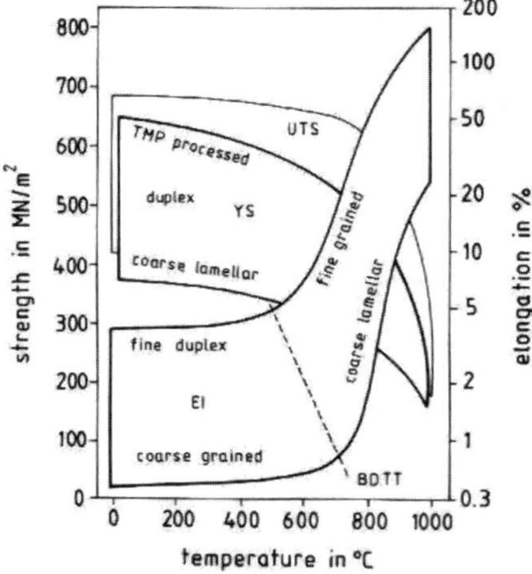

Fig. 17. Temperature dependence of ultimate tensile strength (UTS), tensile yield strength (YS), brittle-to-ductile transition temperature (BDTT), and tensile elongation (EL) for two-phase TiAl alloys with various microstructures produced under various processing conditions, in particular thermomechanical processing.[2, 40]

Further optimization of the ratio of high-temperature strength and ductility is achieved by alloying with further elements, in particular W and Nb and thermomechanical treatments – see e.g. refs.41 – 45.

3.1.3. *TiAl$_3$*

The trialuminide TiAl$_3$ crystallizes with the tetragonal D022 structure which is common to various other trialuminides. However, TiAl$_3$ is a line compound and is inherently brittle. Additiional alloying allows for preparing ternary trialuminides with compositions corresponding approximately to Al$_{66}$Ti$_{25}$M$_9$ with M = Cr, Mn, Fe, Co, Ag, Cu which show the cubic L12 structure.[46] Since a reduced brittleness is expected for cubic structures with sufficient numbers of slip systems for plastic deformation in accordance with the von Mises criterion[29], these trialumindes have been regarded as promising for materials developments and are subject of ongoing research.[47 – 50]

3.2. Nickel aluminides and related phases

3.2.1. *Ni$_3$Al*

The nickel aluminide Ni$_3$Al with cubic L12 structure is the familiar strengthening γ' phase in the Ni-base superalloys[51] and thus is the best known and most studied intermetallic. Its enormous strengthening effect in the Ni-base superalloys results from its anomalous temperature dependence of the yield stress, i.e. the yield stress of Ni$_3$Al increases with increasing temperature up to a critical temperature, and only above this temperature the familiar normal softening with rising temperature occurs, which was already discussed in Section 2.1. It is noted that the strengthening γ' phase is precipitated coherently with only a small lattice misfit which is accommodated elastically. During creep the interaction between the applied stress and the mismatch stress of the precipitated γ' particles gives rise to the rafting effect, i.e. asymmetric growth – see e.g. Ref. 52.

                                        G. Sauthoff

3.2.2. *NiAl*

The creep behaviour of NiAl with the effects of deviations from stoichiometry were already discussed in Section 2.3.2. The comparatively high diffusivity of NiAl, which is due to the open B2 structure, is disadvantageous for structural applications at high temperatures since it enhances creep, i.e. the creep resistance of NiAl is comparatively low. As in conventional alloys, the creep resistance can be increased through precipitation hardening. Even precipitate particles, which are softer than the matrix, are obstacles for moving dislocations - see Fig. 18 - and improve the creep resistance as is illustrated by Fig. 19.

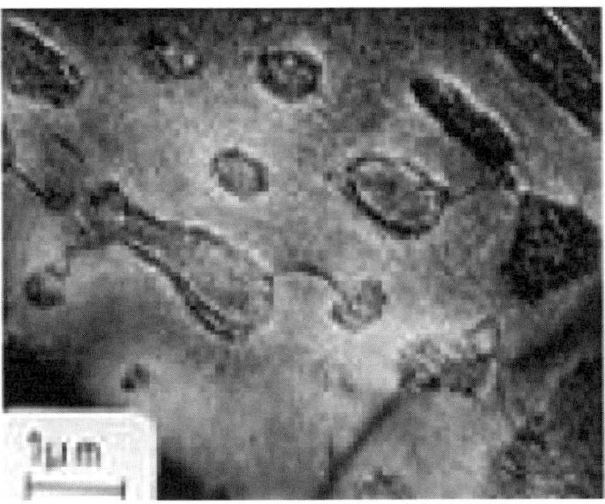

Fig. 18. Transmission electron microscopical micrograph of a two-phase Ni-44 at.% Fe - 28 at.% Al with substoichiometric B2 (Ni,Fe)Al matrix and about 15 vol.% bcc Fe precipitate particles after compressive creep at 900 °C.[54]

Clearly the precipitated bcc Fe particles shift the creep rate curve to higher stresses corresponding to a higher creep resistance. However, the matrix phase is off stoichiometric due to the thermodynamic local equilibrium between precipitate particles and matrix phases, and the decrease of creep resistance due to off-stoichiometry is larger than the increase due to precipitate hardening.

Nevertheless Fe-base alloys with a high volume-fraction of coherently precipitated NiAl particles can be used for developing novel ferritic

Fe-base superalloys with microstructures analogous to those of the familiar Ni-base superalloys as is exemplified by Fig. 20.[56]

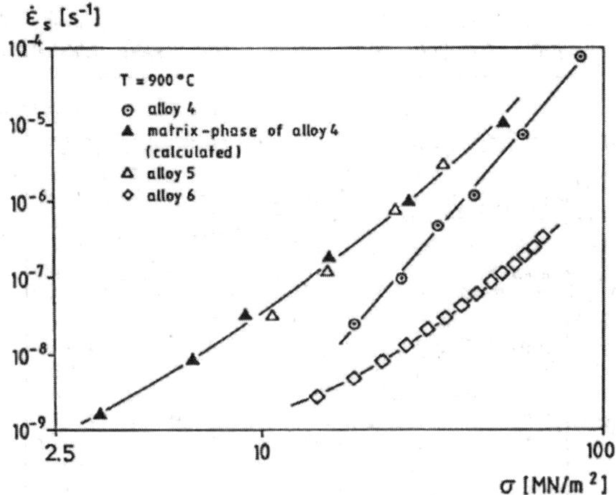

Fig. 19. Secondary creep rate at 900 °C as a function of compressive stress for various NiAl alloys: two-phase Ni-44 at.% Fe - 28 at.% Al with substoichiometric B2 (Ni,Fe)Al matrix and about 15 vol.% bcc Fe precipitate particles (alloy 4), single-phase Ni-10 at.% Fe - 40 at.% Al, i.e. substoichiometric B2 (Ni,Fe)Al (alloy 5) corresponding to the matrix in alloy 4, and single-phase Ni-10 at.% Fe - 50 at.% Al, i.e. stoichiometric B2 (Ni,Fe)Al (alloy 6).[54, 55]

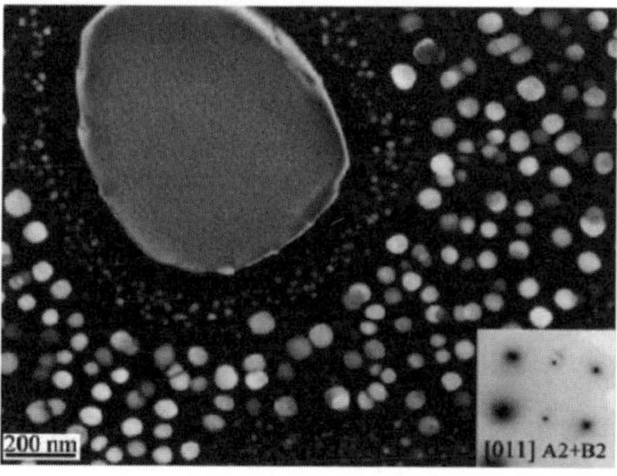

Fig. 20. Dark field TEM image of a ferritic Fe-base alloy with a high volume fraction of (Ni,Fe)Al precipitate after aging at 900 8C followed by air cooling with aged precipitate and cooling precipitates of different size. Note the phase separation within the large (Ni,Fe)Al precipitate.[56]

The achieved creep resistancies indeed compare favourably with those of a respective conventional ferritic steel with only small amounts of NiAl precipitates as shown in Fig. 21 (the data are plotted according to power law behaviour which visualises the presence of the threshold stresses).

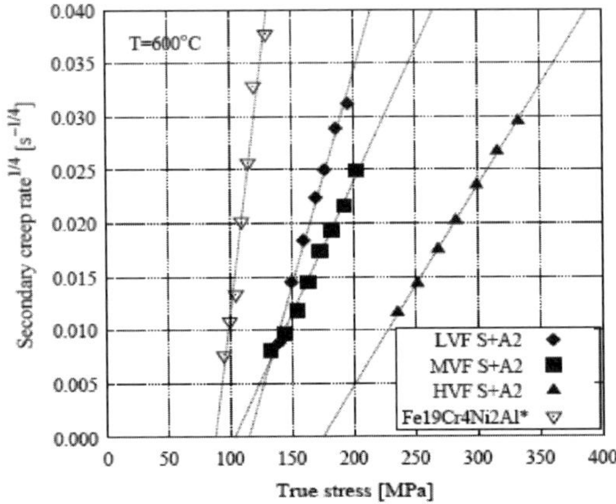

Fig. 21. Secondary creep rate vs. true stress tested at 600 °C for Fe-base alloys with high volume fractions of (Ni,Fe)Al precipitate (the alloys represented by the closed symbols were solution treated at 1200 °C for 24 h followed by aging at 750 °C for 300 h, * conventional ferritic Fe19Cr4Ni2Al steel).[56]

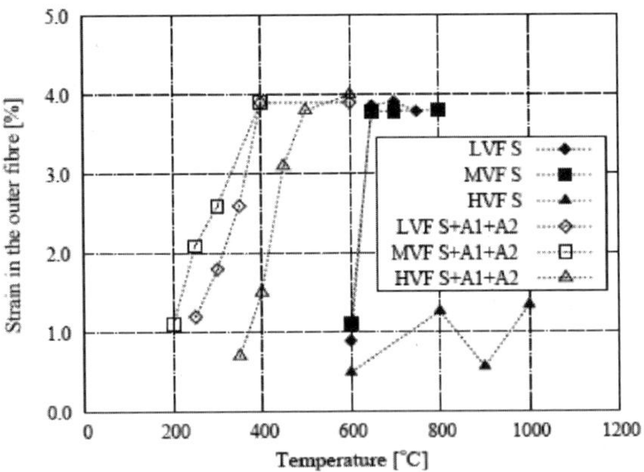

Fig. 22. Strain in the outer fibre of specimens in 4-point bending tests as a function of temperature for Fe-base alloys with high volume fractions of (Ni,Fe)Al precipitate.[56]

However, the increased creep resistancies are accompanied by comparatively high brittle-to-ductile transition temperatures as visible in Fig. 22.

A novel NiAl-base alloy development relies on strengthening the NiAl phase by precipitation of the ternary Laves phase TaNiAl with hexagonal C14 structur.[57, 62] The obtained NiAl-base Ni-Al-Ta-Cr alloy with 45 at.% Al, 2.5 at.% Ta and 7.5 at.% Cr, which is known as alloy IP75, contains a comparatively coarse distribution of Laves phase particles and a fine distribution of Cr-rich particles in a NiAl matrix – see Fig. 23.

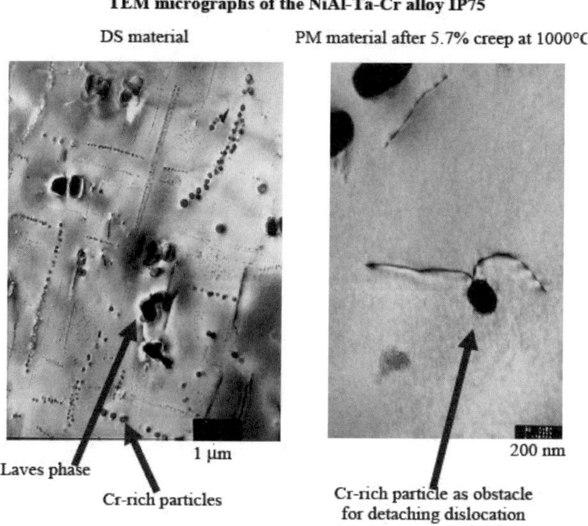

Fig. 23. TEM (transmission electron microscopy) micrographs of the NiAl-base Ni-Al-Ta-Cr alloy IP75 (45 at.% Al, 2.5 at.% Ta and 7.5 at.% Cr) with coarse Laves phase precipitates and fine Cr-rich precipitates after directional solidification (DS, left) and of powder-metallurgically processed materials after 5.7 % creep at 1000 °C (right).[63]

The high temperature strength and creep resistance as well as the oxidation resistance allows for applications up to about 1200 °C. The alloy is brittle, but it is shock-resistant in spite of the brittleness. Both ingot metallurgy and powder metallurgy processing has been established. Figures 24a – c visualize the effects of processing on the deformation behaviour. Combustor liner model panels have been precision cast and tested successfully.

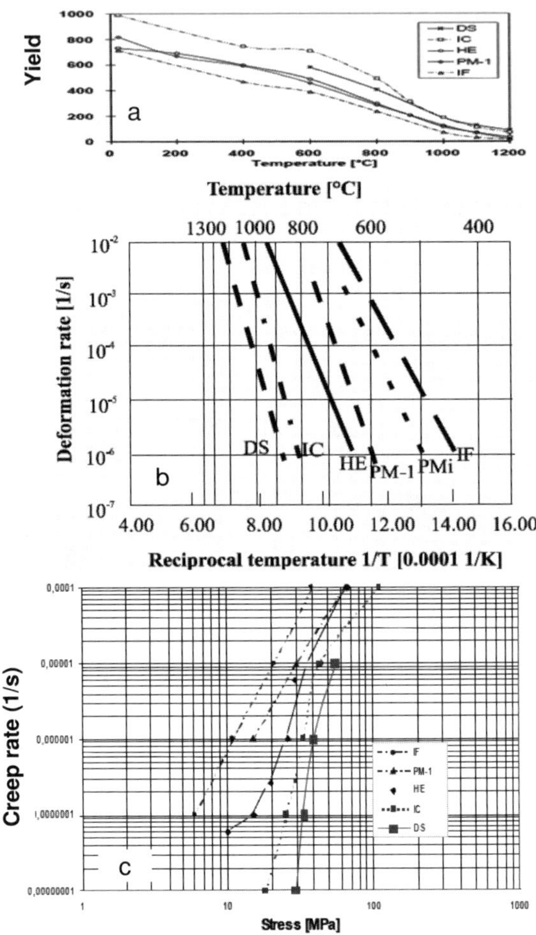

Fig. 24. Effects of processing (DS: directionally solidified, IC: investment cast, HE: hot extruded, PM-1: powder metallurgically processed, PMi: PM-1 alloy with 600 ppm O2, IF: isothermally forged) on yield stress as a function of temperature a), brittle-to-ductile transition temperature (in 4-point bending with 1% plastic strain in the outer fibre) as a function of deformation rate b), and secondary creep rate at 1100 °C as a function of applied stress in compression tests with stepwise loading.[59]

3.3. *Iron aluminides and related phases*

Fe-Al alloys including the iron aluminides are most attractive for high-temperature applications because of their excellent hot-gas corrosion resistance.[64, 65] The ductility of the disordered Fe-Al alloys with Al contents below those of the intermetallic phases Fe_3Al and FeAl decreases with increasing Al content[66] and the aluminides Fe_3Al and FeAl are known as being brittle. However, the yield strength and the ductility of the latter ordered phases varies in a complex way with degree of ordering, i.e. there is a relative ductility maximum with about 8 % elongation at about 28 at.% Al.[67] The yield strength depends very sensitively on the prior heat treatment[68] which is in contrast to the usually observed behaviour of intermetallics. The reason for this special behaviour is the presence of excess vacancies – see Section 2.2 -, which are easily created at high temperatures by thermal activation - the more the higher the temperature is - because of the very low enthalpy of formation and which anneal out at lower temperatures only very slowly because of their high enthalpy of migration.[68, 69] Hardness and yield strength increase with increasing vacancy concentration, and the annihilation of the excess vacancies needs long annealing times at low temperature, i.e. several days at 400 °C.[68]

As in the case of the NiAl alloys, there is a demand for increased high-temperature strength and creep resistance in view of structural applications at high temperatures. This can be accomplished by ternary alloying in order to make use of solid-solution hardening, hardening through enhanced ordering at higher temperatures, precipitation hardening and/ or combinations of the various possibilities.[70]

Figure 25 clearly shows the effect of alloying on ordering. The addition of Ti and V to Fe_3Al enhances ordering, i.e. it shifts the transition from DO3 order to B2 order and thereby the anomalous yield stress peak to higher temperatures.[71]

Of particular interest is the case of Fe_3Al with carbon as ternary alloying element.[72] Fe_3Al has only a very low solubility for C at low temperatures[73], i.e. excess C is precipitated as fine Fe_3AlC particles first on grain boundaries and second within the grains which produces precipitation hardening. Fe_3Al and Fe_3AlC are separated by an extended two-phase field in the isothermal section of the Fe-Al-C phase diagram[72], i.e. the Al content of the Fe_3Al can be varied to vary the degree of order and the amount of Fe_3AlC can be varied between 0 and 100 %.

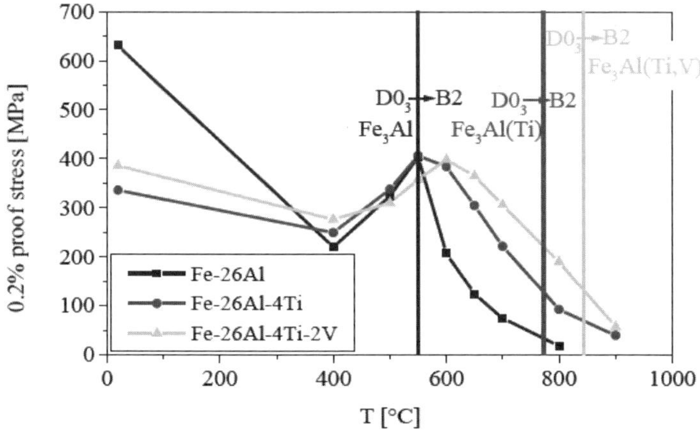

Fig. 25. 0.2% Proof stress (in compression, deformation rate 10 s) as a function of temperature for various Fe_3Al alloys: Fe-26Al, Fe-26Al-4Ti and Fe-Al-4Ti-2V (always at.%).[71]

The yield stress at temperatures below 500 °C increases with increasing C content, i.e. increasing volume fraction of Fe_3AlC, for alloys with about 25 at.% Al corresponding to the stoichiometric composition and about 28 at.%Al, which is expected, whereas the yield stress for the alloys with about 23 at.% Al is not much affected by the variation of the C content (Fig. 26). At higher temperatures the effects of composition variation are smaller and complex due to microstructure effects. However, the ductility also increases with increasing C content, which is in contrast to expectation. Stronger strengthening at temperatures above 600 °C is achieved by additional alloying to prepare quaternary Fe-Al-C-Nb alloys with precipitates of Nb carbide and $(Fe,Al)_2Nb$ Laves phase (Fig. 27).[74]

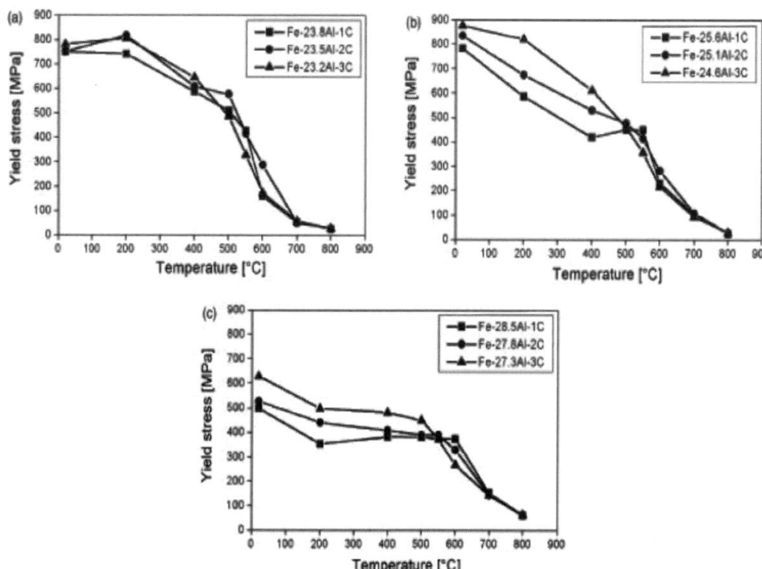

Fig. 26. Compressive yield stress as a function of temperature for various Fe-Al-C alloys.[72]

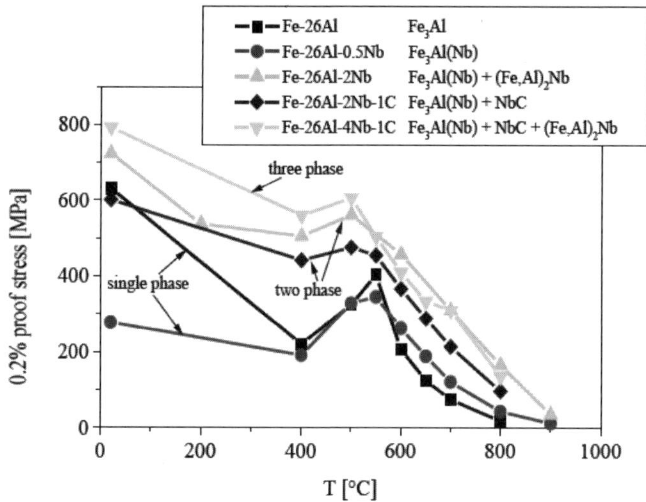

Fig. 27. Proof stress versus temperature for a series of Fe-Al based alloys.

Another case of particular interest is the ternary Fe-Al-Ta system. There is an equilibrium between the Fe-Al solid solution including the disordered bcc phase, the D03-ordered Fe_3Al phase and the B2-ordered FeAl phase on the one hand and the ternary Laves phase Ta(Fe,Al)2 with hexagonal C14 structure on the other hand.[7] Accordingly such Fe-rich Fe-Al-Ta alloys with precipitated Laves phase in a Fe-Al matrix are promising for alloy developments for high-temperature applications since strength, ductility, corrosion resistance and the balance of these can be controlled - and optimized with respect to the envisaged application - by controlling the Al content, which controls the atomic ordering in the matrix, and the Ta content, which controls the amount of Laves phase. Yield stress and ductility have been studied with systematic variation of the Al and Ta contents.[7]

The microstructure studies of the Laves phase-strengthened Fe-Al-Ta alloys have revealed that the expected exclusive Laves phase precipitation occurs in the Fe-Al-Ta alloys with about 25 at.% Al only at temperatures of 800 °C and above first at grain boundaries and second in the grains whereas at lower temperatures an unexpected Heusler phase Ta_2FeAl with L21 structure, which may only be metastable, is precipitated homogeneously as fine particles with cubic orientation and only later Laves phase precipitate particles are observed - see Fig. 28.[7]

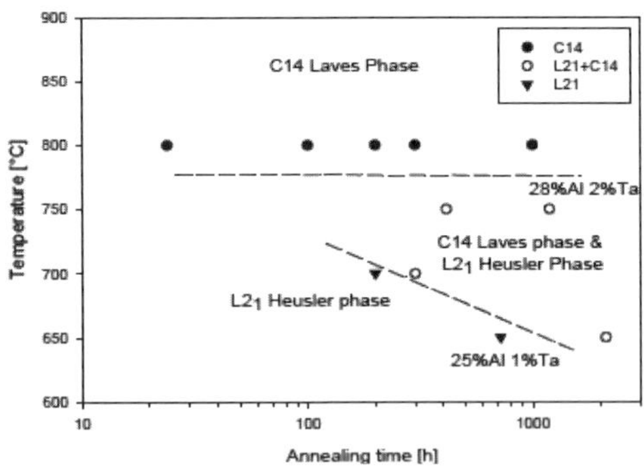

Fig. 28. Tentative temperature-time-transformation diagram for Fe-25at.%Al-2at.%Ta alloys (if not stated otherwise) with data at 700 °C and 800 °C referring to annealing and data at 650 °C and 750 °C referring to creep.[7]

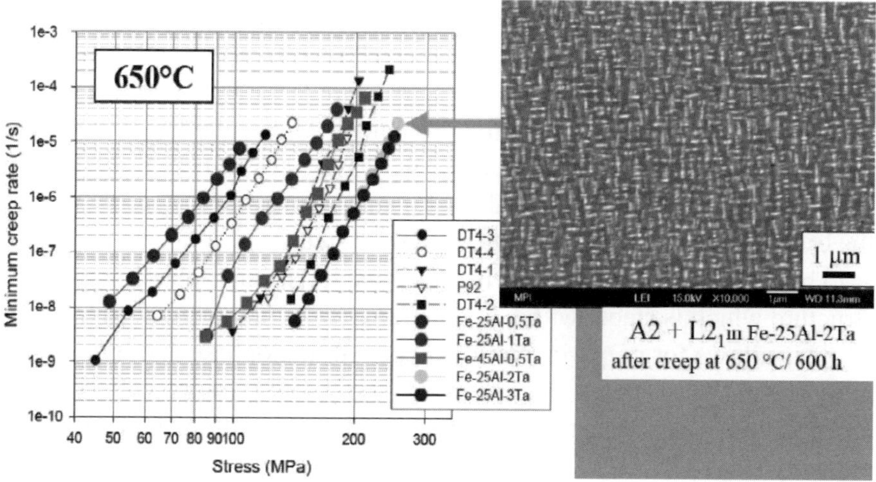

Fig; 29. Minimum compressive creep rate at 650 °C as a function of applied stress for various Fe-Al-Ta alloys in comparison to data for some 12wt.%Cr model steels (DT4-3, DT4-4, DT4-1) and the 9wt.%Cr heat-resistant steel P92 (left) and micrograph (transmission electron microscopy) of the microstructure of the Fe-25at.%Al-2at.%Ta alloy with precipitated Heusler phase (L21 structure) in the bcc Fe-Al matrix (A2 structure) after creep for 600 h at 650 °C.[7]

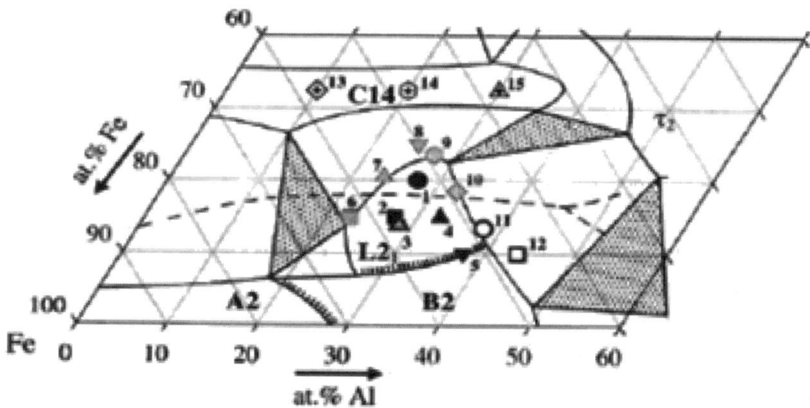

Fig. 30. Isothermal phase diagram of the Fe-Al-Ti system.[75]

The finely distributed Heusler phase particles contribute significantly to improving the creep resistance of these Fe-Al-Ta alloys and this effect is larger than that of the Laves phase precipitate as is visible in Fig. 29. The comparison of the data for the present Fe-Al-Ta alloys with those for the heat resistant martensitic/ferritic 9-12%Cr steels in Fig. 29 clearly shows that the Fe-Al-Ta alloys allow for surpassing the creep resistance of the established martensitic/ferritic 9-12%Cr steels.

The ternary Fe-Al-Ti system contains a Heusler phase with L21 structure which is stable within a wide range of vcomposition (Fig. 30).

This allows for preparing various types of monophase and multiphase alloys with a complex variation of the yield stress with varying composition, microstructure and temperature including a yield stress anomaly (Fig. 31).

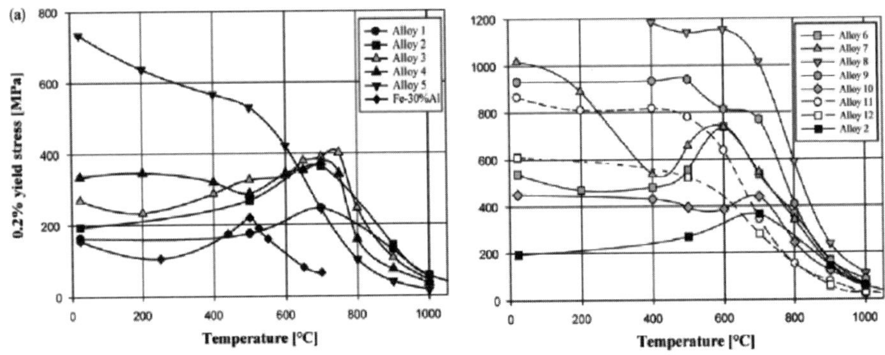

Fig. 31. Compressive yield stress as a function of temperature for various Fe-Al-Ti alloys as indicated in Fig. 30.[75]

4. Conclusions

Atomic order in intermetallic alloys gives rise to superdislocations which may produce yield stress anomalies. Perfect order is possible only at stoichiometry and deviations from stoichiometry produce off-stoichiometry effects due to enhanced point defect contents. Both superdislocations and point defects in addition to the familiar microstructure effects give rise to a complex stress and temperature dependence of the high-temperature deformation behaviour.

This complexity allows for designing alloys for specific structural high-temperature applications. Present materials developments in progress based on titanium aluminides, nickel aluminides, and iron aluminides offer possibilities for adjustment to specific applications.

Besides the intermetallic phases which are subject of present materials developments, there is a multitude of intermetallic phases which are attractive for high temperature applications because of their advantageous high melting temperatures and low densities – see Fig. 32. Indeed a high specific yield stress, i.e. yield stress per unit density, which results from a high yield stress and/or a low density – see Fig. 33 -, is an important criterion for selecting an intermetallic phase for structural materials developments.

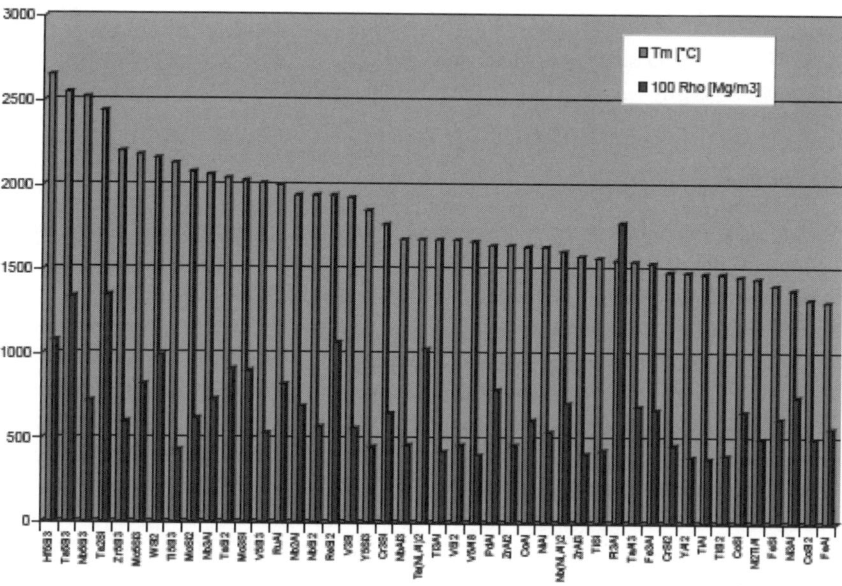

Fig. 32. Melting temperature Tm and density Rho for various intermetallic phases.

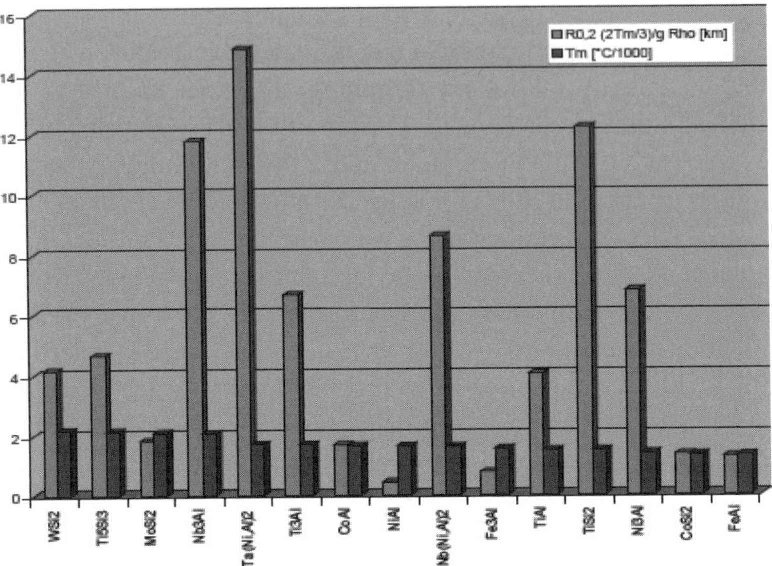

Fig. 33. Specific yield stress (i.e. yield stress R0.2 at 2/3 of the melting temperature Tm divided by the density Rho) for various intermetallic phases.

References

1. *Intermetallic Compounds*, 4 Volume Set, Eds. J.H.Westbrook and R.L.Fleischer, (John Wiley & Sons, Chichester, 2000).
2. G.Sauthoff, *Intermetallics,* (Wiley-VCH, Weinheim, 1995).
3. G.Sauthoff, in: *Ullmann's Encyclopedia of Industrial Chemistry*, Seventh Edition, 2006 Electronic Release(Wiley-VCH, Weinheim , 2006).
4. G.Sauthoff, in: *Landolt-Börnstein* (New Series) Group VIII: Advanced Materials and Technologies, Subvolume VIII/2A: Powder Metallurgy Data, Part 2: Refractory, Hard and Intermetallic Materials, Eds. P.Beiss, R.Ruthardt, and H.Warlimont, (Springer-Verlag, Berlin, 2002)pp. 221-267.
5. G.E.R.Schulze, *Metallphysik* , Akademie-Verlag, Berlin (1967).
6. K.Girgis, in: *Physical Metallurgy*, 3 edition, Eds. R.W.Cahn and P.Haasen, (North-Holland Physics Publ., Amsterdam, 1983)pp. 219-269.
7. G. Sauthoff, in: *Basics of Thermodynamics and Phase Transition in Complex Intermetallics*, Ed. E. Belin-Ferré (World Scientific, Singapore, 2008) pp.147-188
8. G. Sauthoff, *Z. Metallk.* **77** (1986) 654-666.

9. P.Veyssiere, *Mater.Sci . Eng .***A 309,** 44 (2001).

10. Y.Q. Sun, P.M. Hazzledine, in: *Ordered Intermetallics - Physical Metallurgy and Mechanical Properties,* Eds.,C.T. Liu, R.W. Cahn, G. Sauthoff, (Kluwer Acad. Publ., Dordrecht, 1992) pp.177-1 96.

11. G. Sauthoff, in: *Intermetallic Compounds, Volume 2, Basic Mechanical Properties and Lattice Defects of Intermetallic Compounds,* Eds. J. H. Westbrook and R. L. Fleischer, (Chichester:John Wiley & Sons, 2000) pp.41-66.

12. C. de Novion. In: *Intermetallic Compounds, Volume 2, Basic Mechanical Properties and Lattice Defects of Intermetallic Compounds,* Eds. J. H. Westbrook and R. L. Fleischer, (Chichester:John Wiley & Sons, 2000) pp. 131 -155.

13. C. Jiang, L.Q. Chen, Z.K. Liu, *Intermetallics* **14,** 248 (2006).

14. L. M. Pike, I.M. Anderson, C.T. Liu, Y.A. Chang, *Acta Mater.* **50,** 3859 (2002)..

15. C. Herzig and S. Divinski, *Intermetallics* **12,** 993 (2004).

16. D. D. Risanti and G. Sauthoff, to be published.

17. U. Reimann and G. Sauthoff, *Intermetallics* **7,** 437 (1999).

18. D. Y. Seo, J. Beddoes, L. Zhao and G. A. Botton, *Mater. Sci. Eng. A-Struct. Mater.* **329,** 810 (2002).

19. H. Mecking and Y. Estrin. in: *Constitutive Relations and Their Physical Basis*, Eds. S. I. Andersen, J. B. Bilde-Sorensen, N. Hansen, T. Leffers, H. Lilholt, O. B. Pedersen and B. Ralph, (Roskilde:Riso National Laboratory, 1987) pp.123-145.

20. J. Klöwer and G. Sauthoff, *Z. Metallk.* **83,** 699 (1992).

21. H. J. Frost, M. F. Ashby, *Deformation Mechanism Maps* (Pergamon Press, Oxford, 1982).

22. A. Gorzel and G. Sauthoff, *Intermetallics* **7,** 371 (1999).

23. G. Sauthoff, *Z. Metallk.* **80,** 337 (1989).

24. C. Phaniraj, B. K. Choudhary, K. Bhanu Sankara Rao and B. Raj, *Scripta Mater.* **48,** 1313 (2003).

25. M. Kachanov, *International Journal of Damage Mechanics* **3,** 329 (1994).

26. A. Isaac, F. Sket, W. Reimers, B. Camin, G. Sauthoff and A.R. Pyzalla, *Materials Science and Engineering:* **A 478,** 108 (2008).

27. A. Isaac, F. Sket, A. Borbely, G. Sauthoff and A. R. Pyzalla, *Praktische Metallographie-Practical Metal lography* **45,** 242 (2008).

28. H. A. Lipsitt, D. Shechtman and R. E. Schafrik, *Metall.Trans.,* **11A,** 1369 (1980).

29. R.von Mises, *Ztschr.Angew.Math.Mech.,* **8,** 161 (1928).

30. Y.-W. Kim and F. H. Froes, , in: *High-Temperature Aluminides and Intermetallics,* Eds. S. H. Whang, C. T. Liu, D. P .Pope and J. O.Stiegler, (TMS, Warrendale,1990) pp.465-492.

31. Y. Minonishi, *Philos.Mag.,* **A, 63** 1085 (1991).

32. T. Nakano and Y. Umakoshi, *J.Alloys Compounds,* **197,** 17 (1993).

33. D. A. Koss, D. Banerjee, D. A. Lukasak and A. K. Gogia, in: *High Temperature Aluminides and Intermetallics,* Eds. S. H. Whang, C. T. Liu, D. P .Pope and J. O. Stiegler (TMS, Warrendale, 1990)pp. 175-196.

34. F. H. Froes, C. Suryanarayana and D. Eliezer, *ISIJ INT.*, **31,** 1235 (1991).

35. R. G. Rowe, in: *High Temperature Aluminides and Intermetallics*, Eds. S. H. Whang, C. T. Liu, D. P. Pope and J. O. Stiegler, (TMS, Warrendale, 1990) pp. 375-401.

36. D. Banerjee, R. G. Baligidad, A. K. Gogia and J. L. Strudel, in: *Structural Intermetallics 2001* (Proc. ISSI-3), Eds. K. J. Hemker, D. M. Dimiduk, H. Clemens, R. Darolia, H. Inui, J. M. Larsen, V. K. Sikka, M. Thomas and J. D. Whittenberger, (TMS, Warrendale, 2001) pp.43-52.

37. C. J. Cowen and C. J. Boehlert, *Intermetallics* **14**, 412 (2006).

38. F. Tang, S. Nakazawa and M. Hagiwara, *Mater. Sci. Eng.* **A**. **325**, 194 (2002).

39. H. A. Lipsitt, D. Shechtman and R. E. Schafrik, *Metall.Trans.*, **6A,** 1991 (1975).

40. Y. W. Kim and D. M. Dimiduk, *JOM-J.Min.Met.Mat.*, **43**, 40 (1991).

41. J. Lapin, *Intermetallics* **14,** 115 (2006).

42. J. A. Jimenez, M.. Carsi, G. Frommeyer, S. Knippscher, J. Wittig and O. A. Ruano, *Intermetallics* **13,** 1021 (2005).

43. C. T. Yang and C. H. Koo, *Intermetallics* **12,** 235 (2004).

44. M.Takeyama and S. Kobayashi, *I ntermetallics*, **13,** 993 (2005).

45. F. Appel, M. Oehring and J. D. H. Paul, *Adv.Eng.Mater.*, **8**, 371 (2006).

46. K. S. Kumar, *Int.Mater.Rev.*, **35**, 293 (1990).

47. Y. V. Milman, D. B. Miracle, S. I. Chugunova, I. V. Voskoboinik, N. P. Korzhova, T. N. Legkaya and Y. N. Podrezov, *Intermetallics* **9**, 839 (2001).

48. Seung Hyun Lee, Seung Uk Lee, Kyoung Il Moon, and Kyung Sub Lee. *Materials Science and Engineering* **A 382**, 209 (2004).

49. M. Heilmaier, D. Handtrack and R. A. Varin. *Scripta Mater.* **48,** 1409 (2003).

50. C. Brandt and O. T. Inal. *J.Mater.Sci.* **37,** 4399 (2002).

51. M. McLean, *Nickel-Base Superalloys: Current Status and Potential, in: High-Temperature Structural Materials*, Eds. R. W. Cahn, A. G. Evans, and M. McLean, (Chapman & Hall, London,1996) pp. 1-14.

52. K. Zhao, Y. H. Ma, L. H. Lou, Z. Q. Hu, *Materials Science and Engineering*: **A 480,** 205 (2008).

53. S. Divinski, F. Hisker, W. Loser, U. Sodervall and C. Herzig, *Intermetallics*, **14,** 308 (2006).

54. I. Jung and G. Sauthoff, *Z.Metallk.*, **80,** 484 (1989).

55. I. Jung and G. Sauthoff, *Z.Metallk.*, **80,** 893 (1989).

56. C. Stallybrass, A. Schneider and G. Sauthoff, *Intermetallics* **13,** 1263 (2005).

57. W. Kleinekathöfer, A. Donner, H. Meinhardt, M. Hengerer, G. Sauthoff, B. Zeumer, G. Frommeyer and H. J. Schäfer, , in: *Symposium Materialforschung - Neue*

Werkstoffe des Bundesministeriums für Forschung und Technologie (BMFT), Eds. U. Dahmen, I. Gilbert, D. Lillack and S. Runte, (KFA-PLR, Jülich, 1994)pp. 1014-1015.

58. B. Zeumer and G. Sauthoff, , *Intermetallics, 5,* 563 (1997).

59. B. Zeumer, W. Sanders and G.Sauthoff, *Intermetallics, 7,* 889 (1999).

60. M.Palm and G.Sauthoff, in: *Structural Intermetallics 2001 (Proc. ISSI-3)*, Eds. K. J. Hemker, D. M. Dimiduk, H. Clemens, R. Darolia, H. Inui, J. M. Larsen, V. K. Sikka, M. Thomas, and J. D. Whittenberger TMS, Warrendale, 2001) pp.149-156.

61. M. Palm, J. Preuhs and G. Sauthoff, *J.Mater.Process.Technol.*, **136,** 105 (2003).

62. M. Palm, J. Preuhs and G. Sauthoff, *J.Mater.Process.Technol.*, **136,** 114 (2003).

63. M.. Palm and G. Sauthoff, in: Materials for Advanced Power Engineering 2002 (Proc. 7th Liège Conference), II.653 (2002).

64. J. Klöwer, Mater.Corros., **47,** 685 (1996).

65. P. F. Tortorelli and K. Natesan, *Mater.Sci .Eng .***A 258,** 115 (1998).

66. J.Herrmann, G.Inden and G.Sauthoff, *Acta Mater.*, **51,** 2847 (2003).

67. C.G.McKamey, , in: *Physical Metallurgy and Processing of Intermetallic Compounds,* Eds. N. S. Stoloff and V. K. Sikka (Chapman & Hall, London, 1996) pp.351-391.

68. J. L. Jordan and S. C. Deevi, *Intermetallics*, **11,** 507 (2003).

69. J. Wolff, M. Franz, A. Broska, R. Kerl, M. Weinhagen, B. Köhler, M. Brauer, F. Faupel, and T. Hehenkamp, *Intermetall ics*, **7,** 289 (1999).

70. M. Palm, A. Schneider, F. Stein and G.Sauthoff, in: *Integrative and Interdisciplinary Aspects of Intermetallics*, Eds. M. J. Mills, H. Inui, H. Clemens, and C. L. Fu, (MRS, Warrendale, 2005) pp.3-14.

71. F. Stein, A. Schneider and G. Frommeyer, *Intermetallics* **11**, 71 (2003).

72. A. Schneider, L. Falat, G .Sauthoff and G. Frommeyer, *Intermetallics, 13,* 1322 (2005).

73. J. Herrmann, G. Inden and G. Sauthoff, *Steel Research International*, **75,** 343 (2004).

74. L. Falat, A. Schneider, G. Sauthoff and G. Frommeyer, *Intermetallics* **13**, 1256 (2005).

75. M. Palm and G. Sauthoff, *Intermetallics* **12**, 1345 (2004).

CHAPTER 5

METADISLOCATIONS IN COMPLEX METALLIC ALLOYS

Marc Heggen and Mick Feuerbacher

Forschungszentrum Jülich GmbH
52425 Jülich, Germany
E-mail: m.heggen@fz-juelich.de

Metadislocations are novel and highly complex structural defects in complex metallic alloys. This Chapter covers the basic geometric concepts of metadislocations involving different metadislocation types and the relation between the number of phason planes, Burgers vector and the elastic energy. Reactions between metadislocations like splitting, the formation of loops and networks are presented. As metadislocations form phase boundaries between complex phases, they are furthermore pivotal defects mediating phase transformations. We will mainly focus on metadislocations in ε-type CMA phases, however also give an example of a metadislocation in the complex phase $Al_{13}Co_4$.

1. Introduction

The plasticity of complex metallic alloys is a highly interesting field of metal physics. Novel mechanisms of plastic deformation have been reported involving new types of defects. The most prominent among those are the so-called metadislocations, which were discovered 10 years ago by Klein and coworkers in the complex metallic alloy ε_6-Al-Mn-Pd.[1] Since that time several different types of metadislocations were observed in various complex metallic alloys, their microstructure was studied, the reactions and their modes of motion were investigated.

Today, metadislocations are considered the most complex and probably the structurally most fascinating defects in materials science.

161

Without going into any structural details at this point, the electron micrograph in Fig. 1 already points out the enormous size and complexity of a metadislocation. The dashed triangle indicates the direct vicinity of the core of the metadislocation, the volume where structural changes take place upon movement. While a dislocation in a conventional material features a number of atoms of the order of unity in its core, a metadislocation features about 15000 atoms per unit thickness in the direct vicinity of its core.

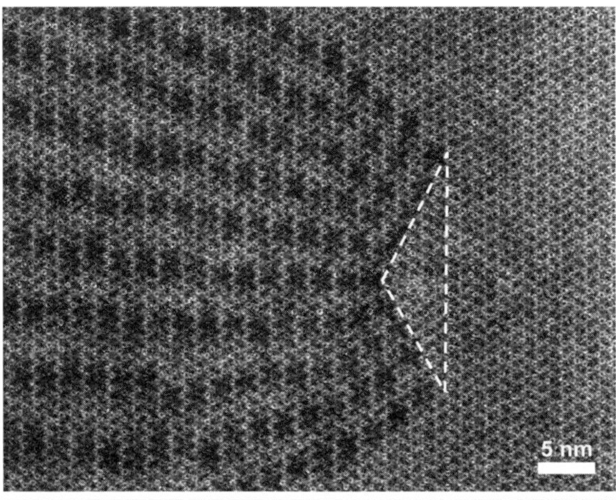

Fig. 1. Electron micrograph of a metadislocation in the complex metallic alloy phase ε_6-Al-Pd-Mn. The volume surrounding the dislocation core contains about 15000 atom.

After a brief introduction of the main structural characteristics of complex metallic alloys in section two, we will discuss the structure of the basic phase where metadislocations are observed in, the ε_6-phase, and the structure of related phases. The concept of metadislocations is introduced in section three. We describe the essential features at the example of a metadislocation with six associated phason planes, in the "basic" ε_6-phase and then proceed to other types. In section four we briefly discuss metadislocation types in other crystal structures, generalize their construction principle and discuss their modes of motion.

2. Complex Metallic Alloys

In recent years complex metallic alloys emerged as a novel field in materials science.[2] These materials are intermetallic phases, which possess a highly complex atomic structure. As a working definition, we can characterize this class of materials by the following features:

- The lattice parameters are large, exceeding ~10 Å. Correspondingly, the unit cells are of large volume and contain many tens up to several thousands of atoms.
- The atoms are locally arranged in a cluster substructure, in which icosahedral coordination plays a prominent role.
- A substantial amount of disorder is present in the ideal structure, i.e. the structure contains *inherent disorder*.

2.1. *Example*: *The structure of ε_6-Al-Pd-Mn and the ε-phase family*

CMA phases frequently exist in close relationship to other phases of similar structure, close or identical composition, and based on the same cluster substructure but with different lattice parameters. We refer to such groups of related phases as "phase families". A prominent example is the ε-phase family, which is based on the phase ε_6-Al-Pd-Mn. The ε_6-phase has a composition around $Al_{74}Pd_{22}Mn_4$ and is a ternary extension of the binary phase Al_3Pd. The structure model for ε_6-Al-Pd-Mn, which will be discussed in the following, was presented by Boudard *et al.*[3] ε_6-Al-Pd-Mn has an orthorhombic structure (space group *Pnma*). The cell parameters are large and amount to $a = 23.89$ Å, $b = 16.56$ Å, $c = 12.56$ Å, and the unit cell contains about 318 atoms. The unit cell is shown in Fig. 2a in a perspective view, and in projection along the [0 1 0] direction in Fig. 2b. The phase exhibits a substructure, which is based on so-called Mackay-type clusters. The central position of Pseudo-Mackay cluster[4] is occupied by a Mn atom and is surrounded by an incomplete Al shell, an Al icosidodecahedron (30 atoms), and a Pd icosahedron (12 atoms). Depending on the occupation of the inner shell, a pseudo-Mackay cluster comprises, about 52 atoms (Fig. 2c). The presence of Mackay-type cluster columns in the ε_6-structure can be seen in the perspective view in Fig. 2b. Here, the most obvious features are central Pd atoms which are surrounded by a ten-fold Pd ring (broken circles). Neighboring rings overlap and share two Pd positions. Figure 2d shows an experimental high-resolution electron micrograph. The micrograph was taken in the high-angle annular dark-field (HAADF) mode which provides image contrast proportional to the square of the atomic number.[5] Hence, the positions of the heavy Pd atoms are visible while the light atoms (Al, Mn) are not visible. Cluster columns, characterized by Pd rings can easily be identified (broken circles) and compared to the structure model in Fig. 2b.

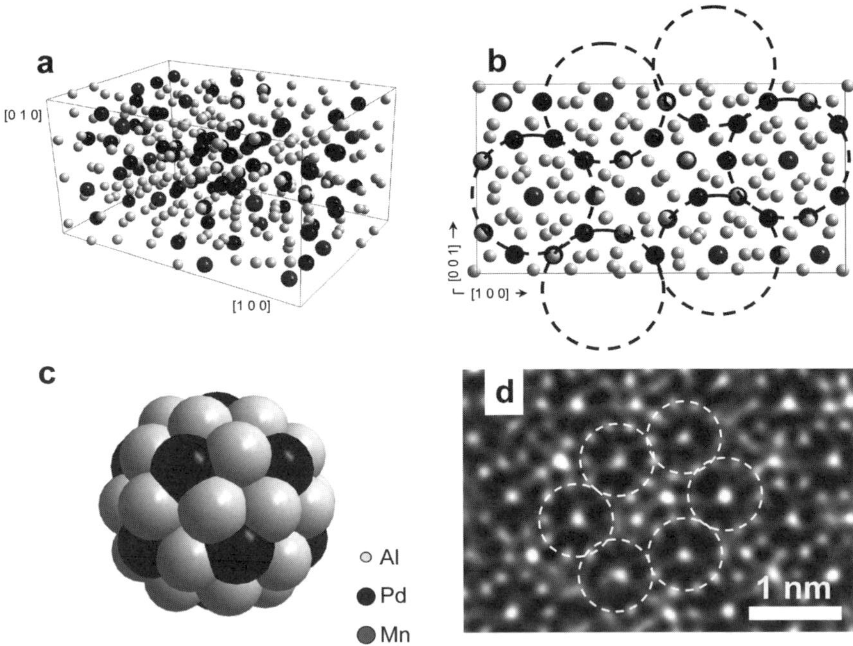

Fig. 2. Structure of ε_6-Al-Pd-Mn. a) Unit cell in perspective view. b) Projection along [0 1 0]. c) Mackay-type cluster. d) Experimental HAADF high-resolution micrograph. Dotted circles indicate Mackay-type cluster columns.

The ε_6-phase forms the basis of the other members of the ε-phase family. A sequence of related orthorhombic phases is referred to as ε_l with $l = 6, 16, 22, 28, 34$ according to the index of the strong diffraction spot (0 0 l), which corresponds to the interplanar spacing of about 0.2 nm occurring in all phases. These phases have identical lattice parameters a and b, and varying c-lattice parameters amounting to 32.4 Å (ε_6), 44.9 Å (ε_{22}), 58.0 Å (ε_{28}), and 70.1 Å (ε_{34}) in the Al-Pd-Mn system.[6,7] Further members of the ε-phase family are monoclinic phases, which can be described as local modulations of orthorhombic phases. These were extensively described and classified by Heggen *et al.*[8] Phases of the ε-family are found in the alloy systems Al-Pd-(Mn, Fe, Rh, Re, Ru, Co, Ir) and Al-Rh-(Ru, Cu, Ni).[6,9]

2.2. *Tiling description*

A highly useful tool for the depiction of structural features, defects, phase boundaries etc. in CMAs is the representation in the form of a tiling. A tiling is a two-dimensional projection of a given structure along a specific direction in terms of an area-filling arrangement of polygons, which correspond to characteristic structural features.

A straightforward way to construct a useful tiling for a given CMA structure is to connect the centers of its characteristic clusters. It is directly obvious from Fig. 2d that the clusters columns in the ε_6-Al-Pd-Mn phase are arranged as flat hexagons, hence the structure may be described in terms of a flattened hexagon tiling. Figure 3a is an electron micrograph of the ε_6- phase at a lower magnification. According to the specific contrast conditions applied, the cluster centers appear as white spots. Connecting these leads to the characteristic tiling shown in the upper right part of the micrograph and in Fig 3b. The tiling consists of flattened hexagons which are arranged in two different orientations. Neighboring hexagon rows have alternating orientations and equally oriented hexagons are stacked in rows along [0 0 1].

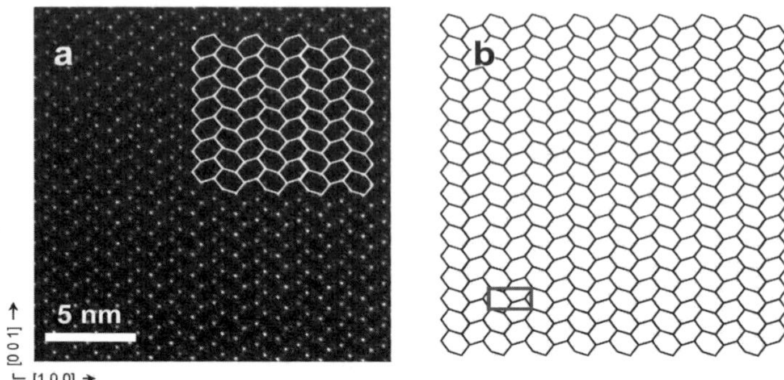

Fig. 3. a) Electron micrograph of ε_6-Al-Pd-Mn along the [0 1 0] direction with superposed hexagon tiling. b) Hexagon tiling.

The phases related to ε_6-Al-Pd-Mn, i.e. the other members of the ε-phase family can be similarly described by tilings. Three examples are depicted in Fig. 4. Figure 4a shows the ε_{28}-phase. For its description, in addition to the hexagon tiles, two other tiles are required – a banana-shaped nine-edge tile and a pentagon. The latter are alternately arranged along the [1 0 0] direction and form (0 0 1) planes regularly stacked perpendicular to the [0 0 1] direction, forming a face-centered lattice. The unit-cell projection is shown in light grey. Figure 4b shows the ε_{16}-phase, which can be described utilizing banana-shaped polygons and pentagons only. Figure 4c shows the ξ-phase, which is represented by a parallel arrangement of only one type of hexagon.

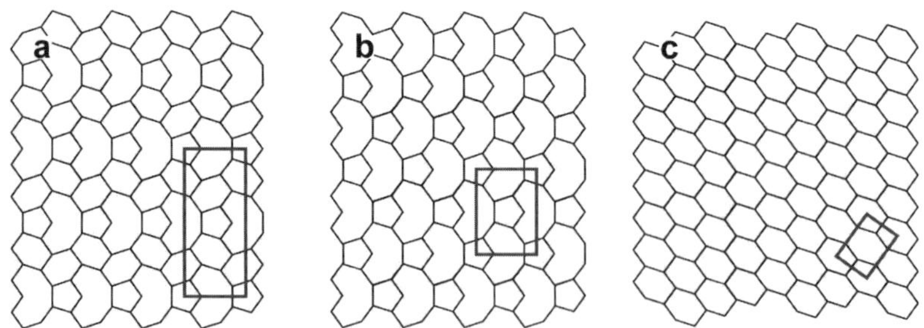

Fig. 4. Tiling representation of a) the ε_{28}-, b) the ε_{16}-, and c) the ξ-phase.

2.3. *Phason lines and phason planes*

In the previous section the tiles to describe ε-type structures were introduced. In particular the combination of a banana-shaped nonagon and a pentagon is a decisive structural element which will be discussed in more detail in this section. A combination of a banana-shaped nonagon and a pentagon represents a linear defect along the [0 1 0] direction, which is referred to as a phason line. If phason lines are present at a certain density in a ε_6-structure, they tend to align along the [1 0 0] direction and form (0 0 1) planes, which are referred to as phason planes.

Figure 5a is a high-resolution HAADF image of a phason plane in ε_6-Al-Pd-Mn and Fig. 5b shows the corresponding tiling representation. Since each individual phason line connects two neighboring hexagon columns with their alternately oriented version, the phason plane is a (0 0 1) mirror. Accordingly, phason planes can be considered as twin boundaries or inversion boundaries in the ε_6-structure.

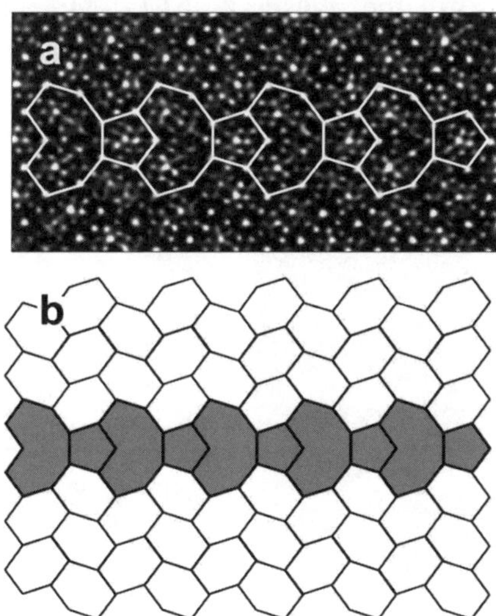

Fig. 5. Phason planes in ε_6-Al-Pd-Mn. a) High-resolution HAADF image. b) Corresponding tiling representation (grey).

Figure 6 shows a micrograph of an ε-type Al-Pd-Mn sample which can be described in terms of flattened hexagons (white) and banana-pentagon tiles (dark grey). Here, phason planes are not strictly oriented along the [1 0 0] direction but show a wavy course. Indeed, single phason lines can easily move along the [0 0 1] direction by means of so-called phason flips[10] and can create deviations from the strict alignment as a (0 0 1) plane which is seen in Fig. 5. In some areas of Fig. 6 the phason planes show (1 0 1) translations between neighbouring phason lines (arrows) and a varying number of hexagon tiles between the phason planes, ranging from 0 to 3 (examples are indicated by circles). The unit-cell projection (rectangle) shows that locally the ε_{28}-phase is found. Over large areas of the micrograph, however, a disordered ε_{28}-type structure is observed and intermixing with other phases is found.

Figure 6 is a typical example of intermixing of structural units of related ε-phases. It demonstrates that phason lines and phason planes assume a dual function in ε-type phases. On the one hand, they are structural defects e.g. in the phases ε_6-Al-Pd-Mn and ξ-Al-Pd-Mn. On the other hand, phason lines and phason planes can arrange regularly and become structural elements of an ideal structure, for instance of the ε_{16}- or ε_{28}-phase (Fig. 4a and b).

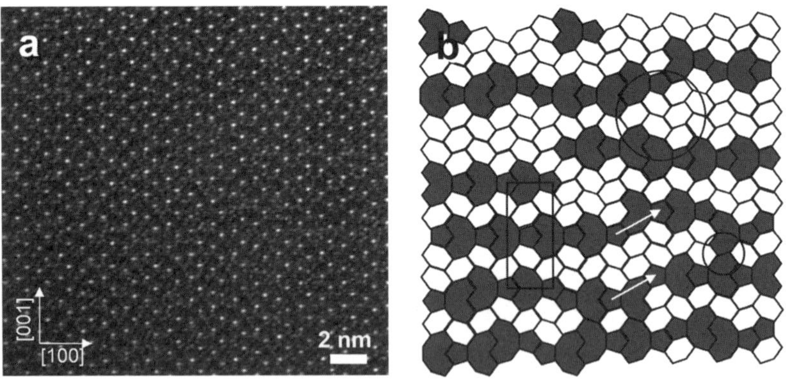

Fig. 6. a) Electron micrograph of an ε-type Al-Pd-Mn sample along the [0 1 0] direction. b) Tiling description.

In the next section metadislocations are introduced. The presence of metadislocations in an ε-type phase is in general connected to the introduction of additional phason planes to the material. Hence, metadislocations are connected to structural variations of the surrounding material and to the local appearance of new phases around them.

3. Metadislocations in the ε-Phase Family

3.1. *Basic features and tiling description*

Figure 7a shows a micrograph of a metadislocation in ε_6-Al-Pd-Mn. The metadislocation shown is imaged end-on, i.e. the viewing direction in the electron microscope is parallel to the line direction. At the right-hand side of the micrograph six phason planes are present which are terminated at the centre of the micrograph. Figure 7b analyses the structure in terms of a tiling. Besides the phason elements (grey tiles) an additional polygon is used (dark grey tile) which represents the core of the metadislocation.

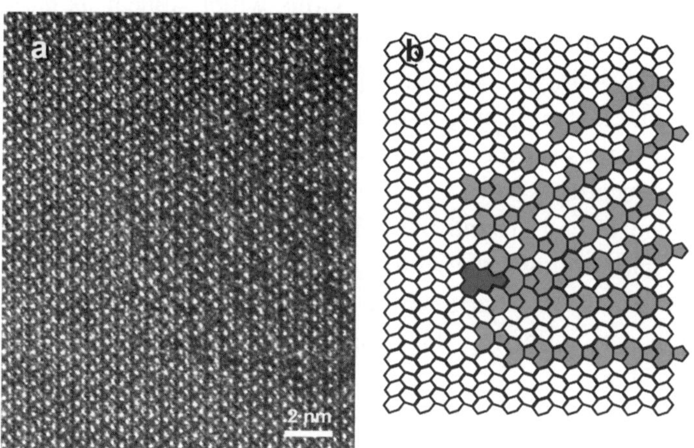

Fig. 7. Metadislocation with six phason planes in ε_6-Al_Pd-Mn. a) Electron micrograph and b) tiling description showing phason elements (grey tiles) and metadislocation core (dark grey tile).

Figure 8a shows a metadislocation embedded in the ε_{28}-Al-Pd-Mn phase at a lower magnification and slightly different contrast conditions than Fig. 7a.

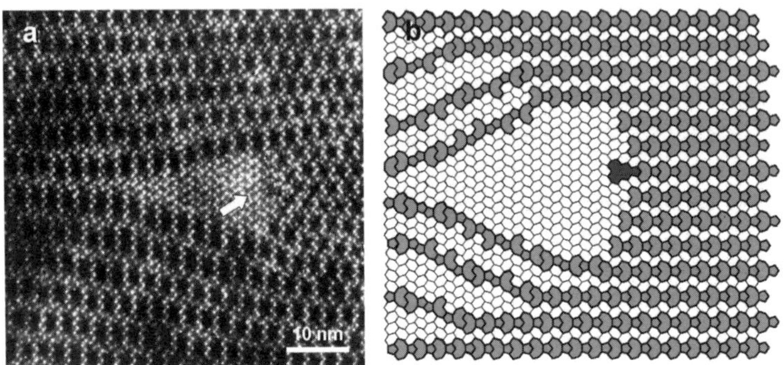

Fig. 8. Metadislocation with six phason planes in ε_{28}-Al-Pd-Mn. a) Electron micrograph (the arrow points at the metadislocation core) and b) idealized tiling description showing phason elements (grey tiles) and metadislocation core (dark grey tile).

The ε_{28}-Al-Pd-Mn phase is recognizable by the periodic stacking of phason planes in the outer image regions which appear as dark contrast features. At a distance from the metadislocation, the phason planes are approximately straight and perpendicular to the [0 0 1] direction. The metadislocation itself consists of a core (arrow), which is associated to six phason planes on the right-hand side. On the other side of the core, the metadislocation shows a region of pure ε_6-structure. The surrounding phason planes bend around the core and, with increasing distance from the latter, narrow down the width of the pure ε_6-region, such that it takes a roughly triangular shape. An enlarged and slightly simplified tiling description, not reflecting the exact curvature of the phason planes, is shows in Fig. 8b.

The principle features of a metadislocation with six phason planes are shown in the idealized tiling in Fig. 9a. Here, the regular stacking of the phason planes represents the ideal ε_{28}-phase. By this means, the metadislocation can be embedded into the ε_6-phase creating a slab of ε_{28}-phase on its right-hand side, i.e. the six phason planes, or into the ε_{28}-

phase, creating a slab of ε_6-phase on the left hand side, which may be "dissolved" by phason flips as shown in Fig. 8.

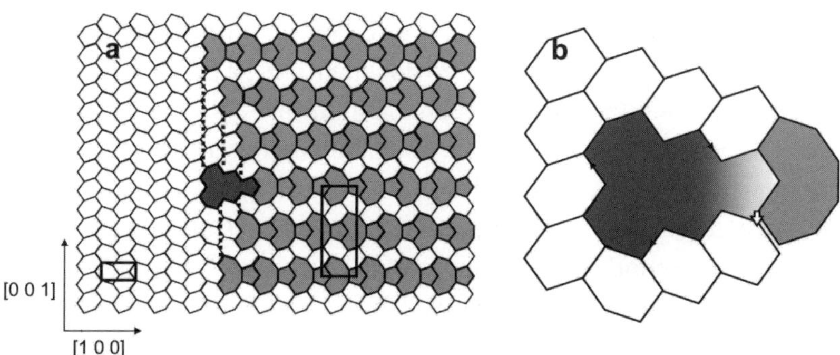

Fig. 9. a) Idealized Metadislocation with six phason planes. b) Burgers circuit around the core using perfect tiles. The white arrow represents the Burgers vector.

Above and below the metadislocation core (dark grey tile) in Fig. 9a two areas can be observed which are represented by a parallel arrangement of hexagons. The parallel arrangement of two hexagons represents stacking fault with (100) plane in the ε_{28}-phase (dotted lines). These stacking faults connect the phason planes and the metadislocation core.

Looking at Fig. 9a, a metadislocation appears to be a complex jigsaw puzzle; however, its "dislocation-nature" may not be directly obvious. Looking a schematic representation of an edge dislocation in a simple material (compare Fig. 3 Chapter 1, this volume), the strain of the material which corresponds to the Burgers vector of the dislocation is directly obvious. The Burgers vector of a metadislocation, however, is much smaller than the lattice constant. This is why its strain is not directly obvious from the tiling description in Fig. 9a.

3.2. *Burgers vectors and metadislocation series*

The Burgers vector of a metadislocation can be determined, in analogy to simple materials, by performing a Burgers circuit. A closed circuit is performed around the core and transferred to the ideal unstrained structure. In Fig. 9b the circuit around the core of the metadislocation, represented by arrows, is transferred to an unstrained tiling using perfect tiles. Here, a gap opens up and the vector connecting the finish and start point of the right-hand side circuit is the Burgers vector $\vec{b} = -c/\tau^4 [0 \quad 0 \quad 1]$. With the lattice constant $c = 12.56$ Å for ε_6- and ε_{28}-Al-Pd-Mn we obtain a Burgers vector length of 1.83 Å. Like mentioned before, the Burgers vector of the metadislocation is a fraction of the lattice constant, i.e. the metadislocation is a partial dislocation in the ε_6- and ε_{28}-structure. The length of the Burgers vector is of the same order as typical values found for simple metals, such as 2.9 and 2.6 Å in aluminum and copper, respectively. Hence, the elastic strain energy of the metadislocation is energetically acceptable, while the energy of a corresponding perfect dislocation would be unphysically high.

In ε-Al-Pd-Mn, five different Burgers vector lengths for [0 0 1] metadislocations ranging from 0.70 to 4.80 Å are experimentally found.[11] Each Burgers vector length is associated to a certain number of phason planes. Figure 10 displays corresponding TEM micrographs, all taken along the [0 1 0] lattice direction.

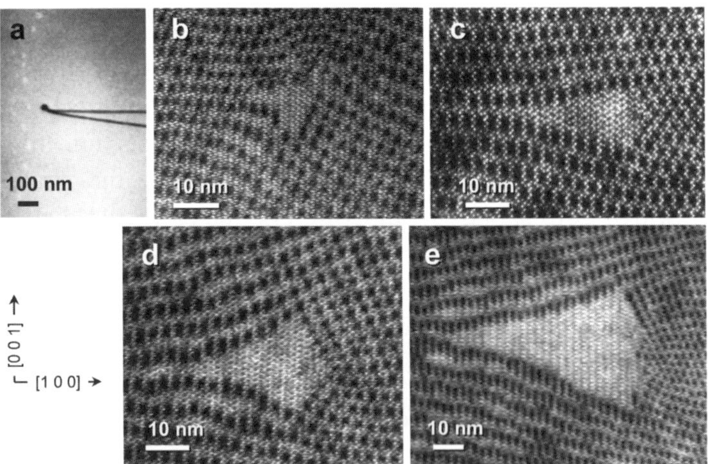

Fig. 10. Metadislocation series. TEM micrographs of metadislocations with a) two, b) four, c) six, d) ten, and e)16 phason planes.

Figure 10a is a low-resolution micrograph of a metadislocation with two associated phason planes in ε_6-Al-Pd-Mn. The metadislocation is seen on the left-hand side of the micrograph and terminates two phason planes, imaged as dark lines on the light-grey background, which stretch out to the right-hand side. The corresponding Burgers vector length is 4.80 Å. Figure 10b is a micrograph of a metadislocation with four associated planes. The associated phason planes are not perpendicular to the [0 0 1] direction but are slightly tilted downwards. The corresponding Burgers vector length amounts to 2.96 Å. Figures 10c, d and e show metadislocations with six, ten and 16 associated phason planes, with Burgers vector lengths of 1.83, 1.13 and 0.70 Å, respectively.. All members of the metadislocation series, along with the corresponding Burgers vector lengths are listed in Table 1.

Regarding the series of [0 0 1] metadislocations, the following facts are noteworthy:

- The numbers of associated phason planes in the series take twice the values of the Fibonacci numbers 1, 2, 3, 5, and 8.
- The Burgers vector lengths are related to each other by factors of τ.
- The sequences of associated phason planes and Burgers vector lengths are opposed: with increasing number of phason planes, the Burgers vector lengths are decreasing.
- Experimentally, it is found that the different members of the metadislocation series are not evenly distributed but their number distribution shows a broad maximum. In deformed and undeformed ε-Al-Pd-Mn, most metadislocations found have six associated phason planes. Metadislocations with ten planes are observed almost as often, while those with four phason planes are found considerably less frequently. Metadislocations with 16 phason planes are even less frequently found and species with two planes were only observed in very few occasions despite an extensive number of experimental investigations. The frequency of appearance of different types of metadislocations considering their energy is discussed in section 3.4.

Table 1. Experimentally observed metadislocations with [0 0 1] Burgers vectors in ε_6- and ε_{28}-Al-Pd-Mn. p: Number of associated phason planes. h: Number of associated hexagon planes according to the modified Volterra process. $|\vec{b}|$: Burgers vector length in terms of c-lattice constant and absolute value.

p	h	$|\vec{b}|/c$	$|\vec{b}|/\text{Å}$
2	3	$-\tau^{-2}$	-4.80
4	5	$+\tau^{-3}$	+2.96
6	8	$-\tau^{-4}$	-1.83
10	13	$+\tau^{-5}$	+1.13
16	21	$-\tau^{-6}$	-0.70

3.3. *Construction principles*

Metadislocations are associated to a certain number of phason planes. Constructing metadislocations in terms of a tiling will enable the reader to comprehend the relation between the number of associated phason planes and the Burgers vector length.

A conventional Volterra process for the construction of an edge dislocation, as employed in structurally simple crystals, involves cutting the crystal on a plane, which ends in the volume of the crystal, inserting or removing of an atomic halfplane, and subsequent elastic relaxation of structure. For metadislocations a similar process will be applied, which comprises the removal of a slab of material of several nanometers in thickness. Subsequently, the gap is filled by another slab of structurally different material, which is closely related to the bulk material.

Figure 11 illustrates this construction principle for a metadislocation with four associated phason planes in the ε_6-phase. From the ideal ε_6-structure (a), five hexagon planes (grey) are removed (b). Then the gap is filled by four phason planes (c). This slightly overcompensates the gap and hence leads to overlap of the tilings (arrow) since the four phason planes are slightly "thicker" than the five hexagon planes. Then the structure is elastically relaxed (d), which leads to the presence of elastically distorted hexagon tiles and a central area that cannot be filled by regular tiles (dotted lines), in which the dislocation core is located.

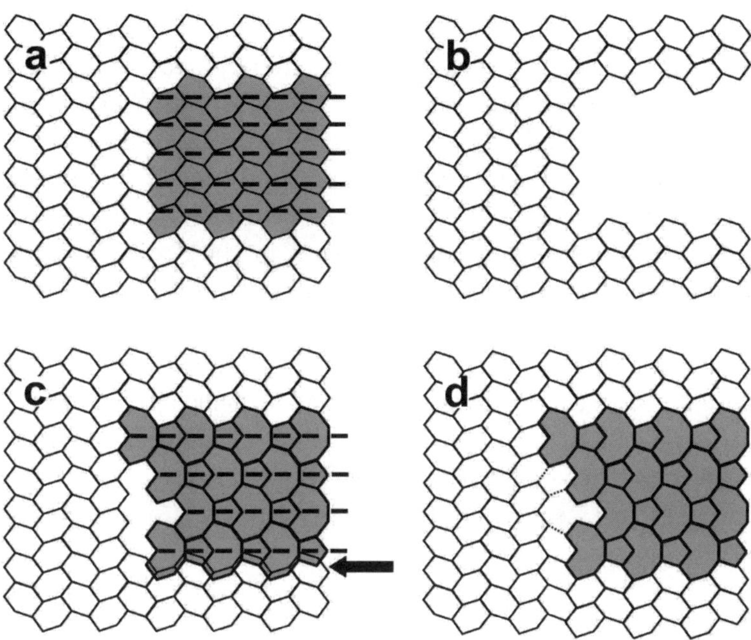

Fig. 11. Construction of a metadislocation with four associated phason planes in the ε_6-phase by a modified Volterra process (see text).

The Burgers vector of the so-constructed metadislocation is obtained by balancing the removed hexagon against the inserted phason planes. With $d_h = c$ and $d_p = c\tau^2/2$ beeing the "thickness" of row of hexagons and a phason line according to Fig. 11 b we obtain the Burgers vector $-5d_h + 4d_p = c/\tau^3$. Analogously, we can construct the metadislocation with six associated phason planes by removing eight hexagon and inserting six phason planes, which yields $-c/\tau^4$. This exactly corresponds to the Burgers-vector length of a metadislocation associated to six phason planes. The negative sign is due to the net removal of a thin layer of matter.

The process can be generalized as

$$b_p = c\left(\frac{\tau^2}{2}p - h\right)$$ (1)

where h is the number of hexagon planes removed and p is the number of phason planes, the metadislocation is associated with. Each member of the metadislocation series is represented by a duplet of numbers p and h which constitutes its Burgers vector. The full series of metadislocations as experimentally observed is listed in Table 1 along with the corresponding parameter h. As discussed in the previous section the experimentally observed metadislocations correspond to double Fibonacci numbers $p = 2, 4, 6, 10, 16$. The numbers of corresponding hexagon planes is also a series of Fibonacci numbers $h = 3, 5, 8, 13, 21$. This construction correctly reproduces the sequence of Burgers vectors lengths, as well as the opposing sequence of numbers of associated phason planes. The sign of the Burgers vectors alternately changes along the series, as the inserted phason planes alternately overcompensate or partially compensate the removed hexagon planes.

3.4. *Elastic energy*

Generally, the modified Volterra process, introduced in the last section, is not limited to Fibonacci and double Fibonacci numbers. In principle, it can be performed with p/2 and m being any natural number. In this section we discuss the corresponding elastic strain energies estimated as $E \approx \mu b^2$ (see Chapter 1, in this volume). With a shear modulus of 30 GPa[12], we obtain about 10^{-9} J/m for a metadislocation with six phason planes, which is comparable to that of a Shockley partial in copper (1.5 10^{-9} J/m). On the other hand, a perfect [0 0 1] dislocation in the ε_6-phase would have a very high elastic energy of 4.7 10^{-8} J/m.

Figure 12 shows the elastic energies for experimentally observed and hypothetical metadislocations with $p/2$ and h being natural numbers. The greyscale of each box represents the logarithm of the elastic energy of the metadislocation corresponding to a (p,h) doublet, where the darkest grey corresponds to the lowest energy. We find a valley of low-energy values along a line, which is approximately diagonal in Fig. 12.

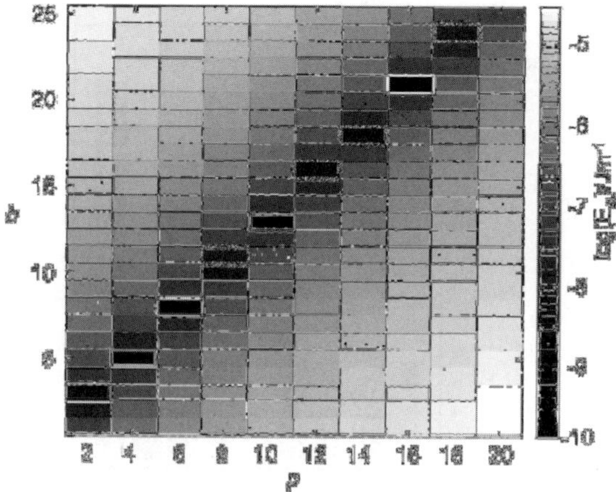

Fig. 12. Elastic strain energy of metadislocations represented by the parameters p and h. Experimentally observed metadislocations are marked by boxes with solid white outline, other hypothetical metadislocations are marked by boxes with broken white outline.

Experimentally observed metadislocations are indicated by boxes with solid white outline. The boxes with broken white outline represent metadislocations which also possess relatively low energy, but which were not observed experimentally (for example such with 8, 12, or 14 associated phason planes). In Fig. 13 a the energies of those metadislocations represented by boxed (p,h) doublets in Fig. 12 are shown in a bar graph. The grey bars represent the elastic energy of hypothetical metadislocations and the black bars represent the energy of those experimentally observed. Among the latter, the metadislocation with two associated phason planes apparently has the highest elastic energy, even exceeding that of the hypothetical metadislocations with 12 and 14 phason planes. The metadislocations with higher numbers of associated phason half planes seem to be energetically favourable. This is due to the smaller Burgers vectors of the latter, which directly leads to a smaller elastic line energy. However there are additional energy factors which have not been accounted for in the above consideration, the most evident of which is the cost of the phason planes associated with the metadislocation.

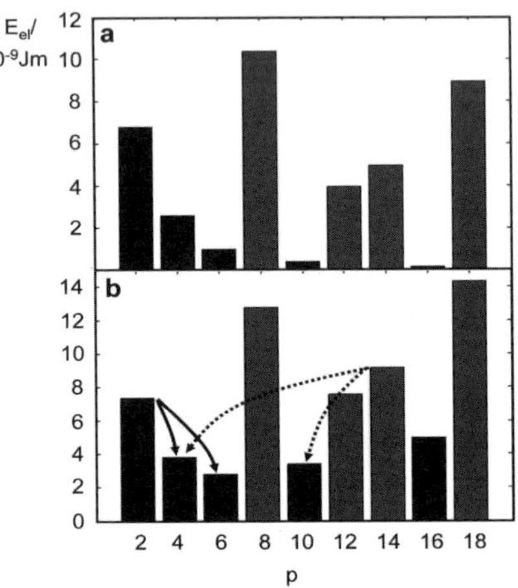

Fig. 13. Energy of metadislocations represented by boxed *p,h* doublets in bar-graph representation. Black bars represent the energy of metadislocations which were experimentally observed, grey bars represent other hypothetical metadislocations. a) Elastic energy contribution. b) Elastic energy plus fault-plane contribution. The solid and dotted arrows indicate metadislocation splitting and dissociation, respectively.

In Fig. 13b, a corresponding contribution was taken into account, in a first approximation, as an additive energy term proportional to the number of associated phason planes. Due to the lack of experimental data, the energy contribution of the additive term was adjusted such that the total energy reflects the experimental number-density distribution. Metadislocations with six phason planes, which are experimentally most frequently observed, have the lowest energy. Less frequently observed metadislocations with 16 and two phason planes have the highest elastic energies among the experimentally observed metadislocations. We furthermore find that the total energy of metadislocations with 12 and 14 phason planes is very close or even smaller than that of the metadislocation with two phason planes. The fact that they are nevertheless not experimentally observed may be due to reactions: metadislocations with 14 phason planes, for example, may dissociate into metadislocations with 10 and 4. This process is indicated in Fig. 13b by dotted arrows. This figure shows that the energy of two metadislocations with 10 and 4 phason planes is lower than that of one metadislocation with 14 phason planes at constant total number of phason planes. On the other hand, metadislocation splitting increases the number of phason planes and is indicated by solid arrows in Fig. 13b. This process is discussed in detail in the following chapter.

3.5. *Metadislocation reactions*

In this section different reaction between metadislocations are discussed. We will introduce the formation of loops and the splitting of metadislocations into two or more "partials". Figure 14 is a TEM image of two metadislocation loops in ε_6-Al-Pd-Mn. In the center of the image, two metadislocations are seen, which share six associated phason planes. Obviously, the two metadislocations terminating the phason planes are segments of a loop, which has a habit plane close to (0 0 1). In the lower right corner a smaller example of a metadislocation loop with ten associated phason planes is seen. Since metadislocation loops have [0 0 1] Burgers vectors, they are pure edge loops.

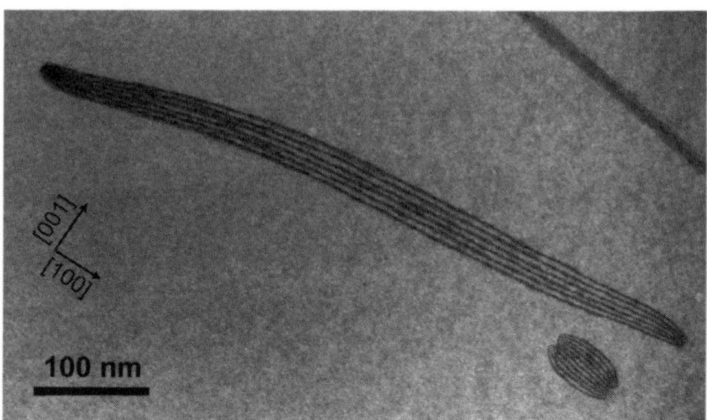

Fig. 14. Metadislocation loops in ε_6-Al-Pd-Mn sharing six and ten associated phason planes.

Splitting of perfect dislocations into partials is a phenomenon which is commonly observed in various crystalline materials. Since metadislocations are partial dislocations *per se*, it may seem surprising that splitting of metadislocations into partials having smaller Burgers vectors is often observed, as well.[13, 14]

Figure 15a shows a micrograph of two metadislocations associated with ten (1) and six phason planes (2), respectively. They are mutually connected by six phason planes.

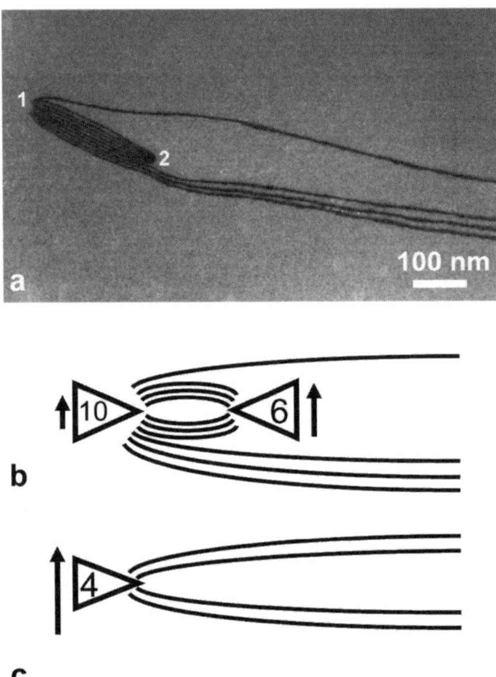

Fig. 15. Two metadislocations associated with ten (1) and with six phason planes (2), which are mutually connected by six phason planes. a) TEM micrograph. b) Schematic representation in terms of triangles and lines, representing metadislocations and phason planes. The number in the triangle denotes the number of associated phason planes, the arrow indicates length and direction of the Burgers vector. c) Metadislocation with four phason planes.

Four phason planes are remaining; one is visible in the upper part and three are visible in the lower part of the micrograph. The situation is schematically depicted in Fig. 15b using triangles and lines representing metadislocations and phason planes, respectively. The number of phason planes associated to a metadislocation is denoted in the corresponding triangle, and an arrow indicates the length and direction of its Burgers vector. According to the opposite line directions of the metadislocations in Fig. 15b, their Burgers vectors are oriented in the same direction and add up to (1.13 - (-1.83)) Å = 2.96 Å (Table 1), which corresponds to the Burgers vector of a metadislocation with four phason planes.

This situation can be interpreted in terms of metadislocation splitting: A metadislocation with 4 phason planes has split into an equally oriented metadislocation with 10 phason planes and an inversely oriented metadislocation with 6 phason planes. The process of metadislocation splitting reduces the local elastic strain energy near the metadislocation cores. According to Frank's rule [15] $b_1^2 > b_2^2 + b_3^2$, where b_1 is the Burgers vector length of the initial dislocation and b_2 and b_3 are those of the products, the splitting is energetically favourable. Fig. 13 a shows that the elastic energy of a metadislocation with four associated phason planes is higher than the sum of those for metadislocations with 6 and 10 phason planes. On the other hand, splitting requires the creation of six local phason planes. This introduces additional fault-plane energy, which has to be balanced against the energy gain due to splitting.

Figure 16 shows a more complex case of splitting into three metadislocations. A metadislocation with ten phason planes (1) is connected with two metadislocations, each with six phason planes (2 and 3). Two phason planes leave the metadislocations at the right hand side (4 and 5). Two additional phason planes seen in the micrograph (6 and 7) are not connected to the metadislocation arrangement. The net Burgers vector of the arrangement is (-1.13 − 2 × 1.83) Å = -4.80 Å which is equal to the Burgers vector of a metadislocation associated with two phason planes. This example discloses another unique feature of metadislocations: The associated phason planes can be connected to two or more other metadislocations. By this way metadislocations can associate large and complex network structures.[14]

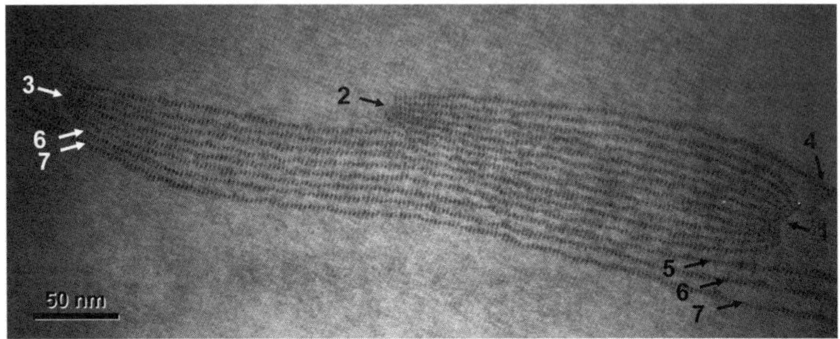

Fig. 16. Complex splitting into three metadislocations (see text).

4. Metadislocations in different complex metallic alloys

In the previous chapters metadislocations in the phases ε_6- and ε_{28}-Al-Pd-Mn were discussed. In the following section it is shown that different types of metadislocations exist in other CMA phases. To start with, we discuss metadislocations in monoclinic ε-phases. These are closely related to the orthorhombic ε-phases, and so are their metadislocations. In section 4.2, metadislocations in the phase $Al_{13}Co_4$ are discussed.

4.1. *Metadislocations in ξ- and monoclinic ε-phases*

Figure 17a shows a micrograph of a metadislocation in the complex alloy $Al_{72.0}Pd_{22.8}Fe_{5.2}$. The metadislocation core is located in the lower left part of the image. It is associated with three planar defects extending to the upper right (dark contrast), which can be identified as phason planes. The surrounding host structure is formed by a parallel arrangement of flattened hexagons and hence can be identified as ξ-phase (Fig. 4c). In the basis of the ξ-structure, the associated phason planes are oriented approximately parallel to the (0 0 1) planes. Figure 17b shows a slightly simplified tiling representation of the metadislocation. The host structure is given by parallel hexagons, representing the ξ-phase seen along the [0 1 0] direction. A Burgers circuit in terms of edges of the hexagon tiles around the metadislocation core (section 3.2) reveals a closure failure of $1/2\tau^4[1 \quad 0 \quad 1]$ in terms of the ξ-lattice, which corresponds to the metadislocations Burgers vector. The Burgers vector length is 1.79 Å.

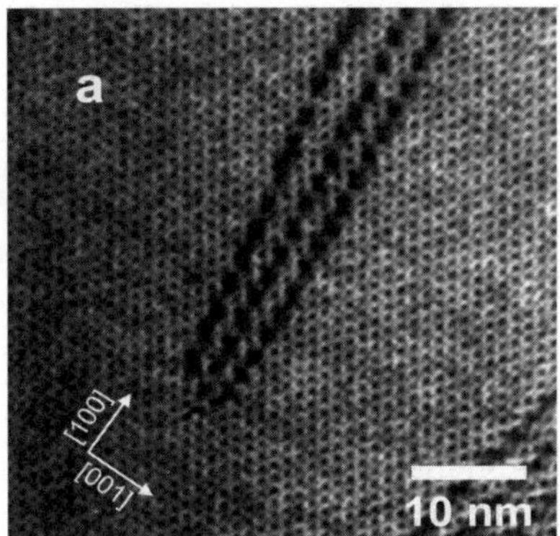

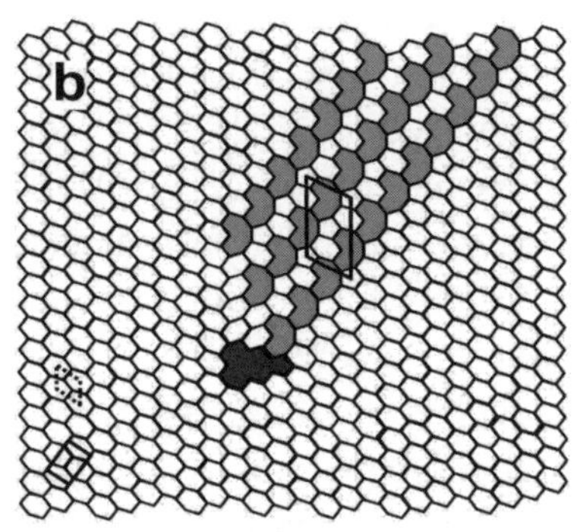

Fig. 17. Metadislocation in $Al_{72.0}Pd_{22.8}Fe_{5.2}$ associated with three planar defects in a ξ-host structure. a) TEM micrograph. b) Corresponding idealized tiling representation.

Comparing the tiles representing the metadislocation cores in the ξ- and of the metadislocation with six associated phason planes in the ε_6-structure (Fig. 17b and Fig. 7b), it is obvious that the same polygon is used. In both cases the Burgers vector length, in terms of the edge length a_0 of the hexagons, is given by a_0/τ^3. The difference in Burgers vector length for ξ-Al-Pd-Fe and ε_6-Al-Pd-Mn (1.79 and 1.83 Å, respectively) is merely due to the slightly different lattice parameters of these phases.

In complete analogy with the case of metadislocations in ε_6-Al-Pd-Mn (section 3), a series of metadislocations with other numbers of associated phason planes do exist.

Figure 18 shows a micrograph of a metadislocation with five associated phason planes. This metadislocation is not embedded in the ξ-

phase but in a phase having phason planes as structural features. It can be identified as the monoclinic (1,-1) variant of the ε_{22}-structure with additional local modulations on the lower right-hand side and the ε_{16}-phase at the top-part of the micrograph.[8] The phason planes surround the metadislocation forming an inner almost triangular region of pure ξ-phase. In this sense this metadislocation is directly comparable to metadislocations in the ε_{28}-phase (see Fig. 8a). The metadislocation core is represented by a dark grey tiling in Fig. 18 b and it corresponds to the core of a metadislocation with 10 phason planes in the ξ'-phase.[11] The Burgers vector is $1/2\tau^5\begin{bmatrix}1 & 0 & 1\end{bmatrix}$ in terms of the basis of the ξ-lattice, corresponding to a length of a_0/τ^4. Analogously, metadislocations with one, two, and eight associated phason planes can be constructed. As in ε_6-Al-Pd-Mn, the Burgers vector lengths of the metadislocations in the sequence are related by factors of $-\tau$, and with increasing number of associated phason planes, the Burgers vector length decreases. However, in the ξ-structure, the sequence of associated phason planes is the Fibonacci sequence 1, 2, 3, 5, 8, …, whereas in ε_6 the sequence is 2, 4, 6, 10, 16, …, i.e. a series of double Fibonacci numbers.

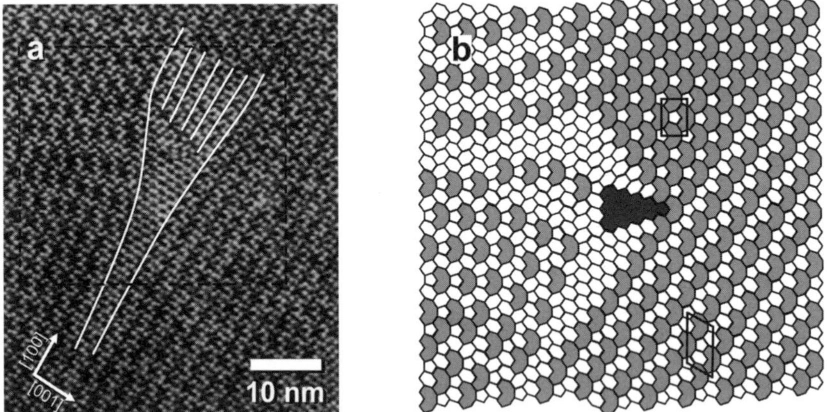

Fig. 18. Metadislocation with five phason planes in a modulated ε_{22}-type structure. a) TEM micrograph. b) Corresponding idealized tiling representation of the boxed region.

4.2. *Metadislocations in orthorhombic Al₁₃Co₄ phases*

o-$Al_{13}Co_4$ is an orthorhombic phase with space group *Pmn2₁* and lattice parameters a = 8.2 Å, b = 12.3 Å, and c = 14.5 Å.[16] The main structural feature are pair-connected pentagonal-prismatic channels extending along the [1 0 0] direction.[16] Within the (1 0 0) plane, the structure can be matched by a tiling consisting of regular pentagons and rhombs[17] as depicted in Fig. 19 a, where the rhombs are arranged in an antiparallel manner. The edges of the tiles correspond to atomic columns, which are predominately occupied by cobalt atoms. The orthorhombic unit cell is superimposed onto the tiling.

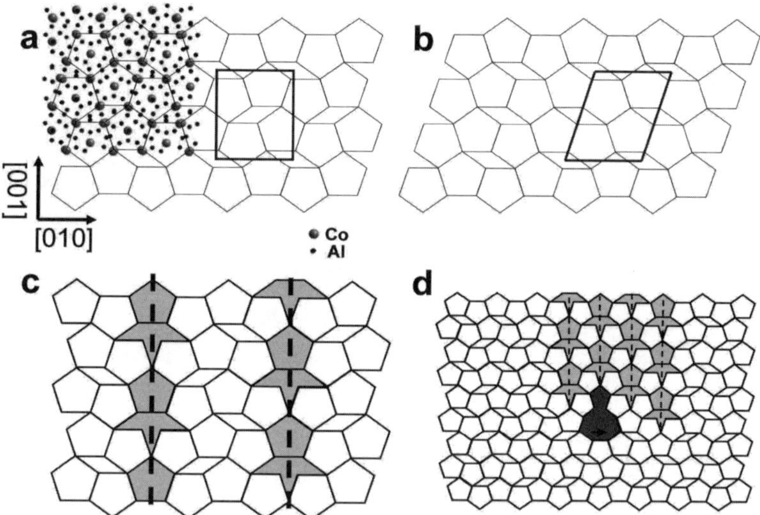

Fig. 19. a) Projection of the unit cell of the o-$Al_{13}Co_4$ phase along the [1 0 0] direction and corresponding tiling description. b) Tiling representation of the m-$Al_{13}Co_4$ phase. c) Phason lines (grey) forming (0 1 0) planes (broken lines). d) Metadislocation with four phason planes. The dark grey tile represents the metadislocation core.

The structure of monoclinic m-$Al_{13}Co_4$ is homeotypic to that of monoclinic $Al_{13}Fe_4$, and structurally closely related to that of o-$Al_{13}Co_4$[16,18-20] The tiling description of m-$Al_{13}Co_4$ (Fig. 19b) is characterized by a parallel arrangement of the pentagon and rhomb elements. Phason lines (light grey tiles in Fig. 19c) can be constructed by introducing an additional tile, a boat shaped heptagon. The phason lines can arrange along the [0 1 0] direction, forming (0 0 1) planes (Fig. 19c). The phason planes can be used to construct a new type of metadislocation in $Al_{13}Co_4$.[17] Figure 19d shows a hypothetical metadislocation with four associated phason planes. Its core is given by the dark grey tile. The Burgers vector can be calculated as $\vec{b} = b/\tau^3 [0 \quad 1 \quad 0]$. In analogy with the case of ε-type structures, a sequence of metadislocations can be constructed.

In experimental investigations of plastically deformed $Al_{13}Co_4$ crystals, however, planar faults possessing (0 1 0) habit planes are not observed and phason planes, as depicted in Fig. 19c, are not present. Instead, planar defects with (0 0 1) habit planes are found in high density (Fig. 20a).[21]

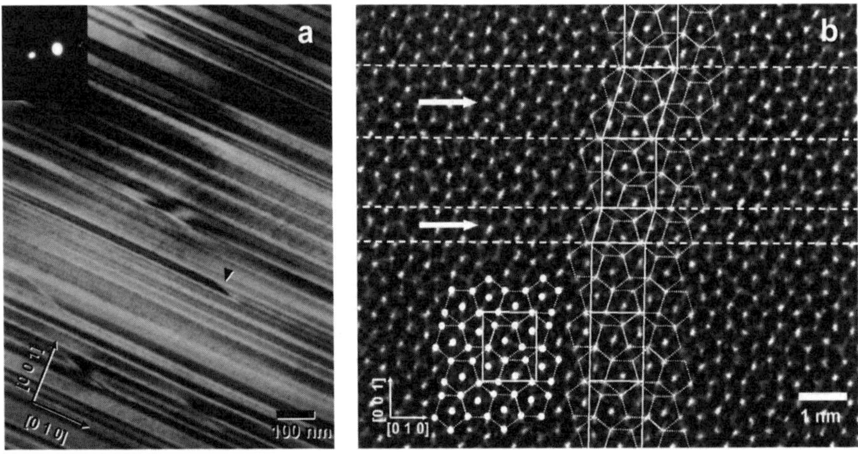

Fig. 20. (a) Bright-field TEM micrograph showing a high density of planar defects (parallel stripes) in a deformed o-$Al_{13}Co_4$ sample. The arrowhead points at a metadislocation terminating a planar defect. (b) High-resolution HAADF micrograph of planar defects in edge-on orientation (arrows) with superposed tilings for the o- and m-$Al_{13}Co_4$ structure.

In the TEM micrograph, the planar defects are seen as parallel stripes covering the entire image area. The arrowhead in Fig. 20a points at a metadislocation terminating a planar defect. Most of the observed metadislocation line directions are parallel to [1 0 0]. Contrast-extinction experiments reveal that the metadislocation Burgers vectors and the displacement vector of the associated planar defects are both parallel to the [0 1 0] direction. This shows that metadislocation glide on (0 0 1) planes is the basic mechanism of plastic deformation in $Al_{13}Co_4$. Figure 20b is a HAADF micrograph of planar defects in edge-on orientation (arrows). Bright dots in the micrograph correspond to atom columns which are preferentially occupied by cobalt atoms and can be matched by the o-$Al_{13}Co_4$ tiling. As in the T-phase, the planar defects are slabs of the structurally related phase represented by a parallel tile arrangement. In this case, the latter correspond to m-$Al_{13}Co_4$.

Figure 21a shows a high-resolution TEM micrograph of a metadislocation (central part of the image) in end-on orientation. The metadislocation terminates a (0 0 1) planar defect, i.e. a structurally modified slab of m-$Al_{13}Co_4$, extending to the right-hand side of the image. The outline of the o-$Al_{13}Co_4$ unit cells is depicted by white rectangles. Rhomboids representing the unit cell of the monoclinic $Al_{13}Co_4$ phase are used to cover the structurally modified slab. Figures 21b and 21c show two alternative models of the metadislocation-core tile. Which tiling corresponds to the real structure cannot be decided on the basis of the available TEM micrographs. For both examples, the Burgers vector can be calculated as $\vec{b} = b/\tau^3 [0\ 1\ 0]$ which is the same as for the predicted metadislocation in Fig. 19d. The way that these metadislocations are associated with planar defects is however different. The experimentally observed one is not associated with phason planes but with a slab of monoclinic m-$Al_{13}Co_4$ phase.

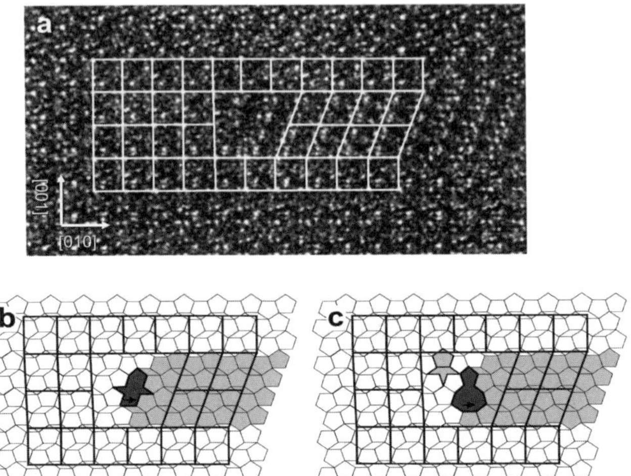

Fig. 21. Metadislocations in o-$Al_{13}Co_4$. a) High-resolution TEM micrograph. The metadislocation terminates a (0 0 1) slab of m-$Al_{13}Co_4$ extending to the right-hand side of the image. White rectangles and rhombs represent the unit cell of the o- and m-$Al_{13}Co_4$ phase, respectively. b) and c) Alternative tilings of the metadislocation core structure.

4.3. *Comparison of different types of metadislocations*

In this section different types of experimentally observed metadislocations are compared and their mode of motion is analysed. Figure 22 depicts metadislocations in ε_6-Al-Pd-Mn (a), ξ-Al-Pd-Fe (b) and in $Al_{13}Co_4$ (c). Their cores are represented by dark grey polygons and their associated planar defects are shown in light grey. Their Burgers vectors are indicated by arrows.

The Metadislocations in Figs. 22a and b are associated to six and three phason planes which have (001) habit planes. The metadislocations move along the - [1 0 0] direction. Their Burgers vectors are $1/\tau^4[0\ 0\ 1]$ and $1/2\tau^4[1\ 0\ 1]$, i.e. the ε_6-type metadislocation moves by pure climb and the ξ-type metadislocation moves by a mixed glide and climb process. Fig. 22 c shows a metadislocation in o-$Al_{13}Co_4$, which is associated to a (0 0 1) planar defect, i.e. a slab of m-$Al_{13}Co_4$.

This metadislocation moves along the [0 -1 0] direction which is parallel to its Burgers vector, i.e. the $Al_{13}Co_4$–type metadislocation moves by pure glide. This mechanism resembles that of a shear transformation as observed in Laves phases, where moving dislocations introduce structurally modified slabs.[22-24] In the C14 Laves phase Cr_2Hf, for instance, dislocation motion creates a slab of the cubic C15 Laves phase within the C14 matrix.[24]

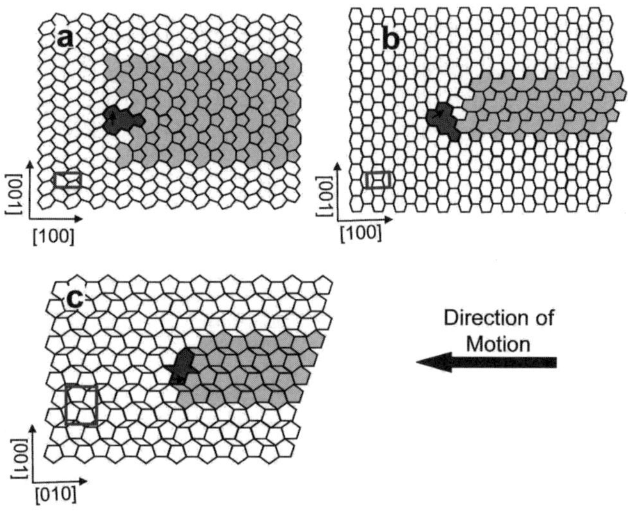

Fig. 22. Tiling representation of experimentally observed metadislocations comparing their modes of motion in ε_6–Al-Pd-Mn a), ξ-Al-Pd-Fe b) and in $Al_{13}Co_4$ c).

The broad variety of constructible and observed metadislocations suggests a definition independent of their mode of motion and the type of associated planar defect: A metadislocation is a line defect with a Burgers vector corresponding to a $1/\tau^n$ (n = 1, 2, 3,...) fraction of the corresponding lattice constant, mediating a local transformation to a phase that is structurally related to the matrix and accommodates the irrational Burgers vector to the lattice. The metadislocation core region comprises structural features of both phases. This definition includes the experimentally observed dislocations in several CMAs like $Al_{13}Co_4$, ε_6-

and ε_{28}-Al-Pd-Mn[1, 10, 13, 14, 17], ξ- and ε_{22}-Al-Pd-Fe[25], and several other ε-type phases.[8, 26]

References

1. H. Klein, M. Feuerbacher, P. Schall and K. Urban, *Phys. Rev. Lett.* **82** 3468 (1999).
2. K. Urban and M. Feuerbacher, *J. Non Cryst. Sol.* **334 & 335** 143 (2004).
3. M. Boudard, H. Klein, M. DeBoissieu, M. Audier and H. Vincent, *Philos. Mag.* A **74** 939 (1996).
4. L. Beraha, M. Duneau, H. Klein and M. Audier, *Philos. Mag.* A **76** 587 (1997).
5. P. D. Nellist and S. J. Pennycook, *Ultramic.* **78** 111 (1999).
6. M. Yurechko, B. Grushko, T. Ya. Velikanova and K. Urban, in: *Phase Diagrams in Materials Science,* Ed. T. Ya. Velikanova (Materials Science International Services, Stuttgart, Germany, 2004).
7. H. Klein, Phd thesis, Inst Nat Polytech de Grenoble, France, 1997.
8. M. Heggen, M. Engel, S. Balanetskyy, H.-R. Trebin and M. Feuerbacher, *Phil. Mag.* **88** 507 (2008).
9. S. Balanetskyy, B. Grushko and T.Y. Velikanova, *Z. Kristallogr.* **219** 548 (2004).
10. H. Klein, M. Audier, M. Boudard, M. de Boissieu, L. Beraha and M. Duneau, *Phil. Mag.* **73** 309 (1996).
11. H. Klein and M. Feuerbacher. *Phil. Mag.* **83** 4103 (2003).
12. M. Feuerbacher, H. Klein and K. Urban, *Phil. Mag. Lett.* **81** 639 (2001).
13. M. Heggen and M. Feuerbacher. *Mat. Sci. Eng.* **400-401** 89 (2004).
14. M. Heggen and M. Feuerbacher, *Phil. Mag.* **86**, 935 (2006).
15. D. Hull and D.J. Bacon, *Introduction to Dislocations* (Pergamon Press, Oxford, 1984).
16. J. Grin, U. Burkhardt, M. Ellner and K. Peters. *J. All. Comp.* **206** 243 (1994).
17. M. Feuerbacher and M. Heggen. *Phil. Mag.* **86** 985 (2006).
18. R.C. Hudd and W.H. Tailor, *Acta Crystallogr.* **15** 441 (1962).
19. Y. Grin, U. Burkhardt, M. Ellner and K. Peters, *Z. Kristallogr.* **209** 479 (1994).
20.] B. Grushko, R. Wittenberg, K. Bickmann and C. Freiburg, *J. Alloys & Comp.* **233** 279 (1996).
21. M. Heggen, D. Deng and M. Feuerbacher, *Intermetallics* **15** 1425 (2007).
22. C. W. Allen and K. C. Liao, *Phys. Stat. Sol.* A **74** 673 (1982).
23. P. M. Hazzledine and P. Pirouz, *Scripta. Met.* **28** 1277 (1993).
24. M. F. Chisholm, S. Kumar and P. Hazzledine, *Science* **307** 701 (2005).
25. M. Feuerbacher, S. Balanetskyy and M. Heggen, *Acta Mater.* **56** 1849 (2008).
26. M. Engel and H.-R. Trebin, *Phil. Mag.* **86** 979 (2006).

CHAPTER 6

COLD WELDING DUE TO IMPACT AND FRETTING UNDER VACUUM. CONSIDERING SCALING FOR APPLICATIONS IN SPACE MECHANISMS

Andreas Merstallinger and Muriel Sales

Austrian Institute of Technology AIT
2444 Seibersdorf, Austria
E-mail: andreas.merstallinger@ait.ac.at

A common failure mode seen during the testing and operation of spacecraft is termed "cold welding". European laboratories also refer to this as either "adhesion", "sticking" or "stiction". This chapter is intended to give an introduction to the effect of cold welding in general. It surveys basic theory, continues to a test method and equipment developed by AIT. Finally, findings and results on "cold welding" between two contacting surfaces under condition of both impact and fretting will be given. Results will be drawn from "engineering surfaces" which may be considered as bare metallic, inorganically coated, or organically coated metals and their alloys.

1. Introduction

1.1. *Cold welding failures on spacecrafts*

Spacecraft subsystems contain a variety of engineering mechanisms which exhibit ball-to-flat surface contacts. These may be periodically closed for several (thousands of) times during on-ground testing and operational life of the spacecraft. These contacts are usually designed to be static, but in reality they are often subjected to impact forces. Other static contacts are closed without impact, but will be subjected to fretting during the launch phase, during deployment of arrays but also in service life of the spacecraft. In the latter case, the origin may be vibrations of the spacecraft caused by gyros or motion of antennas. In most cases metals are used in the construction of these mechanisms, preferably light metal alloys, but these are strongly prone to adhesion. Impacts and fretting occur also in terrestrial applications, but the main difference in space is the missing atmosphere (oxygen).

On ground it is unusual to witness any adhesion between either metallic interfaces independent whether they are subjected to impact or fretting. This is because the surface are re-oxidised after each opening. Hence, the next closing is made on new oxide layers. In space, the oxide layers are broken irreversibly. Therefore, the following closing is metal-metal-contact thereby enabling welding effects. These effects may in open literature also be referred to as sticking, stiction or adhesion. Regarding ESA space mechanisms and standardisation the relevant source is[1], using the item "separable contact surfaces"

An **impact** during closing can eventually degrade the mechanism's surface layers whether they are natural oxides, chemical conversion films or even metallic coatings. This can dramatically increase the tendency of these contacting surfaces to "cold-weld". Fig. 1 shows an example for such a mechanism. The anchor was actuated from it's resting position (middle) electromagnetically, and impacted on both end stops. Finally, an anomaly from the flight model of this mechanism on a satellite was reported, that the anchor kept blocked on the left side (as shown in Fig.1). In technical terms, the adhesion forces were higher than the separation forces available from this mechanism's spring. Ground simulation of the mechanism in a vacuum chamber led to an adhesion force in the range of 0.3 N. This could be confirmed by impact testing at AIT.[2]

Another even more dangerous effect is **fretting**: vibrations occurring during launch or during movement of e.g. antennas in space, can lead to small oscillating movements in the contact, which is referred to as "fretting". This lateral motion causes even more severe surface destructions compared to impact. It may lead to cold welding effects similar to bonding techniques. Adhesion forces may increase to values higher than the closing forces. A failure due to cold welding after fretting occurred on the Galileo spacecraft.[3] The high gain antenna was shaped like an umbrella. Their ribs were locked for launch. The deployment of the high gain antenna could not be fully reached due to cold welding (1991). Investigations have shown that fretting during transport and lift-off caused cold welding of ribs in the launch lock position.

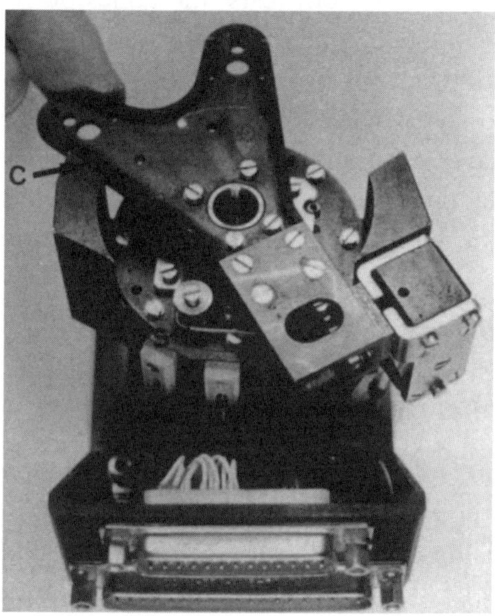

Fig. 1. Mechanism of a satellite: anchor was actuated from its resting position (middle) electromagnetically. It impacted on both end stops. Finally, an anomaly from the satellite was reported, that the anchor kept blocked on the left side. It was "cold welded".

1.2. *Objective to setup test method*

"Cold welding" was diagnosed to be the cause of some spacecraft mechanism failures in the late 1980's and early 90's. It was clear that some laboratory testing was needed in order to assess the effect of different surfaces making "static contact" under vacuum.

This was done by constructing two dedicated equipments: the "impact facility" and the "fretting facility". Both were developed at AIT and have been used to investigate several combinations of bulk materials and coatings for their tendency to "cold-welding". The test philosophy is based on repeated closing and opening of a pin-to-disc contact. In an impact test, in each cycle, the contact is closed by an impact with defined energy (no fretting applied). During a fretting test, the contact is closed softly (without impact), and while being closed, fretting is applied to the contact. For both tests, the adhesion force, i.e. the force required to re-open the contact, is measured at each opening. Basic studies[4] were carried out to show the influence of the main parameters impact energy and static load (contact pressure). These first results have been used to set up a standard test method with fixed parameters.[5] In the following, a survey on basics will be given. This is followed by sections with findings by the authors relevant to space applications.

2. Scientific Background to Cold Welding Effect

2.1. *Basics scientific: Definitions science*

The term adhesion (force) herein refers to the force which is necessary to separate two bodies (pin and disc). The adhesion coefficient is calculated by division of the adhesion force by the load. The test itself is designated to "cyclic static adhesion test", meaning the measurement of the adhesion force after each soft (= quasi-static) loading and unloading for several thousand cycles. If loading is done with an impact, the term 'impact cyclic adhesion test' will be used.

The adhesion force herein is no property of one material but of the whole (tribo-)system, consisting mainly of the two contacting bodies and their environment. The contact is made between a sphere and a flat disc. The pressure distribution within the contact zone may be calculated by Hertzian theory, presumed the maximum pressure does not exceed the elastic limit, given e.g. by the criterion of Tresca.

In a microscopic scale the contact is made between single asperities of the rough surfaces. During loading the asperities are deformed and their number increases until the real area is big enough to carry the load. This real area depends only on the load, on material properties (Young's modulus, elastic limit/hardness), but not on the nominal, i.e. macroscopic, contact area.

If both surfaces are clean, the asperities in contact may adhere to a measurable extent. Atomically spoken the atoms of the both bodies are close enough to form a metallic or covalent bond. The effect is preferably seen between metallic bodies, but in general to a lower extent, adhesion due to covalent binding is also reported.[6]

The following Table 1 indicates how the bulk material is hidden by several layers, which thereby inhibit adhesion, i.e. reduce it to an immeasurable value:

To find high adhesion forces, at least some hundred atomic layers have to be removed, i.e. the adsorption and chemisorption layers. This may be achieved by e.g. sputter etching or sliding. In both cases the material's structure in the vicinity of the contact differs from that of the bulk.

Table 1. Structure of surface.[7]

Range of Depth	Description	e.g.
1 - 10 Å	adsorbed layers	water vapour, …
10 - 100 Å	chemisorbed layers	oxides, …
100 - 50000 Å	plastically deformed zone	machining, grinding, …
> 50000 Å	bulk material	

2.2. *Effect of the space environment and space mission on adhesion*

Table 2 shows a survey of the environmental characteristics of satellite orbits. Satellites themselves own a higher local atmosphere.[8]

The main tribological difference to earth is the lack of atmosphere, which results in the following processes:

- adsorption layers desorb,
- contaminants outgas,
- oils and greases decompose and sublimate,

- sputter etching by particle impingement (outer sides of space vehicles),
- chemisorbed layers, e.g. oxides, if removed, do not reform.

In a technical sense the term cold-welding refers to the blocking/sticking of mechanisms like eject devices or switches. This kind of failure is achieved, if adhesion forces exceeds the resetting force of the springs used for re-opening which often refer to only a few hundreds of Millinewtons.

Table 2. Characteristics of the space environment, satellites own a higher local atmosphere.[9]

Altitude km	Pressure (common) mbar	Pressure (local) mbar	Composition	Comment
400	10^{-7}		$N_2, O_2, ...$	Shuttle,
800	10^{-9}		O, He, O^+, H	Lowest satellite orbits
2000	10^{-11}	10^{-6}		e.g. Meteosat
6500	10^{-13}		H^+, H, He^+	
22000	10^{-13}	10^{-8}	H^+, He^{2+}	Telecommunication

The chance of cold-welding increases, if
- metals are soft and have the same crystal structure,
- contact pressure is high,
- oxide films may easily be damaged/removed. (Oxides are generally more brittle than metals and therefore more likely to crack and exposure underlying metal if it is too soft to provide a firm support under load,
- contact surfaces become clean (sputter etching, sliding, fretting[a], ..),
- low temperature, affects ductility and toughness,

---

[a]Fretting refers to a kind of friction process but with oscillatory movement within the contact plane with amplitudes in the range of a few to a few hundred microns. (Compare e.g. OeNORM M 8120/Part2.)

- use of lubricants is restricted, due to the vicinity of sensitive elements, like optical and electronic components or high reflecting surfaces, which can be contaminated by condensed layers.

For the sake of completeness it has to be noted, that also the ground activities as well as the launch (high accelerations, vibrations resulting in fretting) influence the further operational life in orbit.[10]

3. Cold Welding Test Method

3.1. *General description of test method*

The test method reported herein is based cyclic contacts. A pin is pressed onto a disc for several thousand times. At each opening the force required to separate pin and disc is measured. This force is referred to as "adhesion force" of this cycle. The adhesion force is plotted as function of cycles. Comparison of different materials is based on the maximum value of adhesion found during a whole test.

To enable comparison of cold welding tendency between different material pairings, the following testing philosophy was set-up at AIT (it is described in detail in an in-house specification of AIT[5]): the parameters static load and impact energy are fixed for each pairing with respect to elastic limit (EL) of the contact pairing. Hertz' theory is used to calculate to contact pressure in the ball-to-flat contact. Using the yield strength of the softer material, the "von MISES-criterion" defines an elastic limit (EL): if the load (contact pressure) exceeds this EL, plastic yield would occur. Similarly, for the impact energy a limit (W_Y) can be deduced above which yielding occurs[5, 11] (the formulae can be seen in *section 3.3.*) Based on parameter studies[12, 4], an AIT-standard was defined and approved by ESA: the static load is selected to achieve a contact pressure of 40, 60 and 100%EL. An impact test is started with a static load, which achieves 40% of EL. After 10000 cycles, the load is increased to achieve 60%EL. After, another 5000 cycles 100%EL are applied. The impact energy is kept constant to 40 times the W_Y. This stepwise increase of load enables to continually obtain data during one test run. (From the point of possible irreversible plastic deformations, loads may be increased but must not be decreased. In the latter case, work hardening of material might have increased hardness, and therefore the actual contact pressure is lower than calculated.) For fretting tests, only one static load (related to 60%EL) is applied for 5000 cycles. Fretting parameters are a stroke of 50µm at a frequency of 200Hz. Uncoated specimen are freshly ground to a surface roughness of Ra<0.1µm before testing.[5] The contact is closed for 10 seconds and then opened for 10 seconds, too. At impact the base pressure of vacuum was less than 5.10^{-8} mbar, i.e. surfaces are not recovered during opening. During fretting test, a base pressure of 5.10^{-7} mbar is sufficient, since the change from oxidative wear to adhesive wear is in a range of 0,1 to 10^{-3} mbar. (The devices are described below, a detail of the fretting test equipment is shown in Fig. 3.)

The European Co-operation for Space Standardisation (ECSS) has released specifications on contact surfaces. In the ECSS - E-30 Part 3A, section 4.7.4.4.5 "Separable contact surfaces"[1], following main requirements are stated:

b) Peak Hertzian contact pressure shall be below 93% of the yield limit of the weakest material. (This refers to a contact pressure of 58% of the elastic limit, EL.)

d) ... the actuator shall be demonstrated to overcome two times the worst possible adhesion force ..

Therefore, results obtained from cold welding tests acc. to the ARCS-in-house-specification[5], can be used to address the necessary opening forces for actuators in mechanisms. (Both, impact and fretting test are done at 60% EL.)

3.2. *Description of cold welding test devices*

This very specialised and unique equipment enables to simulate cyclic closed contact, like e.g. in relays, or at end stops, and to measure the forces necessary to re-open the contact, i.e. the adhesion forces. This effect is herein referred to as "Cold Welding", but other terms may be e.g. stiction. Two facilities cover the most dangerous types of contacts: Impact during closing and Fretting within the closed contact. The latter facility may also be adapted to fretting wear tests.

It is the aim of the test to provide designers and engineers with data on adhesion force which have to be considered in design of mechanisms, i.e. to assess on one hand cold welding on bare metal contacts and on the other hand the anti-adhesion performance of coatings. If no specific parameters are demanded, tests are done according to an in-home-specification, to enable comparison of results for different material pairings and the collection of data into a data-base. There was undertaken a lot of effort in the past to be able to offer Special Competence in Cold Welding Effects.

The **impact facility** (Fig. 2a), one for loads up to 100 N enable the measurement of adhesion forces under vacuum between contact points after cyclic contacts, varying from static contacts (long-term compression) to slow closing cyclic contacts (static adhesion) and to impacts with impact energies up to 0.02 J.

The facilities were designed and developed in house. They consists of an UHV-system (10^{-8} mbar after bake out at 130 °C) with an ion getter pump and an air damping system for vibration free measurements. Emphasis was given to the universality of the test set up covering a wide range of impact energies, contact pressures and contact times. The contact is made between a ball and a flat disc. The ball is mounted on a pushrod, which is driven by an electromagnet. A low friction loading system enables an accuracy of 1 mN (0.1 g) for the adhesion force measurement which is made directly above the pin in the vacuum chamber using a piezo force transducer. The transducer measures the impact force as well as the adhesion, i.e. the force needed for the separation of the two materials. Cyclic loading may be done either slowly ("static") or by an impact ("dynamic") with defined energy which is determined by the mass of the pushrod and its velocity at impact as measured by a distance sensor. The impact energy, the impact force, the contact duration, the load during contact and the separation is controlled by a computer. By variation of the ball radius - typically to 2-20 mm - the contact pressure can be adjusted to the yield strength of most materials. Optionally the contact surfaces can be cleaned in situ by glow discharge before the test. The surface roughness is characterised by stylus profilometry.

Above test methods are adequate to detect the affinity of cold welding in an early state. They are capable to assess the statistical spread with increasing contact cycles in order to see if there is a tendency toward cold welding or a single catastrophic failure. A known failure of a space mechanism was successfully reproduced and studied in the laboratory. General tendencies of cold welding were assessed within a master thesis.[13]

The fretting tests facility (Fig. 2b and Fig.3) enables to investigate the influence of fretting on the tendency to the cold welding of materials, i.e. after a definite number of fretting cycles the adhesion force between contacts is measured.

Fig. 2a. Impact facility (inside view).

Fig. 2b. Fretting facility: device (Upside) (inside view below).

The loading mechanism is analogous to the above "Cold welding - impact test device". The loading and adhesion forces are measured by the z- direction of a 3-axis piezo transducer mounted directly below the disk in vacuum. With the x-direction the friction force due to the fretting movement is measured. The fretting movement (sine, triangle or square wave) is generated by a piezo actuator for frequencies between 0 and 300 Hz and amplitudes up to 100 µm. This lateral movement is transduced to the pin via a spring plate and controlled at the contact by a triangulation sensor.

This facility enables for the first time the simulation of high frequency vibrations resulting e.g. from bearings combined with the measurement of adhesion force. In most cases wear occurs, which is measured by 3D-optical profilometry.

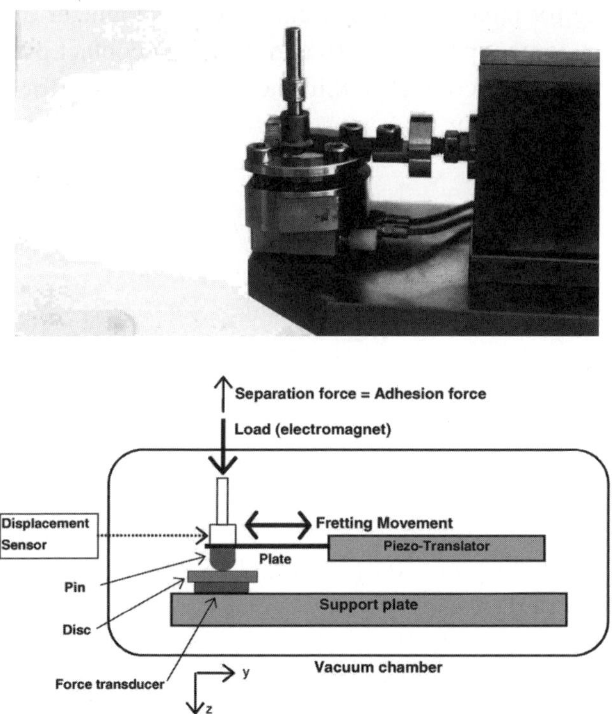

Fig. 3. Fretting device: Detail showing the fixation of pin (upper rod) and disc (mounted directly on a force transducer). Right side: functional principle for generation of fretting movement.

3.3. Hertz theory

3.3.1. Identification of materials

For comparable testing of materials, mechanical data is necessary (Y, E, v, H), values are typically be taken from literature (E, v), from tensile test (Y) and from hardness test (H). Calculation of yield strength from Vickers hardness test shall be avoided[b]. In case of use, Y shall be accompanied by the index 'HV': $Y_{HV} = H_V / 2.6$.

3.3.2. Hertz theory of elastic contact

The following formulas are used to calculate the diameter of contact (a), the maximum contact pressure (p0) and the mean contact pressure acting in a spherical-shaped contact within the elastic regime (for theory refer to e.g. ref. 11, page 93f):

$$\frac{1}{E^*} = \frac{1-v_1^{\,2}}{E_1} + \frac{1-v_2^{\,2}}{E_2} \tag{1}$$

$$a = \left(\frac{3PR}{4E^*}\right)^{1/3} \tag{2}$$

$$p_0 = \frac{3}{2}\,p_m = \frac{3P}{2\pi a^2} = \left(\frac{6PE^{*2}}{\pi^3 R^2}\right)^{1/3} \tag{3}$$

Units are [Pa]-1, [m] and [Pa] for Eq. 1 to 3, respectively. The elastic limit is related to the onset of yield (which occurs not at the surface but inside the softer material). Both criteria, Tresca and von Mises yield the same relation: (see ref 11, page 155):

---

[b]However, this is an often used practice in tribology, since the contact refers to a similar stress situation.

$$p_0 = 1.60Y \tag{4}$$

$$P_Y = \frac{\pi^3 R^2}{6E*^2}(p_0)_Y^3 \tag{5}$$

Thus, the load (Py) necessary to cause onset of yield follows

$$P_Y = \frac{\pi^3 R^2}{6E*^2}(1.60Y)^3 = 21.17\frac{R^2 Y^3}{E*^2} \tag{6}$$

Using the yield criterion, Hertz theory also offers to calculate the (kinetic) energy (WY) necessary to cause yielding (see ref. 11, p. 361):

$$W_Y = 53.4808\frac{Y^5 R^3}{E*^4} \tag{7}$$

For definition of the impact velocity at ARCS the mass of 0.48 kg is used:

$$v = \sqrt{\frac{2W}{m}} = 2.041\sqrt{W} \tag{8}$$

Units are [Pa], [N], [N], [J] and [m/s] in Eqs. 4 to 8, respectively.

3.3.3. *Calculation of test parameter "elastic limit"*

The main objective is to achieve comparability between tests performed on all pairings of materials with different mechanical properties, the following main parameters are chosen to be similar for all tests:

 Contact pressure relative to the elastic limit (EL),

 Impact energy (W) relative to the critical energy (WY).

By use of Hertz theory and the TRESCA criterion the critical static load which is related to the onset of yield, i.e. 100 % EL, can be calculated:

$$P_Y = 21.17\frac{R^2 Y^3}{E*^2} \tag{9}$$

The critical impact energy (WY) which is related to the onset of plastic yield at low impact velocities may be calculated according to the dynamic Hertz theory:

$$W_Y = 53.4808 \frac{Y^5 R^3}{E*^4} \tag{10}$$

Units are in [N] for both equations.

4. Scientific Background to Cold Welding Effect

4.1. *Scaling of cold welding effects*

4.1.1. *Survey and description studies*

A) Atomic scale

Basically adhesion force in the atomic scale can be determined by Atomic Force Microscope (AFM) or Scanning Tunnelling Microscope (STM).[14, 15] But as this thesis focuses on a more macroscopic scale, this technique will not be discussed further.

B) Microscopic scale (Loads in mN-range)

Hartweck[16] investigated the adhesion of two crossed steel cylinders with definite coverage of N, C, P, O, S or H up to one monolayer. The aim was to simulate the cohesion at grain boundaries in bulk steel. The surface composition was determined by auger electron spectroscopy (AES) and the real contact area by contact resistance measurement. The loads were in the range of 10^{-3}-10^{-5} N.

The adhesion force per real area as function of the coverage showed a distinct maximum between 0.1 and 1.5 monolayers depending on the size of the atom of the adsorbed element (values of 400-2000 N/mm^2). Benard[17] notes that the maxima may refer to a methodical error in the resistance measurement. Buckley[18] studied adhesion with loads of 0.2 mN between single crystals of iron in definite orientations in clean and contaminated state (hydrocarbons, H$_2$S). Adhesion was found to be 0.1 -

4 mN.. He also studied adhesion after sliding (0,7 cm, 0.001 cm/s) between single crystals of Cu, Co, W, Ni and poly-crystalline Cu.[19] The ball-on disc system was loaded with 0.5 N for 10 seconds. In situ cleaning was done by outgassing and an electron gun. Sliding increased adhesion. Adhesion was correlated to solubility, orientation/lattice planes/lattice structure.

Gane *et al.*[6] investigated adhesion of clean/contaminated combinations of metals and covalent materials at loads of 1 - 100 mN. *In situ* cleaning of the crossed cylinders was done by sputtering, loading by a balancing arm. They proposed a theoretical calculation of the adhesion force:

$$F_a = k * A_h * U$$

k shape factor of order 2^{20}

A_h Hertz-Area, proportional to (load)$^{2/3}$

U Tensile strength

Vedamanikam and Keller[21] investigated the correlation between adhesion and sliding friction force. The system consisted of two crossed cylinders, one mounted on a balance arm, which were loaded with a few grams. In-situ cleaning was done by heating and Ar-sputter etching. More recently Takahashi[22] supplied a surface analysis apparatus with a mechanism to measure the adhesion force between a ball, radius 0.23 mm, and fixed with an electroconductive paste to a lever, and a flat block mounted on the standard sample holder. One test between an Au-ball and a silicon wafer is reported. After loading with 350 μN for 15 seconds an adhesion force of 280 μN was measured.

C) Macroscopic scale (Loads in N-range)

Kellogg[23] investigated a lot of typical space materials (metals and ceramics with and without coatings) to qualify them for flight reactors, i.e. nuclear fission reactors for satellites. Therefore the test temperatures were 370-700 °C and the loading durations 70 - 10000 h (typ.: 1000-2000 h). The loads were 5- 18 N with an accuracy of 450 mN using a flat-to-flat geometry. Adhesion coefficients for metal-ceramics-combinations were measured up to 0.5 (Load 7N, Beryllium vs. Al_2O_3)

at temperatures of 370- 700 °C which is even higher than for the metal-metal-couple Be-Be: 0.13.

One of the most recent studies has been done by Lewis[24] at the European Space Tribology Lab (ESTL) and is called long term static adhesion test. Herein the adhesion force after a loading duration of up to one year and at contact pressures in the range of the yield strength (loads 100 - 300 N with ball-on-flat) was measured. The materials selected, Al-alloy, Ti-alloy, AISI 52100, SS 440 C, Vespel SP3 (a polyimid made by Dupont), partially with hard or solid lubricant coatings, are the most commonly used ones for space applications. Although the test should have been finished in end 1994 the results are not yet published.

Maillat[25] submitted a set of common materials to an impact cyclic adhesion test. (Refer to section 3). In his experiment the loading arm of a vacuum-pin-on-disc-tribometer was moved up and down by an eccentric which was driven by an electric motor. For static load a dead-weight was used. The accuracy was 50 mN due to vibrations and noise. (Contrary to all other test set-ups, which used ion pumps, a turbo pumping unit with bellow and separate ground plate was used[c].) The adhesion force was observed by oscilloscope, approximately 30 of 10000 cycles were recorded. An impact consists of several rebounds. Plots of the force show negative values, which were interpreted as "dynamic" adhesion forces[d]. The test parameters were: loading duration 10 s, resting time between two cycles 2 s, vacuum $<3.10^{-7}$ mbar at beginning, max. 10000 cycles, impact speed 0.1 m/s, geometry sphere on flat (r = 2 mm). The maximum pressure at impact was set to 300 % of the elastic limit. Among other materials Al AA 7075, stainless steel (17Cr-7Ni-1Al) partly with TiC (CVD)- or MoS_2-coatings, Densimet (sintered W) and TA6V (Ti-6Al-4V, IMI-318) were selected.

---

[c]A similar construction was used in this thesis for evacuation, but prooved to be too noisy for adhesion test: vibrations cause junctions to break continuously during unlaoding and not all togehter at end.

[d]In similar tests, done by Merstallinger and Semerad[26], no dynamic adhesion could be confirmed.

The adhesion forces, except from Ti, were in the range of the experimental error (50 mN). The maximum adhesion coefficients were less than approximately 0.005, except Ti: 0.02. The maximum coefficient of dynamic adhesion is in the range of 0.04, the highest value (0.1) was found for Al. SEM/EDX investigation show material transfer from discs to pin (Densimet to steel pin: W-chips, also for TA6V-disc to stainless steel pin: Ti-plates).

Recently a study including impact cyclic adhesion of magnesium alloy versus itself, partially with a hard coating, and a Densimet stainless steel combination was finished. (For description of test device see section 3.) The test geometry was ball-on-disc. Using impact energies of up to 210^{-2} Nm, i.e. contact pressures in plastic range, adhesion forces up to 60 mN (coated Mg-alloy versus metallic Mg-alloy), 280 Nm (metallic Mg-alloy versus itself) and 100 mN (Stainless steel versus Densimet) were found, i.e. adhesion coefficients[e] of 0.0055, 0.0255 and 0.0091, respectively.[26]

Rossitto and Ghersini[27] set up an experiment to fly with spacelab in orbit. The measurement of adhesion after impact of a sphere on a flat was intended, but due to data failure no adhesion could be recorded.

Another on-orbit-experiment investigating cold-welding (micro-welding) of metallic materials, was "FRECOPA" flown on the Long Duration Exposure Facility (LDEF) by NASA and finished 1992. After 70 months on low orbit in space closed static contacts, which were disassembled on earth, showed no signs of adhesion.[28]

Although the following study does not investigate the adhesion in the sense of separating two bodies perpendicular to the contact planes, it may include some facts which may also contribute to the understanding of the behaviour in adhesion tests:

Pepper[29] studied the shear strength of clean and contaminated metals (Ag, Cu, Ni, Fe) against sapphire ((0001) Al_2O_3). The sliding distance of 5 µm is less than the diameter of the Hertzian contact area and therefore the main process is similar to adhesion tests: breaking of the microjunctions.

---

[e]For definition refer to section 2.1.

For the clean metals shear strength increased with the stability of the oxide of the elements. This is explained by the Chemical-Bond-Hypothesis (CBH): Formation of a chemical bond between metal-atom and sapphire-oxygen. After exposure to O_2, Cl_2 or C_2H_4 the shear strength increased (O_2) or decreased (Cl_2). For the increase, again element specific, between the oxygen-exposed-metals and sapphire the Intermediate Layer-Hypothesis is proposed: formation of a complex-oxide-layer between the metal and the sapphire.

D) Macroscopic scale (Very high loads, tensile test devices)

Another technique to obtain clean metal surfaces is to fracture a specimen under UHV using a tensile test devices. Adhesion then may be measured by loading it and fracturing it again. The cohesion (adhesion) coefficient is defined as the second separation force divided by the load.

Conrad and Rice[30] used this fracture technique for measurement of adhesion of clean and contaminated metals (Cu, Ag, Al, Ni, Brass) to themselves. The contact was loaded for up to 900 s with loads equal or less than the first fracture force. The adhesion coefficient for the bare metals varied from 0.8 to 1, for brass a value of approximately 0.5 was found. The adhesion coefficient could also be correlated to the load. The second fracture did not occur preferentially along the first fracture line.

Gilbreath[31], using the same fracture technique, investigated the adhesion of pure metals (Al, Mg, Ti, ..) and alloys (SS 17-4, brass). A load of 450 N per 27 mm^2 area was applied for different durations. The adhesion coefficient decreased from 0.95 for Pb (fcc), to 0.85 Al (fcc), 0.37 Mg (hcp) down to 0.1 for Bi (rhom). He stated an increase of adhesion with temperature, longer contact time or vibratory loading[f], but a constant value for contact pressures of 0.1 to 60 MPa for Pb, Cu and Mg.

---

[f]The direction of vibrating was not described properly. Vibration in plane of the surfaces, designated to "fretting" are known to increase adhesion.

Hordon[32] also used a tensile probe, but did not fracture specimen. The two flat samples, after evacuation and bakeout, were cleaned by a wire brush. The loads achieved pressures of 1 to 25 times the yield strength of the tested material. Metals and alloys were selected to show the influence of lattice structure (fcc, bcc, hcp) on adhesion. Adhesion was found to be negligible without wire-brush-cleaning and also in the elastic range. The slope of adhesion as function of the load decreased with

    1) increasing hardness, i.e. higher Young's modulus, and

    2) with increasing load.

Winslow *et al.*[33] investigated adhesion after contact pressures of 80% of the yield strength (with maximum of 700 N/mm^2) at temperatures of 300 - 500 °C. The loads were applied by a hydraulic ram for durations of 10 seconds to 19.5 hours. No in-situ cleaning was done. The adhesion coefficients decreased from 1.5 (!) for Cu-Cu and 0.5 for Al AA 2014 to zero for e.g. SS 304 or Ti-alloy (Ti6Al4V) at 500 °C.

Rittenhouse[10] notes that a motion in direction of the contact planes, i.e. some kind of sliding or fretting, was either not prohibited or not defined by the author. But he[10] further assumes that this may also apply to practical applications.

4.1.2. *Adhesion coefficients survey*

Although a variety of studies concerning adhesion/cold welding were found, a comparison seems rather difficult. Tab.3 shows a survey of adhesion tests for Cu, mainly OFHC, and Al, or Al-alloys. Adhesion coefficients range from zero up to values higher than 10 (!). Adhesion is not an effect predominately acting in metal-metal-contacts, but was also found between metals and covalent materials.[6] One problem in comparing is the lack of material or geometric data, which makes it impossible to recalculate contact pressures from loads or vice versa: in the low load range (<1N) adhesion commonly is compared to the load, in the high load range (>50N) to the contact pressure. Therefore in Table 3 the missing data is estimated and written in italics.

In Fig.4 the adhesion coefficients listed in Table3 are plotted against the load. At an increase of seven orders of magnitude of the load, the adhesion coefficient even seems to decrease, i.e. the adhesion force is limited which corresponds to the fact, that the contact pressure becomes stable, if the full-plastic regime is achieved: ~300 % of the yield strength. The maximum contact pressures range from 50 % to ~300 % of the yield strength over the whole load range.

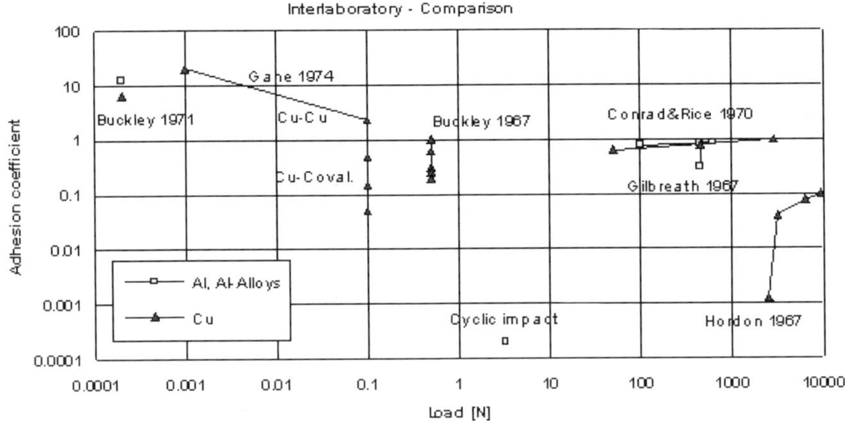

Fig. 4. Interlaboratory comparison of adhesion between Cu or Al another materials.

4.1.3. *Scaling for space*

The literature survey shows that adhesion is strongly depending on the "scale". On the other hand, the effect of cold welding is for space applications an important risk to mechanisms. Typical loading forces of mechanisms on spacecrafts are in the range of Newtons, roughly spoken from 1 to 50N. Engineering standards require contact pressures not higher than 2/3 of the elastic limit defined by Hertzian theory. Hence, the focus on experimental testing, discussion and conclusions will be in this loading regime.

Table 3. Interlaboratory comparison of adhesion between Cu or Al and other materials. (Due to the lack of material or geometric data it is impossible to recalculate contact pressures from loads or vice versa: missing data is estimated and written in italics.[13])

Material		In-Situ-	Adhesion	Contact			Temp.	Dur-	Geo-	Device-Type	Study:	
Pin	Disc	Cleaning	coeff.	Load *	Pressure *			ation	metry		Date	Author
				N	MPa	%Y	°C					
Aluminium												
Al AA 7075		-	0.0002	3.2	*1923*	300	RT	10 s	b-f	Cyclic impact	1992	Maillat, CSEM
Al (99.99%)		-	0.84	450	*16.67*		RT	3 s	frac	As-fractured	1967	Gilbreath
Al 2024 (T4)		-	0.31	450	*16.67*		RT	3 s	frac	- " -	1967	Gilbreath
Al		(Air!)	1.8		?		RT	?	f-f	Twist-Compress	1963	Sikorski
Fe (110)	Al (111)		12.5	2E-4	?		RT	?	b-f		1971	Buckley
Al AA 1100		-	0.77	100	130	*107*	RT	5 min	frac	As-fractured	1970	Conrad&Rice
Al AA 1100		-	0.85	620	560	*463*	RT	5 min	frac	- " -	1970	Conrad&Rice
AA 2014 (T6)		-	0	*17615*	445	80	150	20 h	f-f	Tensile	1966	Winslow
AA 2014 (T6)		-	0	*396*	10	80	300	20 h	f-f	- " -	1966	Winslow
AA 2014 (T6)		-	0	*1979*	50	80	300	2.7 h	f-f	- " -	1966	Winslow
AA 2014 (T6)		-	0.39	*1979*	50	80	300	20 h	f-f	- " -	1966	Winslow
AA 2014 (T6)		-	0.5	*1267*	32	80	500	1.3 h	f-f	- " -	1966	Winslow
Copper												
Cu OFHC		-	0.78	450	16.667		RT	3 s	frac	As-fractured	1967	Gilbreath
Cu		(Air!)	0.7		?		RT	?	f-f	Twist-Compress	1963	Sikorski
Cu (100)	Cu (100)	e-gun	1.02	0.5	*180*	*50*	RT	10 s	b-f	Pin-on-Disc	1967	Buckley ASTM
Cu (110)	Cu (110)	e-gun	0.61	0.5	*290*	*60*	RT	11 s	b-f	(balance)	1967	Buckley ASTM
Cu (111)	Cu (111)	e-gun	0.3	0.5	*375*	*90*	RT	12 s	b-f	- " -	1967	Buckley ASTM
Cu poly.	Cu poly.	e-gun	1	0.5	*270*	*60*	RT	13 s	b-f	- " -	1967	Buckley ASTM
Cu (110)	Cu (100)	e-gun	0.25	0.5	*220*	*50*	RT	14 s	b-f	- " -	1967	Buckley ASTM
Cu (111)	Cu (100)	e-gun	0.2	0.5	*240*	*50*	RT	15 s	b-f	- " -	1967	Buckley ASTM
(110)Fe	(111)Cu		6.5	2E-4	?		RT	?	b-f		1971	Buckley
Cu OFHC		-	0.62	50	20	*0.12*	RT	5 min	frac	As-fractured	1970	Conrad&Rice
Cu OFHC		-	0.98	2850	1280	*7.53*	RT	5 min	frac	- " -	1970	Conrad&Rice
Cu		Ar-sputt.	20	0.001	*33*	*13*	RT	min ?	c-c	Balance	1974	Gane
Cu		Ar-sputt.	2.3	0.1	*335*	*127*			c-c	- " -	1974	Gane
Cu	TiC	Ar-sputt.	0.5	0.1					c-c	- " -	1974	Gane
Cu	Sapphire	Ar-sputt.	0.15	0.1					c-c	- " -	1974	Gane
Cu	Diamond	Ar-sputt.	0.05	0.1					c-c	- " -	1974	Gane
Cu		Abrasion	0.00125	*2520*	50.4	80	RT	5 min	f-f	Tensile	1967	Hordon
Cu		Abrasion	0.04	*3150*	63	100	RT	5 min	f-f	- " -	1967	Hordon
Cu		Abrasion	0.075	*6300*	126	200	RT	5 min	f-f	- " -	1967	Hordon
Cu		Abrasion	0.1	*9450*	189	300	RT	5 min	f-f	- " -	1967	Hordon
Cu		-	0	*990*	25	80	300	2.7h	f-f	Tensile	1966	Winslow
Cu	-	-	0.03	*990*	25	80	300	20 h	f-f	- " -	1966	Winslow
Cu		-	0.04	*218*	5.5	80	500	10 s	f-f	- " -	1966	Winslow
Cu		-	0.22	*218*	5.5	80	500	100 s	f-f	- " -	1966	Winslow
Cu		-	1.5	*218*	5.5	80	500	20 h	f-f	- " -	1966	Winslow

4.2. Parameters influencing Cold welding

4.2.1. Influence of contact type

4.2.2. Influence of load

Basically adhesion depends on the real contact area. In case of contact pressures in the elastic regime, the contact area, i.e. adhesion, is correlated to the load, by e.g. Hertz's theory for spherical contacts. This is shown by Gane *et al.* for e.g. copper. After onset of plastic flow the slope of the contact area as a function of the load decreases until full plasticity is achieved and the contact area does not increase any more.[32] (Fig. 4.) A total conformity of the two bodies is hardly to achieve, even in the full plastic range, the surfaces asperities remain.[34, 35] In case of hard covalent materials in contact to themselves adhesion is independent of load (sapphire[6], diamond[6], theoretical[11]).

Two studies report a constant adhesion coefficient for all pressures: for Cu and other metals for loads from 2 to 2000 N (0.06~60 MPa),[31] (see also Table 3.), and Conrad and Rice[30] explain a decrease found for brass by work hardening. (Other materials showed an increase in the elastic regime).

4.2.3. Influence of loading velocity, i.e. impact energy

Crook and Hirst[36] found adhesion for indium after soft loading to be three times higher than for impact loading. (These tests were not done under vacuum!) Cyclic impact tests[26] showed that an increase of impact energy results in increasing adhesion, which was found to be due to a more severe destruction of the surface layers. (Surfaces were not cleaned in-situ, in order to simulate condition of the materials in orbit.) An impact itself was found to consist of several rebounds with total separation of ball and disc. An adhesion after each rebound, as reported by Maillat[25], was not found.

4.2.4. *Duration of contact and unloading velocity*

The increase of adhesion due to longer contact durations of up to 15 minutes was found to be linear[18] wheras Gilbreath[31] found a power function correlation with exponents of 0.009 - 0.14 for Zr to Pb_5Sb. In terms of mechanical behaviour an exponent of 0.14 may be explained by "dislocation creep" of materials, see e.g. ref. 17.

Buckley[37] studied the effect of the time taken to break the adhesive junction. Bénard[17] found this correlation to be similar, but with inverse sign, to that found for the contact duration; from the slope, he proposed the fracture mechanism to be similar to the "dislocation creep".[17]

4.2.5. *Cyclic loading*

The only study measuring adhesion after cyclic impact loading[25], provided very few data points, i.e. the adhesion force of less than 100 cycles was determined. The plots show a strong scatter, but an increase of adhesion with the number of cycles seems probable. This tendency, at cyclic impact testing, could be confirmed.[26] It was inferred from SEM/EDX investigations that the increase of adhesion is due to a progressing damage of the surfaces' contamination layers.

4.2.6. *Surface cleanliness*: *In-situ-cleaning and exposure to gases*

Adhesion tests done without any in-situ cleaning, like e.g. sputter etching:
- negligible adhesion at pressures of up to 300% of the yield strength.[32]
- no signs of adhesion, closed static contact, found after on-orbit experiment 'FRECOPA' flown on the LDEF by NASA.[28]
- adhesion coefficients less than 0.02 during cyclic impact test.[25]
- adhesion coefficients less than 0.025 during cyclic impact test.[26]
- considerable adhesion in case of high loads/contact pressures, using tensile test devices, even in air.[38]
- considerable adhesion in case of higher temperatures.[33, 39]

Despite of a missing in-situ-cleaning procedure, adhesion between metals may be observed, if contact pressures are high enough to remove or penetrate oxide layers. On the other hand adhesion is reduced after exposure.

Exposure of clean as-fractured surfaces to oxygen, formation of only one monolayer, reduced adhesion strongly, whilst exposure to nitrogen, up to atmospheric pressure, did not affect adhesion. CO and CO_2 also reduced adhesion, but to a less extent than O_2.[30] Hartweck[16] investigated polished iron-surfaces which where contaminated with N, C, P, O, S and H. The measured adhesion force per real contact area (resistance measurement) was found to show a maximum depending on the size of the contaminating atom. Bénard[17] mentioned the possibility of an error in the resistance measurement which may cause a wrong calculation of the adhesion tension and asks, if the adhesion force as function of coverage also shows such a maximum. Gilbreath[31] reports that adhesion after exposure to gases is reduced to ~50 %, if only physisorption occurs, e.g. H_2, N_2 or Ar. In case of chemisorption, e.g. O_2, air, or ethylene (C_2H_4), adhesion vanished. Gane et al.[6] also observed a reduction of the adhesion coefficient to less than 0.5 after exposure to oxygen. This was not only shown for a metal-metal-contact (Co-Co), but also for a Cu-TiC. In the latter case the difference of adhesion force with and without exposure seem to vanish with increasing load. Exposure to H_2 or Ar did not affect the adhesion force. Buckley[18] investigated the influence of exposure to various hydrocarbons and H_2S on adhesion of Fe (011). After loads of 0.2 mN adhesion forces of 0.1 - 4 mN were found, i.e. adhesion coefficients up to 20! Vanishing adhesion was only found for ethylene oxide, all other contaminants, containing no oxygen, did only reduce adhesion. Pepper[29] investigating the shear strength of a metal in contact with a covalent material, e.g. sapphire, determined that contamination not necessarily reduce shear strength, but may even increase it. This was explained by the ability of formation of chemical bonds between the metal and oxygen of the sapphire or complex-oxides as intermediate layer.

4.2.7. *Correlation to mechanical properties*

The main factor for the amount of adhesion is the real area of the contact. In case of ideally smooth surfaces, the contact area is defined by the Hertz theory. However, in general the surfaces are rough, the contact is made by single asperities coming into contact. With increasing load their number increases as well as they deform, until the real contact area is able to carry the applied load. In most cases this deformation is plastically. So, the real area is correlated to the yield strength, even if the nominal contact pressures are in the elastic regime.

Two restriction have to be mentioned. Firstly, metals commonly tend to work-harden, if the load is applied[35] and secondly, the yield strength of the asperities may exceed that of the bulk material (see Ref. 35, p.21). That may be seen from profiles of rough surfaces which were taken after loading with contact pressures achieving bulk flow. Although the indentation is clearly visible, the asperities were only partially flattened.[34, 35, 40.]

Hordon[32] reports a decrease of the adhesion coefficient with increasing yield strength. The slopes of the plots of the three lattice structures fcc, bcc and hcp were similar. The scatter within the lattice structure is explained by the different ability of the metals and alloys to work hardening. In case of two different materials the behaviour of the system is rather similar to the more ductile material.

Conrad and Rice[30] explain the fact that the adhesion coefficient as function of the yield strength is steady by firstly the real area being inverse proportional to the yield strength and secondly the yield strength depending on the bonding forces. So, although the real area of the contact decreases with increasing yield strength, the adhesion force remains constant, because the binding force increases. A reduction in case of brass was stated to be due to work hardening.

Sikorski[38], compressing and twisting two flat surfaces in air, determined decreasing slopes of the adhesion coefficient as function of the hardness in the order tetragonal - fcc - hcp. The maximum adhesion coefficient is independent of the hardness which is explained by Habig[7] similar to Conrad and Rice.[30]

4.2.8. *Diffusion*

The statements concerning the influence of mutual solubility on the adhesion, or the more technical term "cold welding", appear to form a dispute covering the surveyed literature from the early 50-ies until now.

The most stringent example for an influence is the technique of "diffusion bonding". (For details see Kazakow.[41]) At high temperatures, 1000-1500 °C, in order to enable considerable diffusion velocities, and moderate contact pressures, 20-60 MPa, bond strengths of up to 800 MPa between steel parts were achieved. That would refer to an adhesion coefficient of up to 40.

Alm[42] summarises in his survey over diffusion bonding techniques that the temperature not only increases diffusion velocity but also ductility. That makes higher adhesion not only due to diffusion but also due to an increased real contact area. The temperature should be in the range of 1/2 - 2/3 of the melting point. Hordon[32] reports a slightly higher increase of adhesion for mutual soluble couples than for insoluble at a temperature of 100 to 150 °C.

Cross sectioning of steel after adhesion test against Be at ~540 °C proved its diffusion into the steel. The adhesion coefficient of this soluble couple ranged up to 0.25. However, in adhesion tests between the insoluble couple Be-Al_2O_3 adhesion coefficients of up to 0.5 were obtained by Kellogg.[39]

On the other hand, in the case of room temperature no evidence was found that mutual solubility affects adhesion:

* self diffusion does not appreciably affect adhesion at room temperature.[30]
* mutual solubility is less important to adhesion than ductility and lattice structure, highest adhesion found for insoluble couple (Pb-Au), i.e. also higher than the soluble-insoluble couples mentioned above.[32]
* no correlation between solubility and adhesion, measured after sliding for a distance of 7 mm.[19]
* no correlation between solubility and adhesion.[6]

4.2.9. *Temperature*

Temperature influences adhesion indirectly. On one hand the decrease of hardness/yield strength with temperature causes an increase of the real contact area. Secondly, diffusion is enhanced. But in terms. of diffusion bonding, the temperature should exceed 1/2 of the melting point.[42] Another point to be considered is, that heating leads to desorption of contaminants, but also to the segregation of solved impurities from the bulk to the surface, thereby again contaminating the surface.

Hordon[32] stated an increase of the adhesion, one component made of copper or silver, with temperature rising from 20 to 150 °C. Below 100 °C adhesion increased similarly for soluble and insoluble couples, above 100 °C, higher slopes were found for the soluble couples. Gane[6] compared adhesion of Ge with hardness as function of the temperature. As hardness had fallen to ~50 % adhesion started to increase (T ~ 0.5 T_m) and exceeded unity at ~2/3 of the melting point. At this temperature the hardness had fallen to less than 10 % of the value at room temperature.

The increase of adhesion as function of temperature was well correlated to the ductility of the metal, e.g. Cu, Mg. In case of 17-4 PH steel the onset of adhesion was found at ~340 °C.[31] Bénard[17] summarises that adhesion starts to increase at temperatures of (0.3-0.4)*T_m .

A comparison of adhesion data, obtained by Kellogg[23] at temperatures of 150 to 500 °C, also indicates this tendency for e.g. Cu, Al AA 2014-T6. Due to a change of contact durations at the different temperatures, 370 - 1000 °C, from the data no increase of adhesion with temperature is visible, except probably for beryllium or stainless steel against Micaceram, a mica-bonded mica.

4.2.10. *Lattice structure and binding forces*

The ductility, i.e. the yield strength, depends on the lattice structure, in especially on the number of sliding planes and the energy necessary to activate them. Combining these two factors, ductility decreases in the order fcc - bcc - hcp.[43] This was validated by several studies, e.g.

Buckley[19], Gilbreath[31], Hordon[32], Sikorski[38]. The latter found rhombic materials to show even less adhesion than hcp.

In case of single crystals, the adhesion force is primarily influenced by the crystal planes coming into contact. A decrease of adhesion, with copper, was found in the order (100), polycrystalline, (110) and (111), i.e. minimum adhesion at close-packed planes. Additionally the adhesion is reduced, if the orientations of the contacting planes were not matched.[19]

The macroscopic adhesion between two bodies is, atomically spoken, determined by the same binding mechanisms as the bulk of a material.[29] So, if two bodies come close enough that the electron orbitals of their atoms overlap, a bond is formed.[44] If the bonds formed are of metal or covalent kind and their number is high enough, the macroscopic adhesion becomes measurable.

The lack of electrons in outer orbitals is, according to Sargent[45], the reason for adhesion and hardness to decrease in the order: B-metals - noble metals - transition metals. In transition metals the high percentage of covalent binding causes the binding forces to be strong dependent on the direction and he also assumes that herein a "thermal activation" is necessary to obtain adhesion. Gane *et al.*[6], on the other hand, found no indication for this assumption.

Binding forces also affect the material transfer. If two different materials adhere, the fracture commonly occurs in the one of the two materials that owns a lower binding force, e.g. Au onto Si.[18]

From the studies of the influence of exposure on the shear strength between a metal and a sapphire, Pepper[29] infers two sources of interfacial strength:

- an intrinsic one, called "chemical bond hypothesis" (CBH), meaning a covalent bond between the metal atom and the oxygen of the sapphire, the shear strength was found to correlate well to the free energy of formation of the metal oxide, and
- an extrinsic, called "intermediate layer hypothesis" (ILH), which proposes the formation of a complex oxide as an intermediate layer.

4.2.11. *Correlation adhesion - sliding friction*

The influence of sliding on the adhesion force may be referred to the section 4.2.6 "Surface cleanliness", since sliding may be seen as a kind of mechanical cleaning of the surfaces. Buckley[19] found adhesion to have strongly increased after a sliding for a distance of 7 mm.

A correlation between the adhesion coefficient and the sliding friction coefficient was proposed by Bowden and Tabor.[35] These tests were done with a steel ball sliding on indium.

Loading crossed cylinders with a balance system with loads of a few grams, Vedamanikam and Keller[21] were able to predict the friction coefficient from adhesion data, by use of a correlation proposed by McFarlane and Tabor.[46]

4.2.12. *Rough surfaces*

Adhesion is directly correlated to the real contact area. But this real contact area deviates the more from the geometric contact area, e.g. calculated by Hertz' theory, the more one surface becomes rougher. Since it was realised that the distribution of surface asperity heights often can be regarded as being Gaussian, several models to calculate the real area from profiles were developed. For the elastic contact of a rough and a smooth surface two models may be mentioned: Greenwood and Williamson[47] (GW-model) and Whitehouse and Archard.[48] (Detailed information, also on all other aspects of rough surfaces, may be obtained from Thomas.[49])

Within the GW-model, load and real area were calculated as function of the separation of the two surfaces. McCool[50] provides calculated examples and tabulated functions which allow to express the real area as function of the load. However, their use is restricted to flat surfaces. Johnson[11] indicates the extension to a sphere in contact with a flat surface.

Whitehouse and Archard[48] provided a theory in which the profile of a surface is described by only two parameters: the r.m.s. roughness (σ), i.e. the standard deviation of the height distribution, and the correlation length ($\beta*$) which is a measure for the distribution of the asperity curvatures. Calculations of the real area and the correlation to the GW-model were studied by Archard.[51] More recently and due to enhanced digitising and computing possibilities numerical solutions were developed. A model, which considers not only the contact of counterformal shape but also of conformal shape, was published by Aronov *et al.*[52] The program also distinguishes between the elastic and the plastic regime. The input of the calculations consist of the digitised 3-D-topographies of both contacting surfaces.

Another study, published by Poon and Sayles[53] presents a "numerical elastic-plastic contact model of a smooth ball on a directionally structured anisotropic rough surface". In a second paper, Poon and Sayles[54], this model is applied to real machined surfaces. Herein, real contact areas and pressures were calculated and plotted as function of the ratio $\sigma/\beta*$.The latter models were developed with the aim to understand wear mechanisms in sliding friction from the point of microcontact modelling. Some[g] authors, e.g. Poon and Sayles[55], report a relation between wear and the surface parameters σ and $\beta*$.

4.2.13. *Summary of known activities of other authors*

Apart from Takahashi *et al.*[22] the only known, existing adhesion/cold welding facility was the adapted vacuum tribometer at CSEM. Maillat[25] studied the adhesion during cycling contacts, but found only a few data points and loading was done by impact, achieving peak forces in the full plastic regime. Lewis[56] investigated adhesion after a contact duration of up to one year, without any cycling. For the calculation of the real area of the contact of rough surfaces a numerical solution for a smooth ball on a rough surface was presented by Poon and Sayles.[53] A consecutive calculation for an experimentally obtained ball-on-flat contact provides a correlation between the real area of contact (calculated) to the measured ratio of r.m.s. roughness to the correlation length.[54]

---

[g]The literature review did not include "friction" and/or "wear".

5. Present State of the Knowledge in the Scaling for Space

In this section, some findings in the field of cold welding for space applications are given. As mentioned above, this "scaling" is "Newtons", i.e. the findings were found be use of the cold welding test devices described above with loading forces in the range of 1 to 20N. It covers on one hand general aspects like influence of contact parameters on adhesion. On the other hand, also some more test results related to materials' and coatings' behaviour divided into impact and fretting are shown.

5.1. *Influence parameters on adhesion*

5.1.1. *Contact type*

Surfaces that are exposed to atmospheric conditions are generally covered by physically or chemically absorbed layers. Even in the absence of absorbed water, grease or other macroscopic contaminants there remain surface layers, such as oxide and nitride layers, which are formed under terrestrial conditions on pure metal surfaces and which can be regarded as natural protection layers against cold welding.

Under vacuum or in space environment, once removed by wear, these layers are not rebuilt and the exposed clean metal surfaces show a higher cold welding probability. So, adhesive and tribological behaviour under space environment or vacuum differs significantly from terrestrial conditions and the use of data collected under latter conditions is rather restricted. Secondly, a modelling of the adhesion forces suffers from the unknown degree of real metal-metal contact, which is linked to the destruction of the surface layers which is strongly affected by the contact situation. Moreover, scientific studies are mostly based on atomically clean surfaces. Hence, the value of adhesion on typical surfaces of spacecrafts produced by "normal engineering" are somewhere between the high values seen from atomically clean surfaces and the "too low values" which would be derived from "pure static" contact. In a recent study, two theoretical approaches to calculate adhesion forces were compared with experimentally measured adhesion forces for a fretting contact. It was shown, that the modelling approaches cannot predict actual adhesion forces.[57]

From general experience[58] and as discussed in previous papers[4, 12], contact situations may be classified in three different types: static, impact and fretting. In a cyclically closed and opened contact, the amount of destruction of surface layers increases in this order: static, impact and fretting. As surface layers are destroyed we see an increase of the adhesion forces. Fig. 5 shows three plots of the adhesion force as function of cycles (=openings). The three plots refer to three contact types applied to a pairing of titanium alloy (IMI834) and Stainless steel (AISI440C).[12] In fretting conditions the maximum adhesion force during the whole test was 9.5N (2.5 times the load of 4 N), under impact 0.96 N (load 29 N), whereas in static contact after 25.000 cycles adhesion of less than 0.1 N occurred (29 N load). A theoretical deduction would have given an estimate of 7,7N without any relation to the real contact situation: Hertzian contact area 0,006mm² times yield stress of this Ti-alloy (~1200MPa).

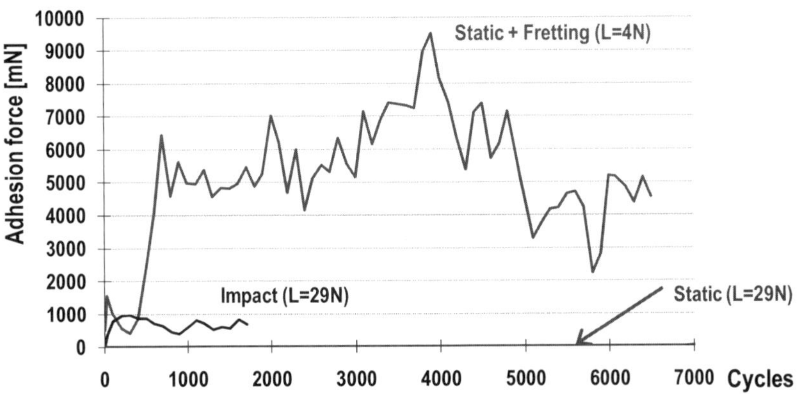

Fig. 5. Adhesion force as function of cycles (i.e. one closing-separation).[12] Comparison of adhesion in static (load 29N), impact (load 29N) and fretting (load 4 N) condition under vacuum. Danger and severity of adhesion increases with contact: static- impact – fretting. (Max. adhesion in static 0.1 N after > 25,000 cycles, in impact 0.96 N, in fretting 9.5 N.)

Summarising, contaminant layers (oxides) are removed under impact and fretting much more quickly as compared to static contacts, and cold welding occurs much earlier than expected. This may not only reduce the life time of a satellite but also can endanger space missions, e.g. any opening or ejection mechanism may fail due to cold welded contacts. A typical opening/closing mechanism can fail, if the adhesion force exceeds the force which is available to open this mechanism, e.g. by a spring. This "blocking" value may be much lower than the applied load. The blocking of the mechanism (Fig. 1) under impact condition was reported with an adhesion force in the range of 0.3 N. This value was confirmed by a first verification study of the impact device.[2]

5.1.2. *Environment*

As mentioned in section 4, the environment has crucial influence on adhesion force.[13] A screening test was performed using a contact pair of technical pure aluminium and copper. Pin and disc were polished to Ra~0,1µm. Contact parameters were a radius of curvature of 12mm and a load of 10N. A static test was performed at a base vacuum pressure of 2.10^{-8}mbar. Increasing the separation time from 2 to 120 seconds between loading cycles leads to an increase of surface contamination. Figure 6 shows the adhesion force as function of the exposure (vacuum pressure times opening time): starting with an adhesion for of 40mN, the adhesion decreases with increasing surface contamination to almost zero. It shows that in a static contact, adhesion is only technically measurable for ultra high vacuum (range of 10^{-8} mbar).

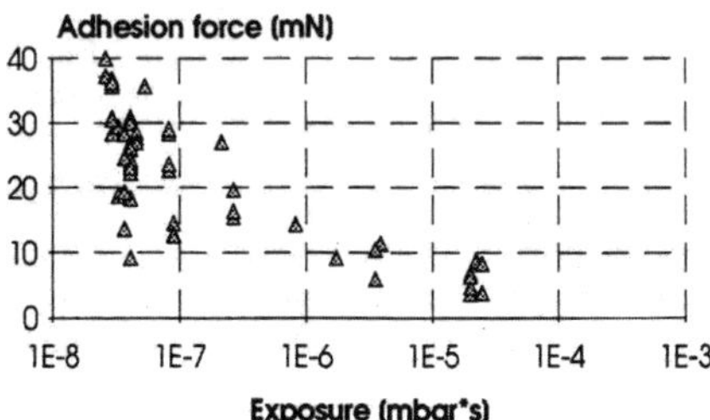

Fig. 6. Adhesion force decreases with exposure to vacuum (i.e. increasing surface contamination).[13]

Following that findings, impact tests are done at this vacuum level. On the other hand, under fretting conditions, much higher adhesion forces are found at vacuum levels up to 10^{-5}mbar. Hence, in case of mechanical surface destructions surface contaminations by residual gas are not that significant (fretting motions leads to scratching of surface layers).

5.1.3. *Roughness*

The influence of typical engineering finishes on adhesion force was screened on a similar contact as before (copper pin on aluminium). Here, the roughness of the aluminium disc was varied by using different types of finishing as indicated in Fig.7. The x-axis in Fig.7 refers to the product for RMS roughness (sigma) divided by the correlation length beta* which was taken as measure for the spatial dimensions.[13] From the data plotted, it can be concluded that medium RMS-roughness between 0,2 and 0,8μm, leads to the lowest adhesion forces. In this range, the contact area is reduced by the asperities and the shape of asperities does not enable too much plastic deformation. In case of higher roughness, (sand blasting) the plastic deformation leads to increased contact area. On the other hand, in case of perfectly polished surfaces initially a relative high contact area is enabled.

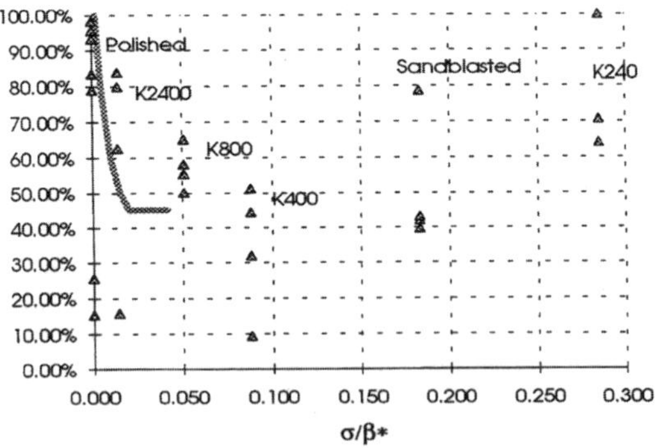

Fig. 7. Relative change of adhesion force with varying roughness.[13]

5.2. Adhesion under impact

5.2.1. Typical space materials under impact

In the following, some results found in impact testing will be compared. Data was obtained from static load of 100%EL. A Table, compiled in Annex explains abbreviations used for materials and their basic properties. A survey of adhesion forces found for a selection of typical (uncoated) space materials is shown in Fig. 8.

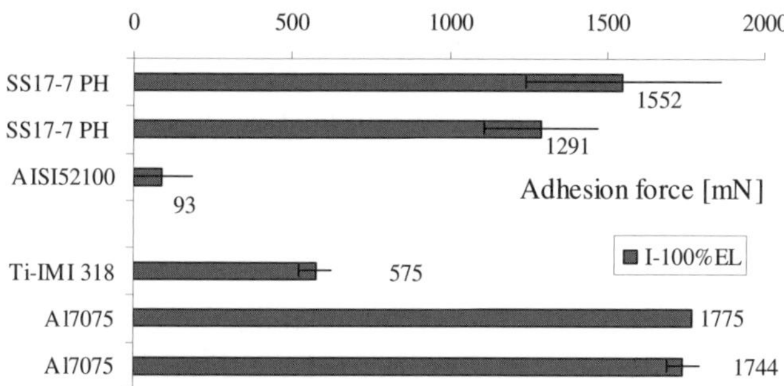

Fig. 8. Adhesion force under impact for materials in contact to themselves. Highest adhesion for stainless steels with Nickel (e.g. SS17-7PH), and Al alloys (Al AA7075). Medium adhesion for Ti-Alloy. Low adhesion for bearing steel AISI 52100 (no Ni).

The measured adhesion forces can be regarded as "high". Especially for stainless steel SS17-7PH vs itself or aluminium alloy (Al7075) vs itself. From a crystallographic point of view, face centred cubic metals are the most prone to adhesion: Fe, Al. This is due to their high ductility. A study[59] on the adhesion of different working materials to a cutting tool made of high speed steel indicated a relation between adhesion force and Ni-content. Regarding standard tests made with different steels to themselves, results show that the standard bearing steel (AISI 52100) has negligible adhesion. For the AISI440C (no Ni) certain adhesion under impact was found. Mixing of steels can reduce adhesion. (Fig. 9.)

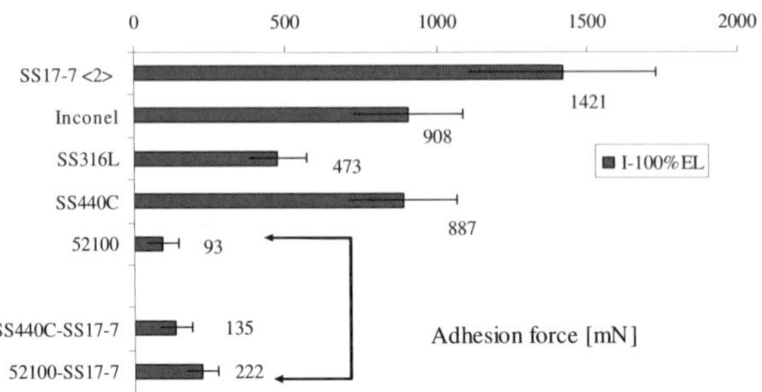

Fig. 9. Adhesion force under impact for different types of steel versus itself: austenitic structure and Ni seem to promote high adhesion: SS17-7ph (7%Ni), AISI 316L (11%Ni), Inconel 718 (52%Ni). No adhesion for AISI52100 versus itself ("52100"). High adhesion of AISI440C has to be proven. Combinations of different steels: adhesion seems to increase in contact to steels with higher tendency to cold welding (arrow).

5.2.2. *Influence of coatings on steel*

One possibility to reduce the risk of cold welding, is to apply coatings onto the contacting surfaces (Fig. 10). Stainless steel discs (SS17-7PH) were coated with TiC and MoS_2 and investigated for their ability to reduce adhesion. These coatings are related to two types of coatings: respectively hard and soft. The efficiency of the first group - **hard coatings** - depends on the load bearing capacity of the underlying bulk: if it is too soft, it is deformed under impact, and the hard coating breaks.[60] Then the underlying metal comes into contact to the metal of the opposite metal surface, and adhesion is found. However, pieces of the hard coating (TiC) are still present. They may be transferred and may act as additional abrasive particles. Hence, adhesion may be decreased in comparison to bare metal surfaces, but since destructed surfaces areas cannot be "re-coated" adhesion still occurs. An example is the **TiC** (2000 HV) on the SS17-7PH (only 441HV): the coating decreased the adhesion force by about four times, after the TiC brake-off it was no longer effective and a marked increase in adhesion force was measured.

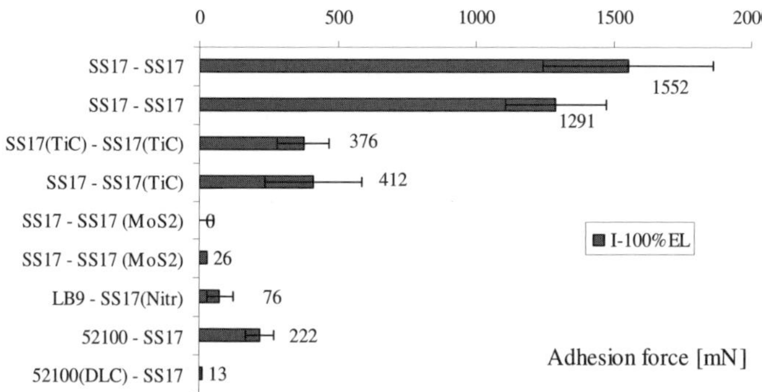

Fig. 10. Adhesion force as function of static load for different coatings on steel: lowest adhesion for SS17-7PH (SS17) with MoS2. Higher adhesion for TiC (coatings were broken). Negligible adhesion between bronze (LB9) and SS17-7PH (LB9-SS17(Nitr)). Low adhesion between AISI52100 and SS17-7PH, can further be decreased by use of DLC coating ("52100(DLC)-SS17").

Hence, a hard coating should be applied on steel types which enable a higher hardness, e.g. AISI440C or AISI52100 (up to 700 HV). This would avoid plastic deformations of the underlying steel substrate, which results in cracking of the coating.

Instead of using a hard coating, a harder steel type may be selected for contact to stainless steel SS17-7PH. By use of steel AISI52100 in contact to SS17-7PH lower adhesion can be achieved (222 mN, Fig.9). This can further be reduced by application of hard coating on hard steel: DLC.[61] (Fig.10.) The hard **DLC-film** did not (visible) peel off and no adhesion during more than 37.000 cycles was measured. Some small amount of steel was transferred from the (un-coated) pin to the DLC-coated disc. Before selecting DLC, great care has to be put on it's composition. Most of conventional DLC coatings are not compatible with vacuum applications. Secondly, a **soft lubricant** coating on SS17-7PH could avoid any adhesion to another SS17-7PH pin. Hence, under impact soft lubricant coatings on stainless steels reveal higher efficiency in prevention of cold welding than hard coatings.

5.2.3. *Influence of coatings on aluminium an titanium*

Hard **finishes on the soft aluminium substrate** showed breaking and removal of the upper layers, but did not enable cold welding. Tests were run up to 50.000 cycles without finding sudden increases of adhesion forces. Figure 11 compares the maximum adhesion forces of Al7075 vs itself (uncoated: Al7075-Al7075) and the influence of selected coatings.

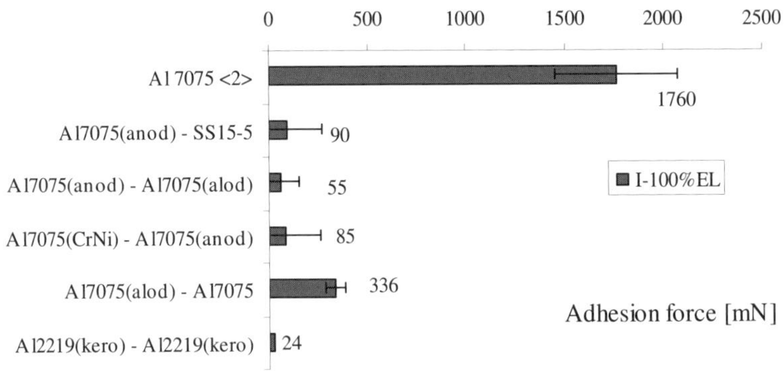

Fig. 11. Maximum adhesion force under impact for different coatings on aluminium (AA7075): negligible adhesion for combined coatings "Hard anodised (anod)", CrNi-plated (CrNi), Alodine 1200 (alod) and Keronite 14, Alodine alone is not sufficient to prevent cold welding (336mN). (For details on Keronite coating refer to Ref. 62.)

No adhesion was found for combinations: Al AA7075 hard anodised versus stainless steel SS15-5PH ("Al7075(anod)-SS15") and Al AA7075 CrNi-coated vs Al AA7075 hard anodised ("Al7075(CrNi)-Al7075(anod)"). However, an Alodine 1200 coating only on the disc is not sufficient to prevent adhesion (Al7075(alod)-Al7075, adhesion force of 336mN). A recently developed coating, named "Keronite"[62], did not show any adhesion, too. But the main advantage of this new coating was that no surface destruction or formation of debris was found.

5.2.4. *MoS$_2$ coatings versus MoS$_2$ composites*

Another strategy to avoid cols welsing was seen in using composite material containing solid lubricants. The investigations included also two composites materials containing MoS_2 particles: Vespel SP3 (Polyimide with 15m% of MoS_2) and a silver alloy "$AgMoS_2$" (with 15v% MoS_2). The polymeric composite "Vespel" shows negligible adhesion against both, stainless steel SS17-7PH and Al AA7075. The silver composite shows some small adhesion 117 mN. SEM-inspection showed the counter-surfaces (partially) to be covered with MoS_2 flakes, which were pressed out of the matrices. This effect is assisted by the fact that adhesion is mainly driven by bonding between two metals. In case of Vespel no metal is present. In case of silver, the very low shear strength enables easy braking of the bonds. Hence, beside coatings also composites provide efficient prevention of cold welding, due to their ability to re-form, i.e. at each impact a new lubrication layer is formed and coating free areas are re-coated. (Fig. 12.)

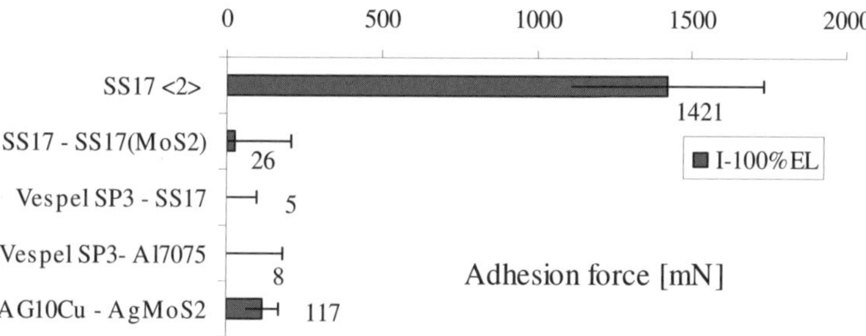

Fig. 12. Adhesion force under impact for different combinations with MoS_2: coatings and composites provide good prevention of cold welding (polymer composite Vespel SP3, Ag10Cu = coin silver, $AgMoS_2$ = silver composite with 15v% MoS_2) .

5.3. Results on fretting

5.3.1. Comparison of impact and fretting contact

As mentioned in the introduction, the fretting movement which is a small sliding, was expected to cause sever surface destructions. This should lead to increased adhesion force, because of creation of reasonable clean contact area. A survey of values of adhesion forces found under fretting and under impact is given in Fig.13.[12, 61, 63-65]

In highest allowed contact pressure at impact (100%EL), typical adhesion forces range up to approx. 2000 mN. Under fretting conditions at even lower contact pressure (60%EL), the adhesion forces exceed these values by factor of up to 10. Stainless steel SS17-7PH versus itself shows adhesion of approx. 1500 mN under impact, but more than 11000 mN in fretting (Fig.13 "SS17-7"). For other metal-metal contacts similar behaviour is found, Al 7075 versus itself.[64] The highest adhesion was found for Inconel 718 (Ni alloy). No adhesion is found for polymer to metal contact, e.g.VESPEL SP3 (polyimide with 15m% MoS_2) versus stainless steel SS17-7 PH.

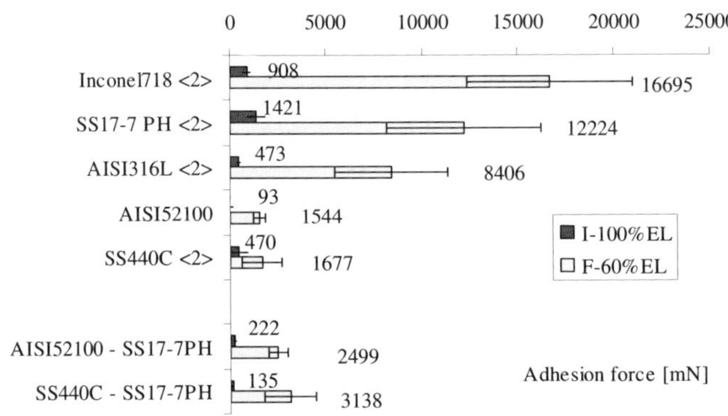

Fig. 13. Comparison of adhesion force under impact (I) and fretting (F) for different steels and Ni-alloys versus themselves: Fretting initiates higher adhesion, for ferrous alloys adhesion decreases with decreasing content of Ni.[59] Combinations of different steels seem to be dominated by the one with higher adhesion: AISI52100 to SS17-7ph. (52100-SS17).

As mentioned earlier, the adhesion of different construction steels to a cutting tool made of high speed steel indicated a relation between adhesion force and Ni-content in fretting conditions.[59] Standard fretting tests[5] done on different steels versus themselves, also show this basic relationship (Fig.13). Adhesion decreases in the order Inconel 718 (52%Ni), SS17-7PH (7%Ni) and AISI 316L (11%Ni), down to AISI 52100 and even lower for AISI440C (no Ni). The bearing steels (AISI

52100 and SS440C) show lowest adhesion under fretting. (The high adhesion of the AISI440C (no Ni) under impact is still under investigation.)

5.3.2. *Coatings on aluminium*

Similar to impact conditions, coatings are expected to reduce risk of cold welding. However, due to the fretting motions it was not clear which coatings would be able to prevent cold welding, i.e. which coatings are not destroyed under fretting conditions. Selected combinations of coatings on **aluminium** were tested under fretting: no adhesion was found between Al 7075 hard anodised and Al 7075 NiCr-plated (see Fig. 14). Al 7075 hard anodised in contact to non-coated SS15-5PH showed negligible adhesion. This is in contrast to the fact, that SEM-images show breakthrough and pealing off of the conversion layer on the Al. But results are in accordance to impact tests done in Ref. 63: despite of a breakthrough of the layer, no adhesion was measured. Coating of only the disc with Alodine, did not prevent from cold welding, medium adhesion of 2036 mN was found. (Fig. 14: "Al7075(alod)-Al7075") Alodine1200 is a chemical conversion coating composed of hydrated aluminium chromate which is very thin ($<1\mu$m). Therefore, the fretting wear resistance is very low and the coating was broken setting free the aluminium metal beneath.

In contrast to this very thin conversion layer, very thick ceramic like coatings may be obtained on aluminium and titanium by Plasma-Electrolytic-Oxidation process (PEO). Such a coating on aluminium may reach thicknesses up to 100μm.[62, 66]

For space applications thickness in a range of 10 to 30μm would be of interest. Such a coating developed by "Keronite" was in a first study screened. It offers not only no adhesion to steel AISI52100, but also did not peel off during fretting.[67] (A later study has confirmed this.[20]) This is an advantage to anodised layers, where debris is produced which could contaminate other areas in the space craft.

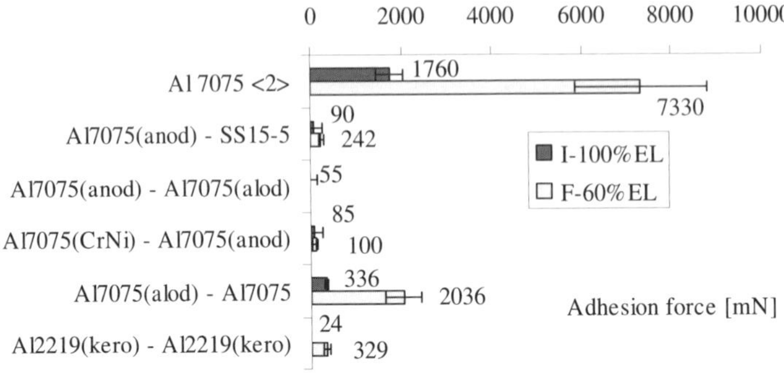

Fig. 14. Adhesion force of aluminium based coatings (I=impact and F=fretting): Adhesion between Al parts is strongly reduced by hard anodising (anod), CrNi-plating (CrNi), no adhesion for (thick) Keronite.[62, 67] A single Alodine coating (alod) is not efficient in prevention of cold welding, because of it's low thickness <1µm.

5.4. *Thin and thick coatings under fretting and thermal cycling*

It can be seen in the results above, that long term fretting testing of hard coatings on softer substrate metals led in most cases to failure of the coating. Examples are TiC on stainless steels[64], anodising of aluminium alloys[61] or several coatings on Ti-alloys.[65] For all these coatings one common parameter is their thickness: it is in a range of a few microns.

A dedicated set of studies was done to investigate if thick coatings are more resistant to cold welding under fretting conditions. The 3G Keronite process enables to achieve thick ceramic coatings on soft metallic substrates. It is based on Plasma electrolytic oxidation (PEO). This is one of relatively new environmentally safe electrolytic coating processes, applicable to light metals like Mg, Al, Ti and their alloys. It represents a rapidly developing sector in surface engineering. The process involves the use of higher voltages than in anodising and the electrolyte usually consists of low concentration alkaline solutions and all variants of this process are considered to be environmentally friendly. The process results in the formation of ceramic layers with thicknesses up to 100µm thick. Therefore, coatings with thickness from 17 to 55µm were applied three common aluminium alloys (AA2219, AA7075 and

AA6082) widely used in spacecraft hardware. They were investigated for their resistance to cold welding under fretting before and after thermal cycling. As counter part conventional bearing steel AISI52100 was used. No adhesion could be found and the coating did not break. Then the coatings were subjected to 20 thermal cycles between -185 and +107°C no cracking/delamination was detected after that. Finally, again fretting test were done: again the coatings survived without showing breakage, debris formation or adhesion.[68]

Hence, 3G Keronite coatings with ~20µm on aluminium in contact to bearing steel do not show breaking of the coating or adhesion in fretting. On the other hand, Keronite coatings with thickness of 6-10µm on Ti-alloy Ti6Al4V was found to be no sufficient to prevent cold welding. In contact to itself or to INVAR breaking was identified and cold welding occurred. However, for Keronite in contact to itself with one additional layer of MoS_2 (PVD) no cold welding was found.[69, 70] This study also considered thermal cycling, even down to liquid helium: no degradation due to thermal cycling was found by metallographic inspection and by second fretting tests. Following these studies, thick Keronite coatings (>17µm) offer really enhance resistance against cold welding under fretting and ARE NOT degraded by thermal cycling. The role of mos2 coatings (resin bonded or PVD) on top of Keronite has still to be investigated in more details: in some cases no additional benefit can be found, but at least no drawback is known yet.[68-70]

5.5. *Surface morphology after impact and fretting*

Surface is strongly changed due to impact and fretting. After impact testing, a SS17-7PH pin shows plastic flow, which can be seen by the piling up at the edges of the pin's contact area (Fig. 15a). On the other hand, fretting of a SS17-7ph steel versus itself shows strong surface destructions due to adhesive wear. Material is torn out of the surface, and pressed back or adheres to the contact partner (Fig. 15b).

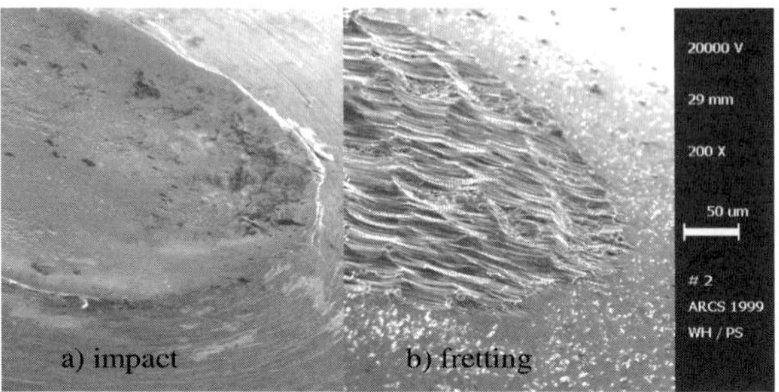

Fig. 15. Surface of a pin (SS17-7ph) after impact and fretting. Impact: only some plastic flow visible by piling up of edges. Fretting: strong destruction of surface, adhesive wear combined with high adhesion forces (Compare to Fig.10 for adhesion forces: "SS17-7".)

As mentioned above, MoS_2 coating on SS17-7PH could not prevent from cold welding. The lubrication effect was lost after 20 cycles (200 seconds fretting, 42000 strokes). Afterwards, adhesion up to 5870 mN was found. Fig. 16 shows strong surface destruction of the MoS_2 coated disc, which is similar to the uncoated. EDX-distribution of Mo taken from the disc shows, that no Mo is present in the contact area after 7000 cycles. Both, pin and disc, show fretting wear scars which are similar to those without MoS_2 coating.

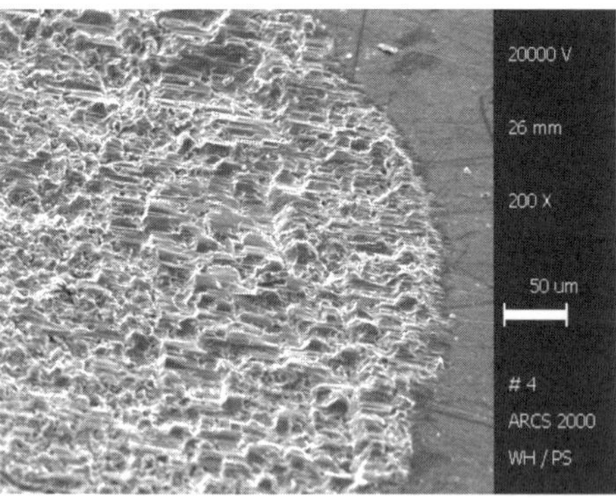

Fig. 16. Surface of disc SS17-7ph with MoS_2 coating after fretting tests: lubrication effect was lost after less than 200 second fretting movement (confirmed by EDAX-mapping: no Mo present in contact area).

In contrast to all coatings investigated until now, the thick Keronite on Al AA2219 was the only one which prevents adhesion and which was not destroyed under fretting conditions. Hard anodising of Al AA 7075 prevented adhesion, but much loose debris was found. (Fig. 17.)

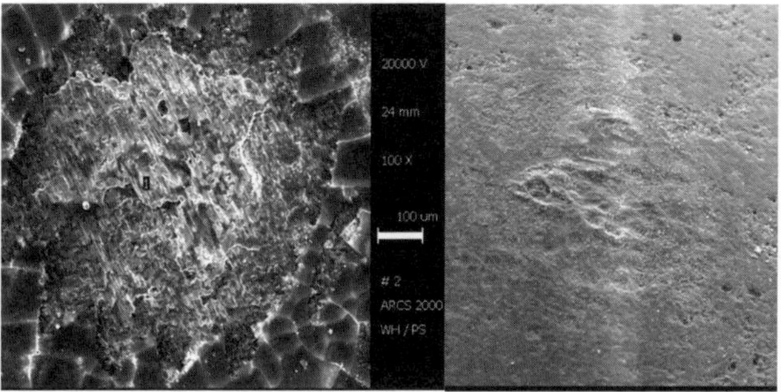

Fig. 17. Comparison of Al-coatings under fretting: Left: Hard anodising on Al7075 was broken. Right: Keronite coating on Al AA 2219 does not show fretting marks. (Compare to Fig. 14 for adhesion forces.)

5.6. *Influence of contact parameters on fretting/theoretical prediction of adhesion forces*

5.6.1. *Background – Influence in impact contacts*

Basic studies[60] by AIT have shown that under impact conditions, an increase of the static load leads to an increase of the adhesion force. Figure 17 shows the adhesion forces found for stainless steel SS17-7ph in contact to itself (ball on flat contact, without any coating). The three bars refer to the measured adhesion under static loads related to contact pressures of 40, 60 and 100% of the elastic limit. It can be seen that the adhesion forces increases with the contact pressure, when impact occurs (no fretting) (see Fig.18.).

                    A. Merstallinger & M. Sales

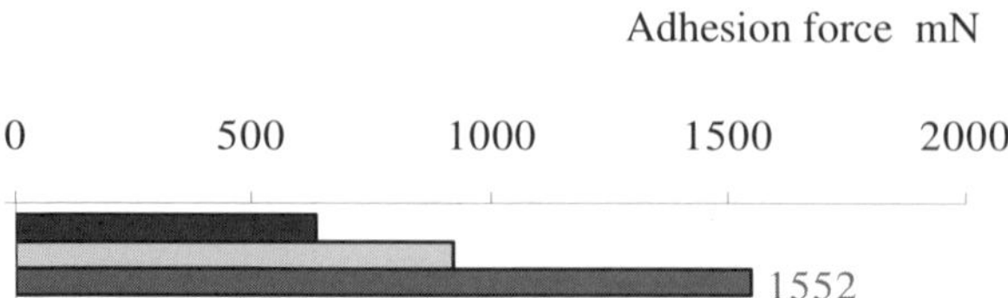

Fig. 18. Adhesion forces (in mN) of steel SS17-7ph versus itself (uncoated) under impact, adhesion increases with static load, i.e. contact pressures 40-60-100%EL.

On the other hand, studies investigating adhesion under fretting have shown strongly severe wear (see above). Impact leads only to some plastic deformations, leading to adhesion forces not higher than 2 N. Fretting, however, causes severe surface damages leading to adhesion forces of several sometimes more than 10 N. Hence, for fretting this influence has to be assessed separately.

5.6.2. *Simulation of contact area — Definition of "REAL CLEAN CONTACT AREA"*

Up to now, the influence of "material" on adhesion forces was discussed. But cold welding does not only depend on material, but also on geometry namely, the contact area. The macroscopically measureable adhesion force is defined by the following basis: ultimate yield strength (of the softer material) times "real clean contact area". Here, the latter is generally unknown, unpredicatable and orders of magnitude smaller than expected. In the following we use following "contact areas":

1. *Nominal contact area*: As typically for tribological contacts in mechanisms, only the nominal contact area is known. In flat-to-flat contacts the "nominal contact area" is obvious, in ball-to-flat-contacts it may be calculated by Hertzian theory assuming the contact to be within the elastic regime.

2. *"Real contact area"*: Due to roughness not the full contact area comes into contact, only the tips will be in contact. Of course this

share between nominal and real contact area is influenced by mechanical and surface properties. In general, this real contact area is by orders of magnitude smaller than the nominal one. Some strategies exist to estimate this "real contact area": Archard has proposed an analytic way to estimate the real contact area: divide the load by the yield strength. Modern computational tools may be used to simulate real contact areas using 3D-topographies and mechanical data. However, this is still only practicable for static contacts. Including motion leads to increased computational efforts.

3. *"Real clean contact area"*: In order to calculate adhesion forces, one more reduction step is necessary: again only a part of the real contact area contributes to adhesion. As mentioned earlier, surfaces are covered by natural contaminant layers. Especially, chemical reaction layers prohibit the cold welding. If the tips of two rough bodies come in to contact, both of these layers must be broken in order to enable a metal-to-metal contact. Only this "single joints between the clean metal surfaces are welded", and adhesion force is the sum of all these single welded joints. Finally, it is this last contribution which cannot be predicted anyhow by simulation.

Hence, it must be anticipated that a theoretical prediction of adhesion forces is not possible. Moreover in case of fretting, wear would have additionally to be considered, since it changes the surface topography and the contact area. In order to prove theoretical and experimental approaches, a study was performed.[71]

5.6.3. *Determination of adhesion forces 3 approaches (models and experiment)*

The objective of this study was to investigate if contact pressure and contact area have an influence on the adhesion force under fretting conditions.

Therefore, a set of 4 test parameters was selected with varying load and pin radius (i.e. curvature of spherical pin tip). Tab.4 shows the test parameters. Three tests were done at load of 1 N with radii of 1, 2 and 15mm. This is related contact pressures of 118, 57 and 19% of the elastic limit. One test was made with similar contact pressure of 58%EL, but using a different combination of radii (10mm) and load 12N. (See Table 4.) For all tests the same material combination was selected: AISI316L in contact or itself without any coating. This is a stainless austenitic steel, which was already tested previously and shows high adhesion forces. Three parallel tests were done for each set of parameters. (For materials properties refer to table in annex. The calculations of contact pressures were done using standard Hertzian theory.[6]).

Table 4. Test parameters: three parallel tests per set.

Tests	Tip Radius mm	Load N	Contact pressure MPa	Contact pressure in % EL	Contact area mm2
F1x	1	1	1272	118	0,0012
F2x	3	1	611	57	0,0025
F3x	10	12	209	58	0,0287
F5x	15	1	627	19	0,0072

A) Theoretical estimation of adhesion forces (Approach 1:"Theory")

Theoretically, adhesion forces could be calculated on following basis: ultimate yield strength times contact area. Whereby, the contact area of ball-to-flat-contacts is calculated using Hertz theory. Following that, the adhesion forces in a ball-on-flat contact should decrease when using a smaller contact area. The values for the stainless steel AISI316L versus itself are shown in Fig. 19. It can be seen that, the (theoretical) adhesion force is directly related to the contact area (Table 4).

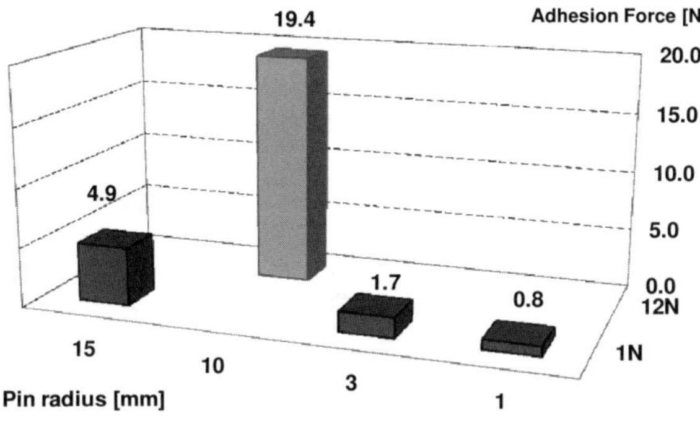

Fig. 19. Adhesion forces: calculated using yield strength times Hertzian contact area.

B) Estimation of Adhesion forces using fretting wear area (Approach 2: "Semi-Theory")

A second way to predict adhesion forces might be based on experimentally derived contact area: use the measured wear contact area after a friction test and multiply it with the yield strength. The results are shown in Fig. 20. Here, the contact area measured after the fretting tests were multiplied with the yield strength.

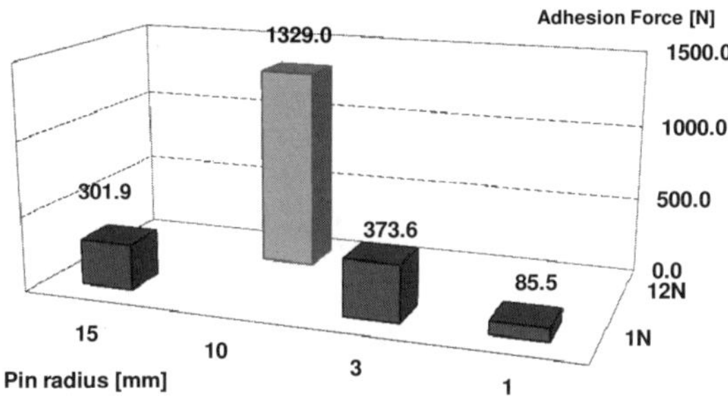

Fig. 20. Adhesion forces: calculated using yield strength times wear contact area, i.e. the contact area measured after the fretting test.

C) Results from fretting experiments (Approach 3 "Experiment")

For each set of parameters three parallel tests were done. Averaged values for each set of parameter are shown in Fig. 21. Adhesion forces are generally high but expected for this material. The only significant difference is seen for the smallest contact area (pin radius 1mm at 1N). Here the adhesion force is slightly lower (6,2N).

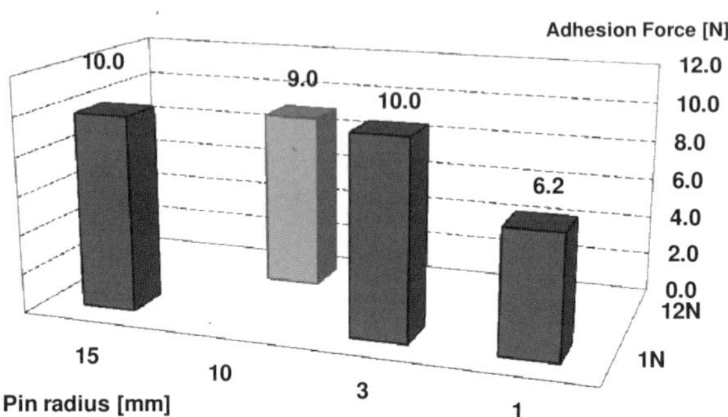

Fig. 21. Adhesion forces: measured values (average value of three parallel test, uncertainty of test method 30%).

On the other hand, fretting wear leads to an increase of the contact area. The contact area of the test with the higher load is remarkably higher than the other tests. Hence, wear is higher for higher loads, even if the contact pressure is comparable: ~58%EL similar for radius of 10mm and load of 12N to radius of 3mm and load of 1N.

5.6.4. *Comparison of modelling and experimental results from fretting*

Adhesion under fretting was derived by three methods:

1. Theory: calculation by yield strength times Hertzian contact area, this would lead to the conclusion that smaller radii are preferable. Neglecting any wear, Hertzian theory gives a smaller contact area. The resulting adhesion force is accordingly lower. (Fig. 19.)

2. Semi-Theory: yield strength times measured wear contact area, i.e. use of the wear contact area which is measured after a fretting test (Fig. 20). The derived values for the adhesion force are similar to approach 1. The conclusion would be similar to the approach 1-Theory: use smaller radius. Even though the contact pressure is

higher than the EL, the wear does lead to a smaller contact area when the tip radius is smaller. However, starting with contact pressure higher than EL is still no issue.

3. <u>Experiment:</u> measured adhesion forces are quite comparable for all parameter sets. They do not show significant influence of initial contact details (Fig. 21). Smaller contact radius seems to be advantageous, the adhesion force is slightly lower (average 6,2N compare to $\approx$ 10N for all other tests).

It is clearly visible that both theoretical and semi-theoretical extrapolations do not fit to the experimental behaviour. The main reason is seen by fact that the definition and determination of the "contact area" is insufficient: the theoretical approaches reveal a "nominal contact area", but the adhesion is related to what shall be referred to as "real contact area".

The real contact area, i.e. the area where metallic bonds actually exist, is much smaller than the nominal one predicted by theory. This is due to the surface roughness and the surface contamination. Both factors reduce the measured adhesion forces by orders of magnitude (compare Figs. 20 and 21). Using theoretical approaches, at a load of 12N the highest adhesion should be found (Figs. 19 and 20.). Even this overall tendency is not found in experimental testing (Fig. 21). Wear due to fretting is levelling out the contact pressure to values in a range of few Megapascals.

Hence, adhesion forces cannot be "modelled" since neither the "real contact area" nor fretting wear are predictable.

6. Summary

1. In the late 1990s devices were developed to measure adhesion forces between metallic contact partners (with and without coatings). This was driven by optimisation of space mechanisms, for which no data on adhesion force on "engineering level" was available. "Engineering level" means in terms of scaling: loading forces in range of 1-50Newtons and surfaces typical for engineering (not atomically

clean). Practically spoken, engineers were looking neither for data from rotation welding machines (loads in range of tons) nor for data from AFM-like tests. This means, that literature showed a huge variation of adhesion force, but no data in the appropriate "scaling".

2. Two test devices were developed. Following first basic studies a test method was fixed to study the cold welding of material interfaces that make contact under conditions of impact and fretting. The method and results enabled to make a step forward in cold-welding effects from "common experience" to measurable numbers, useful for designers of spacecraft applications.

3. Recent studies show, that **theoretical predictions were by far not comparable to experimental** data. The main reason is seen, that the adhesion is driven by the "real contact area" which cannot predicted, e.g. Hertzian theory would predict a "nominal contact area" neglecting surface roughness and surface contamination. Especially, the latter has the main contribution and keeps unpredictable.

4. A wide range of material pairings covering metal-metal (SS17-7 PH versus itself and Al alloy AA 7075 versus itself), metal-polymer (SS17-7 PH versus Vespel SP3), as well as several coatings for steels, aluminium and titanium were investigated under impact and fretting conditions. Some general findings were discussed. More can be obtained form publications or via a WEB based data base: http://service.arcs.ac.at/coldwelddata.

1. Tests have revealed, that the *range of adhesion* forces in uncoated metal-metal-contact with typical engineering surfaces and without coatings *depend on the contact type*:
   * in static contact adhesion forces were below 0,5 N,
   * in impact adhesion force up to 2 N,
   * under fretting adhesion forces in excess of 18 N were found.
   Basic material physics (type of atomic bonds) indicate that there is no technically measureable adhesion between metals and polymers and

ceramics expected A few tests between steel or aluminium and polyimide did not contradict this premise.

2. In order to **avoiding cold welding** polymers or ceramics can be selected, but this may not be suitable for space hardware and their mechanisms. Hence, often metal-metal-contacts cannot be avoided. Then, the first strategy to reduce cold welding risk would be the use of **dissimilar alloy pairs**, e.g. stainless steel versus hard steel (low adhesion likely). The second way is the use of coatings. However, here the type of contact and the substrate material needs to be well known.

3. Under **impact**, hard **coatings on stainless steel** (for instance TiC) may break and, therefore, adhesion is lower but still found. Soft coatings made of solid lubricants (e.g. MoS_2) can repair themselves during impact, i.e. prevention of adhesion is more efficient than for hard coatings. Stainless steels are generally too soft to support a hard coating under impact conditions. Hard anodised **aluminium** can withstand impact. **Titanium** alloy must be coated with hard coatings to resist cold welding, if only impact is expected.

4. Under **fretting** none of the investigated coatings is able to avoid cold welding of stainless steel (SS17-7PH). Also MoS_2 is not effective in fretting, lubrication is lost soon. Hence, main strategy must be the use of different steels (maximum one partner being austenitic). Hard coatings should not be on hard steels. In contrast to steel, hard anodising of **aluminium** prevents adhesion in fretting. However, much lose debris is formed. A thick "Keronite" coating (20µm), which is based on a plasma-electrolytic-oxidation process, is not only resistant to fretting but also avoids debris formation. A test campaign using uncoated **titanium** pin against coated titanium discs, did not give a "general solution". All thin coatings -solid lubricants and hard coatings- were destroyed in the fretting contact. The best combinations still showed medium adhesion after break of the

coating. Combination between titanium and low adhesion steel did also not provide a solution.

References

1. *ECSS-E30 - European Co-operation for Space Standardisation (ECSS):* ECSS - E-30 "Mechanical", Part 3A "Mechanisms" , section 4.7.4.4.5 "Separable contact surfaces", page 32, Pre-Print Version June 2000.
2. A. Merstallinger and E. Semerad, ESTEC Contract No 8198/89/NL/LC, WO 32 (1995).
3. M. R. Johnson, *The Galileo High Gain Antenna deployment anomaly*, California Inst. of Technology, JPL, Pasadena.
4. A. Merstallinger, E. Semerad, B. D. Dunn and H. Störi, I, Zürich, Proceedings ESA SP-374 (1995).
5. A. Merstallinger and E. Semerad, In-house-Standard by ARCS approved by ESA, Issue 2 (1998) and In-home-Standard by ARC Seibersdorf research GmbH, Issue 2, audited in 2003.
6. N. Gane, P. F. Pfaelzer and D. Tabor, *Proc. R. Soc. Lond.*, **A340**, 495 (1974).
7. K-H. Habig, 'Verschleiß und Härte von Werkstoffen', (Verlag München, Wien, 1980).
8. B. D. Dunn, private communication (1995).
9. B. D. Dunn, ESA ESTEC, (1989).
10. J.B.Rittenhouse, ASTM STP 431, 1967, S.88-108.
11. K. H Johnson., *Contact mechanics,* (Cambridge University Press, 1985).
12. A. Merstallinger, E. Semerad and B. D. Dunn *7h European Space Mechanisms and Tribology Symposium,* Proc. ESTEC (Noordwijk (NL), 1997).
13. A. Merstallinger , Doctor theses Techn.University Vienna. (1995).
14. G. Binning, H. Rohrer, Ch. Gerber and E. Weibel, *Phys. Rev. Letters*, **49**, 57 (1982).
15. U. Dürig and O. Züger,*Vacuum*, **41**, 382 (1990).
16. W.G. Hartweck, Dortmund Doctor thesis 1979.
17. J. Bénard , *Adsorption on metal surfaces* (Elsevier, Amsterdam 1983).
18. D.H.Buckley, in New directions in lubrication, materials, wear and surface interactions(Noyes Publications, New Jersey, 1985). pp18-42.
19. D. H. Buckley, ASTM STP 431, pp 248-271, (1967).
20. E. Orowan, J. F. Nye and W. J. Cairns,in: Selected Government Research Reports, Vol. 6, Strength and Testing of Materials, Part 1, , Report Nr.8, p. 127-176 (1952).
21. P.M.Vedamanikam and D.V.Keller, ASLE Preprint 72LC-5B-2 (1972).

22. K. Takahashi, T. Ilyama, N. Katoh and T. Onzawa, *J. Vac. Science Technol.* **A 12** (1994).

23. L. G. Kellogg, ASTM STP 431, pp149-180, (1967).

24. Lewis, to be published

25. M. Maillat, 'Tribology Adaption', M Maillat, Centre Suisse d'Electronique et de Microtechnique SA (CSEM), Switzerland, *Final report Tribology Adaption*, Theme 2, Rapport technique No 552, Poject No. 51.122, ESTEC Contract No 9746/91/NL/PP(SC) Theme 2, (1992).

26. A. Merstallinger and E. Semerad, 'Tribological properties of GA3Z1', ESTEC Contract No 8198/89/NL/LC (1995).

27. F. Rossitto and G. Ghersini *5th Europ. Symp. on Mat. Science and Microgravity* Proc., ESA SP-222. (1984).

28. H. Dursch and S. Spear *'LDEF: 69 Months in Space, Proceedings of the First Post-Retrieval Symposium'*,Langley Research Center, NASA. Part 3, 1565-1576 62-27083 17-99, (1991).

29. S. V. Pepper, J. *Appl. Phys*, **47**, 801 (1976).

30. H. Conrad and L. Rice, ASTM-STP 431, (1967) and *Metallurgical Transactions*, **1**, 3019 (1970).

31. W. P.Gilbreath, ASTM STP 431, pp128-148, (1967).

32. M. J. Hordon, ASTM STP 431, pp 109-127 (1967).

33. P. M. Winslow, D. Horwitz and D. V. McIntyre, *Lubrication engineering* (1966).

34. J. Pullen and J. B. P. Williamson, *Proc. Roy. Soc.* **A327**, S.159 (1972).

35. F. P. Bowden and D.Tabor, *Friction and lubrication of solids'* (Clarendon Press, Oxford, Part I 1954).

36. A. W. Crook and W. Hirst, *Research* **3**, S.342 (1950).

37. D. H. Buckley, NASA T.N., D-7018 (1971).

38. M. E. Sikorski, *Trans ASME* **Ser. F 85**, 279 (1963).

39. L. G.Kellogg, ASTM STP 431, pp149-180 (1967).

40. J. B. P. Williamson and R. T. Hunt, *Proc. Roy. Soc.* **A327**, S.147 (1972).

41. N. F. Kazakov, *Diffusion bonding of materials*, (Engl.Translation, MIR Publishers, Pergamon Press, 1985).

42. G. V. Alm, *Machine Design*, 100, (1968).

43. R. Klinger, TU-Berlin Dissertation (1984).

44. J. Ferrante and J. R. Smith, *Phys. Rev.*, **B19**, 3911 (1979).

45. L. B. Sargent, *ASLE Trans.*, **21**, 285 (1978).

46. J. S. McFarlane and D. Tabor, *Proc. Roy. Soc.* **A202**, 244 (1950).

47. J. A. Greenwood and J. B. P. Williamson, *Proc. Roy. Soc London*, **A295,** 300 (1966).

48. D. J. Whitehouse and F. Archard, *Proc. Roy. Soc. London* **A316**, 97(1970).

49. T. R. Thomas *'Rough Surfaces'*, (Longman, London, 1982).

50. J. I. McCool, *Wear*, **107,** 37 (1986).

51. J. F. Archard, *Tribology International* , S.213-220 (1974).

52. V. Aronov, S. Nair and J. M. Wang, *Trans ASME, J. Tribology*, **116,** 833 (1994).

53. C. Y. Poon and R. S. Sayles, *Trans ASME, J. Tribology*, **116**, 194 (1994).

54. C. Y. Poon and R. S. Sayles, *Trans ASME, J. Tribology* **116**, 850 (1994).

55. C. Y. Poon, 2nd Int. Conf. on Combustion Engines - Reduction of friction and wear, IMechE, London (1989).

56. Lewis 'Static Adhesion in Preloaded Contacts subjected to Long-Term Vacuum Exposure', S. Gill, R. A. Rowntree, ESTL, 5th European Space Mechanisms and Tribology Symposium, ESTEC, Noordwijk Proceedings ESA SP-334,(1992).

57. A. Merstallinger, M. Sales, E.Semerad and B. D. Dunn, *Proc. 13th European Space Mechanisms and Tribology Symposium*ESA-SP-670. (ESTEC Contract No 11760/95/NL/NB, CO65.)(2009).

58. Roberts E. et.al.,,(Eds.) *Space Tribology Handbook*, ESTL, AEA-Technology, 2001.

59. U. Persson, H. Chandrasekaran and A. Merstallinger, *Wear, ***249,** 293 (2001).

60. A. Merstallinger, E. Semerad and B. D. Dunn, *Proc. 8th European Space Mechanisms and Tribology Symposium,* (ESTEC Contract No 11760/95/NL/NB, CO 12, 1998.) (1999).

61. A. Merstallinger, E. Semerad, P. Scholze and C. Schmidt, ESTEC Contract No 11760/95/NL/NB, CO20, Materials Report 3150, (2001).

62. S. Shrestha, A. Merstallinger, D. Sickert and Dunn B. D., Proceedings ISMSE, ESTEC, (2003).

63. A. Merstallinger, E. Semerad, P. Scholze and C. Schmidt, ESTEC Contract No 11760/95/NL/NB, CO21, Materials Report 2663, (2000).

64. A. Merstallinger, E. Semerad and W. Costin, *Influence of steel types on cold welding under fretting and impact*, ESTEC Contract No 11760/95/NL/NB, CO40.

65. M. Sales, A. Merstallinger, W. Costin, G. Mozdzen and E. Semerad, ESTEC contract No 11760/95/NL/NB, Co52 (2006).

66. S. Shresta and B. D. Dunn B.D., *Surface World*, 40 (2007).

67. A. Merstallinger, E. Semerad and W. Costin, ESTEC Contract No 11760/95/NL/NB, CO39, Metallurgy Report No. 3522 (2002).

68. M. Sales, A. Merstallinger, S. Shresta and B. D. Dunn, SMT22 – *22nd int. Conf. on Surface Modification Technologies"* (2008). (ESTEC Contract No 11760/95/NL/NB, CO60.)

69. M. Sales, A. Merstallinger, S. Shresta and B. D. Dunn, *Proc. of the 11th ISMSE: International Symposium on Materials in Space Environment*, (2009). (ESTEC Contract No 11760/95/NL/NB, CO72.)

70. Sales M., Mozdzen G., Merstallinger A., Semerad E., Costin W., COLD WELDING SUMMARY CHART Collection of data of all performed Studies, ESTEC Contract No 11760/95/NL/NB, CO53.

71. A. Merstallinger, M. Sales, E. Semerad and B. D. Dunn, *Proc. 13th European Space Mechanisms and Tribology Symposium*, (2009), ESA-SP-670. (ESTEC Contract No 11760/95/NL/NB, CO65.)

Abbreviations – Properties

US .... ultrasonic (bath)
GD .... glow discharge
HV .... High voltage
SEM .... scanning electron microscope
rH .... relative humidity
R_a .... mean roughness (arithmetic),
R_t .... maximum peak-to-valley-distance

Contact parameters:

H_V .... Vickers hardness
Y .... Yield strength
Y_{HV} .... Yield strength calculated from Vickers hardness
p_m .... mean contact pressure in contact area
p_0 .... maximum contact pressure
P .... load (P_Y refers to load at elastic limit)
R .... radius of curvature (Spherical to flat contact, otherwise refer to e.g. B3)
E .... Young's modulus (of material 1: E_1)
v .... Poisson's ratio (of material 1: v_1)
E* .... reduced Young's modulus
W .... impact energy (W_Y refers to W which firstly causes yield)
m .... mass (herein of pushrod)
v .... velocity

Annexe: Material abbreviations and data to Publications

This Table contains the abbreviations used above. It also gives materials' data used for calculation of test parameters.

Abbreviation	Designation	Composition	Condition	HV daN/m m²	Yield MPa	Poisson	E GPa
Al7075	Al alloy Al AA 7075	2.1-2.9Mg1.2-1.6Cu0.18-0.28Cr5.1-6.1Zn	T7351	170	654	0.33	72
Bronze LB9	Bronze LB9 BS 1400 LB4	Cu-4-6Sn-8-10Pb-2Zn-0.25Fe-0.01Al-0.2Mn-2Ni-0.5Sb-0.1S	AR	160	130	0.34	80
SS15	Stainless Steel SS15-5 PH	14-15.5Cr3.5-5.5Ni0.15-0.45Nb<0.07C2.5-4.5Cu	H1025	393	1000	0.27	196
440C	AISI 440C	Fe-1.01C-0.47Si-0.56Mn-0.014P-<0.002S-17.81Cr-0.27Ni-0.48Mo	Harden	700	2692	0.283	200
SS17	Stainless Steel SS17-7 PH	17Cr-7Ni-1Al	PH	441	1697	0.29	210
Ti834	Ti-IMI 834	Ti5.8-Al4Sn-3.5Zn- 0.7Nb-0.5Mo-0.35Si- 0.06C	AR	334	1285	0.32	112
Ti6AV	Ti-IMI 318	Ti6Al4V	AR	338	850	0.32	105
Vespel SP3	Vespel SP3	85PI-15MoS2	AR	18	68	0.41	2.5
AgMoS$_2$	Ag/MoS2	Ag 15v% MoS2	AR	26	138	0.367	71
Ag10Cu	Ag10Cu	Ag10Cu	AR	150	620	0.367	82.7
Inconel718	Inconel718 / ASTM B 637)	Fe-53.6Ni-18.9Cr-5.3Nb-3Mo-0.98Ti-0.03C-0.13Si-0.12Mn-0.008P-0.001S-0.49Al-0.2Co-0.06Cu-0.004B	AR	348	1338	0,25	211
SS316L	AISI316L	Fe-0.011C-0.41Si-1.42Mn-0.031P-17.3Cr-11.2Ni-2.09Mo-0.05W-0.098Co-0.041V-0.026S	austenitic	175	675	0,28	190
52100	AISI52100 (SKF)	Fe-1C-0.3Si-0.4Mn-0.03P-0.03S-1.6Cr-0.3Ni-0.3Cu	AR	700	2692	0,28	200
AL 2219	AL AA 2219	6.3Cu-0.3Mn-0.18 Zr-0.1V-0.06Ti	T851	138	531	0,33	73,8

CHAPTER 7

MECHANICAL PROPERTIES OF METALS
AT THE NANOSCALE

Ralph Spolenak

*Department of Physics, ETH
Zurich, Switzerland
E-mail: ralph.spolenak@mat.ethz.ch*

The motivation for the topic of this chapter is materials issues in microelectronics and microelectromechanical systems (MEMS). The core of the course is the materials behavior in small dimensions. Focus is laid on scaling of electronic and mechanical properties, thin film mechanics, device reliability and integration issues when dissimilar materials are joined. The chapter is divided into four parts focusing on an introduction to scaling laws and size effects, mechanical properties of thin films, properties of interconnects in microelectronics and finally complex nanostructured geometries. As far as possible, scaling laws are introduced that describe the change in a property as a function of the external dimension. As many of these laws would diverge at extremely small length scales, the limits of scaling laws are discussed. The mechanical properties range from elasticity to strength and fracture toughness, whereas the electrical properties are focused on resistivity and the phenomenon of electromigration. A section on measurement techniques is omitted as these will be the focus of a review article to be published in May 2010. More complex geometries than they are currently found in microelectronics require more sophisticated theories, such as strain gradient plasticity for mechanical aspects, provided they coincide with nanoscale dimensions

1. Introduction to Scaling Laws and Size Effects

Since the advent of microelectronics the need for higher storage capacity, increased processing speed and the need for mobility have driven miniaturization with current feature sizes arriving below 100 nm. Barrier layers have been reduced to below 5 nm and the gate oxide thickness only amounts to several atomic layers. This drive is empirically motivated by Moore's law predicting a doubling in device density every 18 months. However, currently, we are approaching the physical limits and density increase, i.e. feature size reduction may come to a halt. The reasons for this halt are a motivation for the following basic materials principles.

1.1. *Relevance of materials with small external dimension*

Examples for relevant applications can be found in the following sectors: Microelectronics, Microsystems and Nanocomposites

Microelectronic systems are, in general, "2.5 dimensional", as all features are patterned on a flat silicon surface. The feature sizes range from the atomic scale to several microns in film thickness. In addition to the transistor level (built on the silicon surface) up to 12 levels of metallization and dielectric layers are formed with increasing feature size with layer level. Microsystems can be built with a process called surface micromachining, where general microelectronic concepts are used to build mechanical structures in silicon. Similar dimensions apply; however, the progress in miniaturization is not as rapid. In bulk micromaching the entire wafer thickness is used resulting into largest feature sizes on the order of several hundreds of micrometers.

Nanocomposite materials can be found in all materials classes and even in natural materials. For instance the relatively high fracture toughness of natural nacre is attributed to a sophisticated brick-and-mortar microstructure, where the thickness and aspect ratio of the aragonite platelets ("bricks"), which make up 95% of nacre by volume, are optimized for maximum toughness. Energy absorbing mechanisms such as plastic extension of the polymeric "mortar" layer, increased friction due to the roughness of the platelets and an extension of the crack path by the nanocrystalline nature of the platelets further contribute to its toughness. The platelet thickness varies between fifty and several thousand nanometers.[1]

1.2. *Scaling laws – How do external dimensions affect materials properties*

Scaling laws describe how a materials property changes with a change in length scale. This length scale is usually an external sample dimension, but may also be an internal dimension such as a grain size or a domain size in magnetism. In the following we will review several materials properties to see how they are affected by a change in dimension and what the physical origin of the scaling law is.

Let us start out by investigating the mechanical properties of metals. If we consider a classical engineering stress-strain curve we initially observe perfectly *elastic* loading, then the metal starts deforming plastically where usually a 0.2% deviation (at very small length scales this percentage is sometimes increased to allow for a reasonable comparison) from the elastic slope is defined as the *yield strength*. Upon further loading strain localization occurs which eventually leads to material failure at the strain to break, the latter being related to the *fracture toughness* of a material. In general, one can say that the elastic modulus remains constant with change in external dimension until the atoms in a material become countable. The yield strength increases usually with inverse proportionality to the external dimension and fracture toughness decreases with decreasing dimensions.

When one addresses the electrical resistivity as an important materials parameter for the microelectronics industry, one finds that it is not dependent on external dimensions at least until one reaches a length scale of about 100 nanometers. Unfortunately, this is exactly the one that is currently approach in state of the art devices. More on the origins of this limit can be found below.

Another physical property of metals is the phenomenon of magnetism. Let us, for argument's sake, explore the influence of a change in grain size on the coercivity of a material. Down to a critical length scale of again about 100 nanometers the coercivity, the resistance against a change in magnetization, increases with decreasing grain size a so-called magnetic hardening effect. Below this length scale, magnetic softening is observed.

The last example of a scaling law would be scaling in contacts size[2] in the terminal fibrillar adhesive elements of natural species that have one thing in common: They can climb on vertical walls and stick to ceilings. This is true for flies, beetles, bugs, spiders and lizards. Interestingly, the bigger (heavier) the animals get, the smaller their contact elements become ranging from 5 microns down to 200 nanometers (what a coincidence!). This can be explained in very simple terms by exploring contact mechanics.

Let us first assume that every contact element has the geometry of a circular fiber with a half sphere as a terminal element. Then mathematically we just need to consider the contact between this sphere and a semi-infinite half space. This has rigorously been done in the late 19th century by Hertz[3] and adopted to include adhesive forces by Johnson, Kendall and Roberts[4] (JKR) in the 70s of the last century. They arrive at a very simple result, namely that the adhesive force linearly scales with the radius R of the sphere:

$$F_{ad} = -\frac{3}{2}\gamma R \tag{1}$$

where γ is the work of adhesion. If we normalize this force by the projected area of the fiber, we get an apparent contact strength:

$$\sigma_{contact} \propto \frac{\gamma}{R} \tag{2}$$

This immediately tells us that it is beneficial to split up a bigger contact into smaller ones to increase the effective contact strength, a principal well followed in natural systems.

This is illustrated in Fig. 1 where the areal density is plotted against the mass of an animal, i.e. the weight of an animal, which needs to be supported when climbing on a ceiling. Eq. 1 can now be used to derive such a scaling relationship between the areal density N_A of fibrils and the animal's mass m. If one replaces a contact of size R with 4 contacts of size $R/2$ occupying the same area, the pull-off force increases two fold resulting in the following relationship:

$$F_{tot} = \sqrt{n}\,F_{single} \tag{3}$$

If now the force F_{tot} is assumed to be proportional to the mass m, in Eq. 1 we find:

$$m \propto \sqrt{n}\,m^{1/3} \tag{4}$$

assuming that the radius R than scales with the third root of m.

If we now compute the areal density N_A:

$$N_A = \frac{n}{A} \propto \frac{n}{m^{2/3}} \propto \frac{m^{4/3}}{m^{2/3}} = m^{2/3} \tag{5}$$

which comes very close to the experimentally determined slope in Fig. 1. Taking a closer look at Eq. 2, however, shows that if we (or nature, respectively) made features infinitely small, their strength would become infinite. Apart from not being able to making things infinitely small, an infinite strength would also not be physical, but be limited by a theoretical contact strength, determined by the hypothesis of all interfacial bonds breaking simultaneously. This brings us to the next chapter, where the limits of scaling laws are discussed.

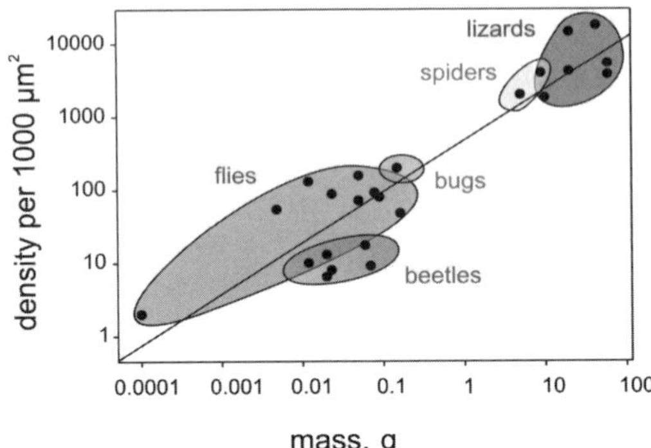

Fig. 1. The areal density of attachment devices is plotted versus the animal mass for different species. The regression curve has a slope of roughly 2/3. This can be justified by the application of the JKR theory. Data in the plot originates from Ref. 2.

1.3. *Limits of scaling*

In the introduction above an argument about scaling laws has been made, in which a change in property is deduced from a change usually in external length scales and the laws that describe that resulting in general statements, such as "smaller is stronger" or "smaller is faster". However, first one has to understand what small actually means. This is obviously a question of a reference. One nanometer is extremely small compared to the diameter of a human hair (100 microns), however it is relatively big compared to the size of a single atom. Therefore for every phenomenon considered, there needs to be a reference length scale that tells us whether something is big or small. In everyday life, for instance, the size of an automobile depends on the parking space in a major city, the width of the streets that we drive upon, or whether we can fit our own body easily on the driver's seat (or not). Below find a list of phenomena and reference scales:

- Car – parking space, size of human body (4 m, 2m)
- Cellular phone – spacing between ear and mouth (10 cm)
- Yield strength of metals – curvature of dislocations (1 $\square$m)
- Young's modulus – atomic spacing (0.1 nm)
- Magnetism – domain size (100 nm)

An insightful review of scaling effects due to internal length scales, e.g. the case of magnetism can be found in a paper by Arzt.[5] In relationship to such length scales, scaling laws, such as in Eq. 2 are defined. They describe a property over a wide range of dimensions, but need to be limited specifically, when they tend to diverge. In the last example of the previous section, the strength of the contact needs to be limited by the theoretical strength of the contact. This theoretical limit can be approximated by dividing a general surface energy by the closest distance between the two surfaces, i.e. an atomic spacing s:

$$\sigma_{ideal} = \frac{\gamma}{s} = \frac{0.01 \, \text{J/m}^2}{10^{-10} \, \text{m}} = 100 \, \text{MPa} \tag{6}$$

where γ has been chosen to be typical of Van der Waals interaction. In the following you will find examples of other scaling laws and their limits.

2. Scaling in the Mechanical Properties of Metal Thin Films and Interconnects

2.1. *Thin film effects*

Already in the 50s of the last century thin films have been found to exhibit higher strength levels than their bulk counterparts.[6-8] This is related to confinement effect that the interfaces respectively surfaces pose on dislocation motion within the thin film. If concepts derived from a classical Orowan mechanism are used, the yield strength of the thin film should obey the following relationship:

$$\sigma_y = \frac{G_f b}{h} \tag{7}$$

where G_f is the shear modulus of the film, b its Burgers vector and h the film thickness. As can readily be seen, the strength diverges for infinitely small thicknesses. Nix[9] and Freund[10] elaborated on this approached by relating the work done by a threading dislocation and the energy release during its advance resulting in an effective shear modulus G_{eff}:

$$G_{eff} = \frac{G_f G_s}{2\pi(1-v)(G_f + G_s)} \ln\left(\frac{\beta h}{b}\right) \qquad (8)$$

where G_s is the substrate shear modulus, v the Poisson ratio of the film and β a constant describing the cut-off radius. The logarithmic term now resolves the divergence for small film thicknesses h, resulting in a well fitting model for single crystalline thin films where the strength is determined by a dislocation propagation mechanism.

If one considers polycrystalline thin film materials, a dislocation is not only hindered in its propagation by the interfaces, but also by grain boundaries. Let us consider the two extreme cases. If the grain size d is much larger than the film thickness h, then the model described above applies, if the grain size is much smaller than the film thickness, the typical Hall-Petch relation applies:

$$\sigma_{H-P} = \sigma_0 + \frac{k_{H-P}}{\sqrt{d}} \qquad (9)$$

where σ_0 is the strength determined by other mechanism and $k_{H\text{-}P}$ is the Hall-Petch constant, which is material dependent. It is the case, where grain size and film thickness are of the same order, where things become interesting. To uniformly plastically deform a thin film it is necessary to have dislocation activity in each single grain. This means that plasticity cannot be carried by few dislocations that travel long distances, but instead many dislocations that travel on average as far as the grain size. Their number should be at least one per grain. Under this scenario it is the nucleation of dislocations rather than their propagation that becomes strength limiting. Generally, the strength will be determined by the nucleation of full dislocations, however, for extremely small grains, the nucleation of partial dislocations may become strength limiting.

For full dislocations the source model[11, 12] becomes:

$$\sigma_{source} = \frac{G_f b}{\pi}\frac{3}{h}\ln\left(\frac{\alpha h}{3b}\right) \tag{10}$$

where the source size is taken as one third of the film thickness and Schmid's factor is assumed to be one half. *A* is a numerical constant of order unity.

The nucleation threshold for partial dislocations[13] is then defined as:

$$\sigma_{source,p} = \frac{2G_f b_p}{h} + \frac{2\gamma_{sf}}{b_p} \tag{11}$$

where the grain size *d* is assumed to be equivalent to the film thickness *h*, b_p is the Burgers vector of the partial dislocation and γ_{sf} is the stacking fault energy.

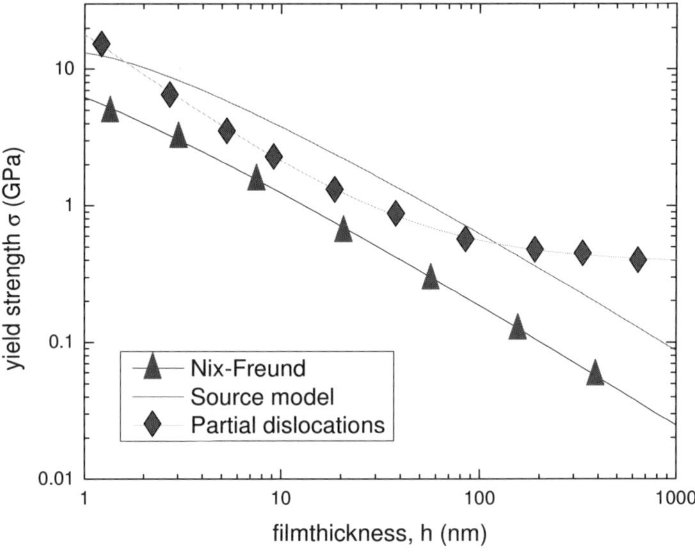

Fig. 2. Yield strength as a function of film thickness for three different models. Note the crossover between the nucleation of full dislocations (source model) and the nucleation of partials at about 100 nm. The case study was done for Cu: G_f = 42 GPa, b = 0.255 nm, b_p= 0.147 nm, γ_{sf} = 40 mJ/m², $G_s = G_f$, n = 0.3, $\beta = \alpha$ = 2.6.

Figure 2 now compares the three mechanisms, where both the Nix-Freund and the source model show the same scaling relationship, but differ in their absolute values, i.e. single crystalline films exhibit lower strength than polycrystalline ones. An interesting observation is the cross-over between the nucleation of full and partial dislocations, which recently has been confirmed experimentally.[14] This cross-over occurs for instance in Cu at around 100 nm and shifts to higher film thickness with decreasing stacking fault energy, e.g. by alloying. The cross-over also results in a change in mechanism and consequently in a change in slope of the scaling relationship.

Although the logarithmic term takes care of unphysical divergence in strength values, there is also a concrete limit to the strength of a material commonly defined as the theoretical strength. Classically the theoretical strength levels scale with the elastic moduli of a material namely with a fraction of the shear modulus G (1/30 to 1/10) for the shear strength and a tenth of Young's modulus for the tensile strength. It is interesting to note that which of these results first in failure depends on the elastic anisotropy of the material.[15] Furthermore, the relationship with elastic properties is a crude approximation of the bonding potential and may sometimes lead to wrong predictions. Secondly, there are view experimental studies that actually unambiguously explore such strength levels, as the absence of any defect (dislocations and dislocation sources for ductile failure and cracks for brittle failure) needs to be ensured. In addition, when investigating nanostructures, also the rate of deformation may play an important role.

2.2. *Time dependent mechanisms*

For most metallic materials room temperature is a sufficiently low a temperature that time dependent effects can be neglected. When we are considering thin films, however, diffusion distances become extremely short. This is accentuated by the often polycrystalline nature of thin films, which in addition exhibit a nanocrystalline nature. Furthermore, surface diffusion may also play a significant part in deformation mechanisms. In the following we will treat three mechanisms as they relate to time dependent mechanical behavior of thin films and nanostructures.

The mechanism usually occurring at the lowest homologous temperature (T/T_m) is thermally activated dislocation glide. Here, dislocations are "assisted" in passing an obstacle by thermal energy. The associated creep rate is defined as follows:

$$\dot{\varepsilon}_{dg} = \dot{\varepsilon}_0 e^{-\frac{\Delta F_{dg}}{kT}\left(1-\frac{s\sigma}{\hat{\tau}}\right)} \qquad (12)$$

where s is Schmid's factor, τ the critical shear stress at 0 K, and ΔF_{dg} the activation energy at zero stress. As it is difficult to perform creep tests for thin films (i.e. constant stress, while monitoring the elongation), uniaxial tensile tests with several different strain rates or performed at different temperatures can be used instead. When the stress plateau is reached, activation energies can be determined by equating the applied strain rate to Eq. 12. A similar concept can be applied when materials are tested at higher temperatures where diffusive mechanisms become rate determining. For Coble creep (grain-boundary diffusion) the following creep equation applies:

$$\dot{\varepsilon}_{gb} = A_{gb}\frac{\delta D_{gb}}{kT}\frac{\Omega}{dh^2}\sigma \qquad (13)$$

where A_{gb} is a constant of about 16, δ the grain-boundary width, D_{gb} the grain-boundary diffusivity and $\Omega\square$ the atomic volume.

In general thin polycrystalline metal films exhibit much stronger rate and temperature dependence compared to their bulk counterparts. This is accentuated if free surfaces come into play as is the case in gold thin films as shown by Gruber *et al.*[16], where creep rates can strongly be reduced by the presence of surface passivation layers.

The above cited mechanism, however, is only feasible for symmetric boundary conditions, i.e. equivalent diffusivities on both top and bottom interfaces of the thin film. If this is not the case a mechanism called

                                    R. Spolenak

constrained diffusional creep (Fig. 3) becomes active.[17,18] Here, a complete relaxation of the thin film by solely diffusive processes is not possible. Different diffusivities on top and bottom interface will result in the formation of wedges at the grain boundaries. The stress state at their tip resembles the stress state in the proximity of a crack tip and results in dislocation propagation parallel to the thin film interfaces. Consequently, this process was labeled as parallel glide and its theoretical treatment experimentally verified.

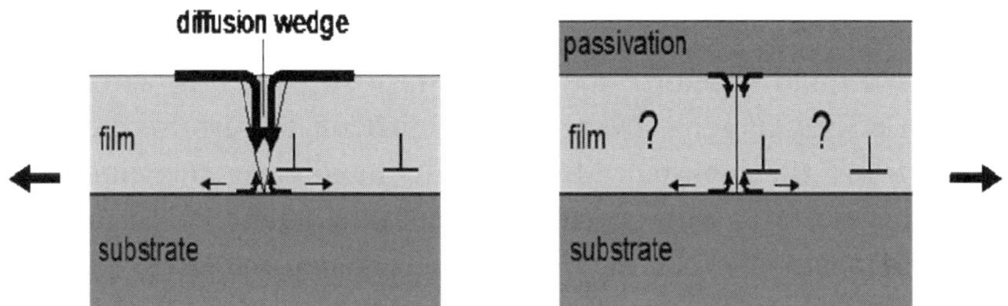

Fig. 3. Schematic representation of the parallel glide process. Provided there is a large difference in diffusivities between bottom and top interfaces, interfacial diffusion will result in the formation of a wedge (left side). This wedge in turn leads to a stress field similar to the one around a crack resulting in the emission of dislocations parallel to the bottom interface. The combination of these mechanisms may result in a complete relaxation of the thin film.

2.3. *Scaling in ductility*

For traditional bulk materials with the exception of nanocrystalline materials high strength comes at the cost of reduced ductility. The same is true for thin metal films. The increase in strength due to the confinement effect as described above results in a reduced energy release rate. For thin films the energy release rate G in a linear elasticity approximation is defined as follows:

$$G \propto \frac{\sigma^2 h}{E} \qquad (14)$$

The proportionality constant is then dependent on the exact details of the crack geometry. An example of this can be found in Fig. 4, where the strength level at the onset of fracture in a Cu/Ta system on polyimide is plotted against the copper thickness (Ta thickness: 10 nm). If one super imposes the strength squared times the thickness, which is a measure of the energy release rate, one sees that with increasing strength the energy release rate drops. This qualitatively supports the argument given above.

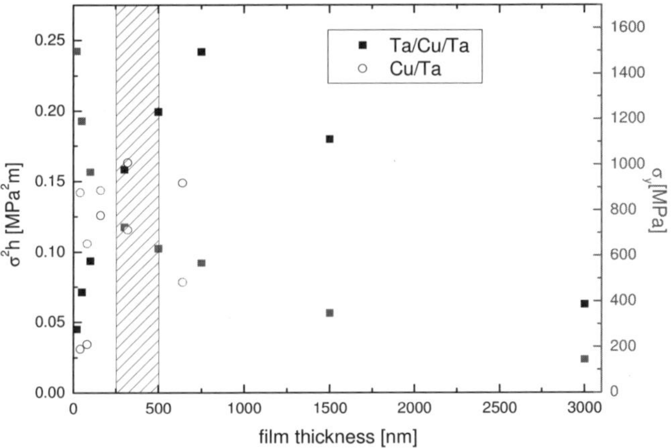

Fig. 4. A parameter describing energy release rate (the product of the strength squared and the film thickness h) vs. film thickness and the strength vs. film thickness. High strength levels resulting from low film thickness also exhibit reduced ductility.

One should bear in mind that the strength of metal thin films usually results in a plateau. With regards to ductility this violates Considère's criterion that the strain hardening rate always needs to be higher than the current stress level. In the thin film case, strain localization, a prerequisite for fracture is usually inhibited by the substrate.

The derivation for Fig. 4, however, is only an approximation for the energy release rate as strain energy only converted into surface energy is considered. More complex analyses involving metal plasticity are presented in Gruber *et al.*[19]

In a very recent study the elongation to brake was studied as a function of film thickness in a system of copper on polyimide substrate.[20] The elongation to break exhibited a steady increase until about a thickness of 500 nm. This rise was attributed to a change in fracture path from transgranular to intergranular at about 200 nm of thickness and a decreasing tendency for decohesion from the substrate. For higher thicknesses (above 500 nm) the elongation to break decreased again, due to a thickness dependent formation of a (100) texture component in the usual surface energy driven (111) component and a consequent strain localization at (100) oriented grains.

3. Scaling in the Properties of Metal Interconnects for Microelectronics

3.1. *Electromigration*

Electromigration is a process of directional material flow induced by an electric current. As the electric current densities in microelectronic devices are very high, electromigration constitutes a main reliability concern for microelectronics. Thus it has been extensively studied for nearly 40 years. It affects materials such as aluminum, copper, their alloys and recently also lead-free solder materials for packaging. The basic mechanism underlying electromigration is the interaction of electrons with defects in a material, the "wind force" on one hand, and the Coulomb interaction between the externally applied electric field and ionized atoms in the material, on the other. For microelectronic materials the "wind force" is dominating. As the electromigration force F_{EM} is weak, it merely biases the diffusion processes in a material and only becomes observable at elevated temperatures ($T_{test} > 100°C$) and high current densities j ($j > 10^5$ A/cm^2) and is defined as follows,

$$F_{EM} = Z^* e\rho j \tag{15}$$

where Z^* is the effective charge that includes "wind" and Coulomb force, e is the elementary charge and ρ is the resistivity of the conductor line.

The effective charge is a material property and even varies for different alloying components within the same alloy. Thus also the presence of an alloying element may significantly reduce the effective charge of the host material. This was the prime motivation for introducing Cu as an alloying element into aluminum interconnects around 1970.[21, 22] It was found that the presence of Cu strongly reduces the effective charge for aluminum and thus resulted in a strongly increased lifetimes of interconnects. The median time to failure (MTF) of interconnects was determined by testing parallel line arrays of identical interconnects, until all of them failed. This was typically carried out at elevated temperatures and current densities. The empirical approach led to the formulation of Black's law[23]:

$$MTF = Aj^{-n}e^{-\frac{E_A}{kT}} \tag{16}$$

where E_A is the activation energy for electromigration and k is Boltzmann's constant and T is the absolute temperature. It is obvious that such an approach led to a rapid optimization of alloy and processing condition, however, provided little insight about the fundamental mechanisms.

When looking at these fundamental mechanisms it is instructive to look at electromigration induced drift rates:

$$v = \frac{D}{kT}F_{EM} = \frac{D}{kT}eZ^*\rho j \tag{17}$$

where D is the effective diffusivity of atoms exposed to the electromigration force. Analyzing Eq.17 one sees that the rate at which electromigration occurs is equally dependent on two parameters, one, how fast atoms can move as described by the diffusivity (note that this term depends exponentially on temperature via one or several activation energies) and second, how strong the driving force is. The diffusivity depends strongly on the diffusion path. According to the geometry and structure of a wire as well as the materials these paths can either be the volume of the material, the grain-boundaries, dislocation cores, interfaces and surfaces. In wide aluminum lines for instance the dominant diffusion path would be the grain-boundaries. In copper, on the other hand, the dominant diffusion path is usually the top interface between the copper line and a glassy dielectric. All these diffusion paths are very similar to the ones found for diffusional creep.

It is important to note that an electromigration induced material transport in itself does not lead to damage regardless whether the rates are high or low. Damage usually accumulates at sites of flux divergence, which can range from differences in surface diffusivities from grain to grain, over a change in diffusion path, i.e. from grain-boundary diffusion to interfacial diffusion, to diffusion barriers as found in so called Blech[24] segments (compare Fig. 5) and dual damascene structures. Blech segments are finite conductor lines deposited on a continuous conductor of significantly lower conductivity. Most of the current flows through the high conductivity segments and causes edge displacement on the cathode side and hillock formation on the anode side.

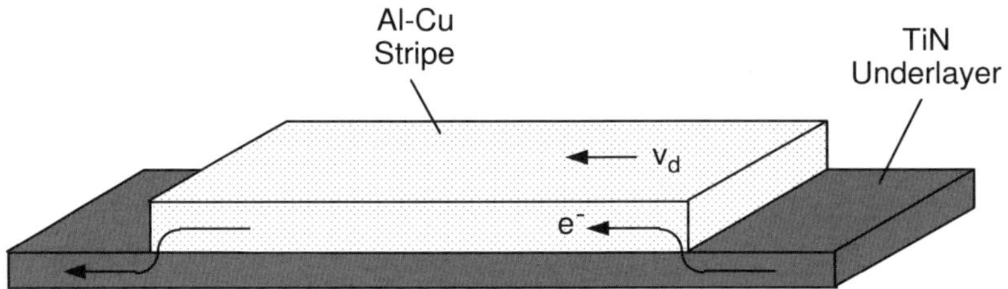

Fig. 5. Schematic describing the geometry of a Blech-type test structure. The drift velocity can be determined by monitoring the displacement of the cathode edge.

The driving force can fundamentally only be altered by a change in material system, e.g. form Cu to Al or the addition of alloying elements, e.g. Cu in Al or Sn in Cu. However, in encapsulated interconnects additional driving forces may arise during the electromigration process. If material is transported in a system of constant volume then also the local stress states will change. Material is usually transported from the cathode to the anode side and thus causes tensile and compressive stress states, respectively. The resulting gradient in stress, subsequently, counteracts the electromigration force and will eventually halt the electromigration process, provided stresses are not released by the formation of hillocks or voids. Consequently, shorter lines exhibit higher stress gradients for given maximal stresses and are less prone to electromigration damage.[24-26] This effect is commonly known as the Blech effect.[24]

In some cases an additional driving force can also originate from thermal gradients and either assist or counteract the electromigration process. The drift rate caused by gradients in stress is given by:

$$v_{\Delta\sigma} = \frac{D}{kT}\frac{\Omega\Delta\sigma}{l} \tag{18}$$

Equating 17 and 18 results in the so-called critical product:

$$\left(jl\right)_c = \frac{\Delta\sigma\Omega}{eZ^*\rho} \tag{19}$$

Thus for a given current density a critical length l_c is defined below which no electromigration damage is expected to occur.

Voids can consequently nucleate only at locations of high tensile stress, which must in turn be a local flux divergence. In addition voids have been found to migrate in the early stages of electromigration so that also the interaction of the motion with the microstructure present becomes important. An overview can be found in Ref. 27.

3.2. Scaling in electrical resistivity

First studies on how the resistivity of metals changes under the influence of an external constraint go back to the 50s of the last century. Fuchs[28] and Sondheimer[29] investigated effects of scattering at interfaces. These are negligible as long as the rate of these additional scattering events is low in comparison to scattering of electrons by phonons. If, however, the distance between to interfaces becomes smaller than the electron mean free path (characterizing the rate of phonon scattering events), then an increase in resistivity can be observed. This increase was expressed as:

$$\rho_{FS} = \rho_0 \left(\frac{3}{8} C (1-p) \frac{1+A_R}{A_R} \frac{\lambda}{w} \right) \tag{20}$$

for the interfacial contribution, where C is a constant of order of magnitude unity, p is the parameter describing the probability of specular scattering of electrons at an interface ($p=1$ implies a 100% probability of specular scattering), A_R is the aspect ratio of the conductor line, λ is the electron mean free path and w is the width of the conductor line. Mayadas and Schatzkes[30] formulated an equivalent model for the scattering of electrons by grain boundaries:

$$\rho_{MS} = \frac{\rho_0 \dfrac{1}{3}}{\left(\dfrac{1}{3} - \dfrac{\alpha}{2} + \alpha^2 - \alpha^3 \ln\left(1+\dfrac{1}{\alpha}\right) \right)} , \alpha = \frac{\lambda}{d} \frac{R}{1-R} \tag{21}$$

where d is the grain size and R is the probability for reflective scattering of electrons ($R=1$ implies a 100% probability for reflective scattering. Both contributions only result in a significant increase in resistivity, if the internal length scale d or the external length scale w become smaller than the electron mean free path. It is also instructive to analyze extreme cases of the scattering parameters p and R. If p goes to zero, one obtains the highest resistivity contribution. If p approaches 1 the Fuchs-Sondheimer term disappears. For the Mayadas-Shatzkes term the scenario is less straight forward as the functional relationship is more complex. For R approaching 1, α becomes very large and the Mayadas-Shatzkes term disappears. In the opposite case (R approaching zero) λ approaches zero as well. In the latter case the resistivity approaches the resistivity value due to phonon and impurity scattering.

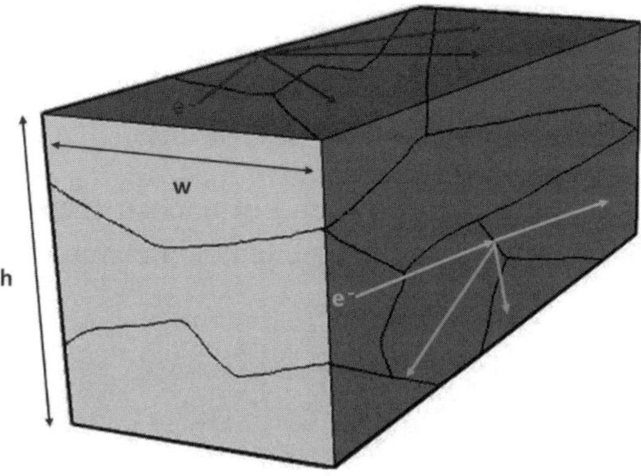

Fig. 6. Schematic describing the geometry of a conducting wire and scattering events at interfaces and at grain boundaries, where *h* is the height of the wire and *w* its width.

Usually both models are used in an additive way. As the electron mean free path of e.g. copper is on the order of 100 nanometers, effect can already be seen in current high performance microprocessors. From a materials point of view the question arises, how one could improve electrical conductivity. The scenario is shown in Fig. 6. As the electron mean free path cannot be increased the following options can be taken:

- Increase of the grain size above the line width w by a thermal treatment or epitaxial film growth.
- Grain boundary engineering to allow only for grain boundaries with low R values.
- Interface engineering to result in high values of p for specular scattering of electrons at interfaces.

All those measures, however, have limited effect in reducing the dimension induced resistivity effect and other approaches such as e.g. the use of metallic carbon nanotubes or possibly graphene (a single atomic layer of graphite) need to be taken into account to ensure further downscaling of interconnect dimension.

Experimentally scaling in conductor lines has been verified several times and most recently by Steinhögl *et al.*[31]

4. Complex Geometries: From Thin Films to Interconnects, Micropillars and Nanoporous Nanoparticles

4.1. *Geometry effects on mechanical deformation*

The following geometries can be found in nanostructures. Each of them has slightly different effects on the mechanical behavior.

- Thin films
- Interconnects
- Pillar geometries
- Nanoparticles
- Nanoporous metals

The first geometry has already been treated extensively, so let us focus on the other geometries. The interconnect geometry predominantly leads to uniaxial stress states and provides fast interfacial diffusion paths on three sides rather than only the top surface.

Pillar geometries, on the other side, have recently become popular[32] for testing size effects, however, do not resemble device geometries. If these pillars are cut from bulk materials by focused ion beam (FIB) systems, samples without grain-boundaries can be made. The defect densities resemble those of molded pillars[33], are, however, always higher than by physical or chemical growth methods or by directional solidification[34, 35] so that theoretical strength is not reached.

In nanoparticle systems the surface-to-volume ratio is further increased, and the small volume results in nearly defect free structures. Ductile and brittle strength levels are close to the theoretical limit.

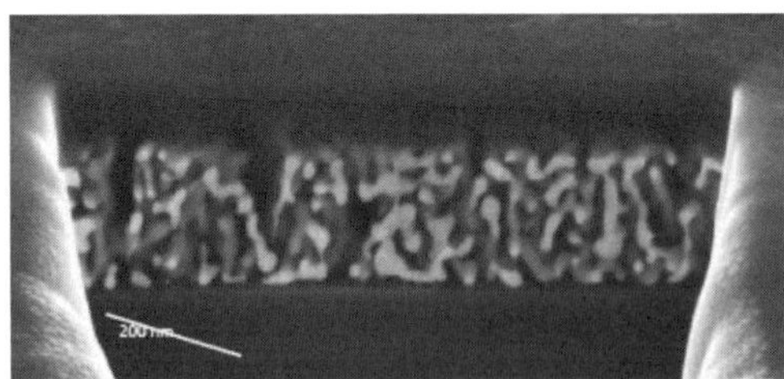

Fig. 7. Micrograph showing a cross-section through nanoporous gold filled with parylene.

Recently, the mechanical properties of nanoporous metallic systems (see Fig. 7) have been investigated and extremely high strength levels have been found, even if those materials are brittle on macroscopic length scales. Local strength levels of single have been determined to approach the theoretical limits by models of porous materials. The explanation for these high strength levels may be attributed to low defect densities, size effects as explained above or strain gradient plasticity, which will be introduced in the following. The latter is due to single ligaments being loaded in a bending mode.

4.2. *Strain gradient plasticity*

If a material is deformed in torsion or in bending, additional dislocations of one sign are necessary to allow for this deformation. An example in bending is described in Fig. 8. These dislocations are frequently also called geometrically necessary or excess dislocations. In principal these dislocations result in an additional hardening term in Taylor's law[36, 37]:

$$\Delta\sigma = CGb\sqrt{\rho_s + \rho_{gnd}} \qquad (22)$$

where C is a constant ρ_s is the density of statistically stored dislocations and ρ_{gnd} is the density of geometrically necessary dislocations. For bulk dimension the density of geometrically stored dislocations ρ_{gnd} is negligible. However, when dimensions are reduced to the micron regime, this term becomes significant. For the example of bending, the strength increase can then be expressed as:

$$\left(\frac{\Delta\sigma}{\Delta\sigma_0}\right) = 1 + \frac{1}{Rb\rho_s} \qquad (23)$$

where $\Delta\sigma_0$ is the strength without strain gradient effects and R is the bending radius.

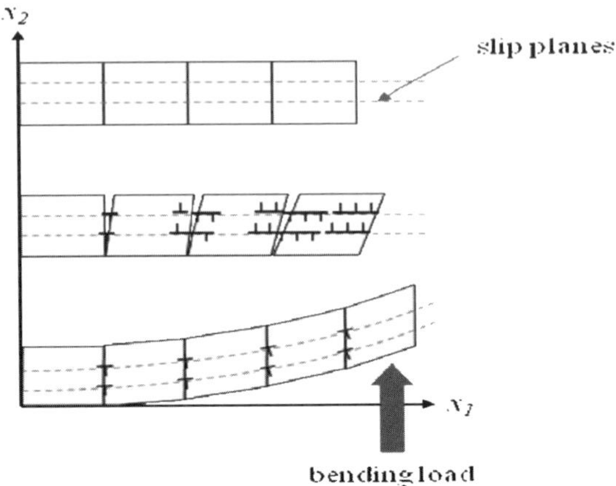

Fig. 8. Schematic of the reasoning behind strain gradient plasticity.

Strain gradient effects can be found in the following deformation modes:

- Bending
- Torsion
- Indentation size effect

Alternatively, one could explain such effects by a length scale that is established over which the stresses applied are higher than the materials strength and then follow classical size effects as explained above.

With regards to application, this point is very important as for the determination of materials properties at the nanoscale, not only the dimension, strain rate and temperature are important (as described above), but also the testing geometry.

4.3. *In-situ testing methods*

Currently, the most common in-situ methods for in-situ mechanical testing of small scale structures are

- Pillar compression
- X-ray diffraction

- Raman spectroscopy
- Scanning electron microscopy (including electron backscatter diffraction)
- Transmission electron microscopy
- Optical microscopy
- Atomic force microscopy

Due to a review in the MRS Bulletin to be published in May 2010, details will be omitted in this chapter.

Acknowledgments

Diana Glück is acknowledged for support with figures.

References

1. F. D. Fleischli, M. Dietiker, C. Borgia and R. Spolenak, *Acta Biomaterialia*, **4**, 1694 (2008).
2. E. Arzt, S. Gorb and R. Spolenak, *Proceedings of the National Academy of Sciences of the United States of America*,. **100**, 10603 (2003).
3. H. Hertz, *Reine Angew Math* **92**, 155 (1881).
4. K. L. Johnson, K. Kendall and A.D. Roberts, *Proceedings of the Royal Society of London Series a-Mathematical and Physical Sciences*, **324**, 301 (1971).
5. E. Arzt, *Acta Materialia*, **46**, 5611 (1998).
6. J. W. Beams, J.B. Breazeale and W.L. Bart, *Physical Review*, **100**, 1657 (1955).
7. J. B. Breazeale, W. L. Bart and J. W. Beams, *Physical Review*, **99**, 615 (1955).
8. A. Catlin and W.P. Walker, *Journal of Applied Physics*, **31**, 2135 (1960).
9. W. D. Nix, Metallurgical *Transactions a-Physical Metallurgy and Materials Science*, **20**, 2217 (1989).
10. L. B. Freund, *Journal of Applied Mechanics*, **54**, 533 (1987).
11. B. von Blanckenhagen, P. Gumbsch and E. Arzt, *Modelling and Simulation in Materials Science and Engineering*, **9**, 157 (2001).
12. B. von Blanckenhagen, P. Gumbsch and E. Arzt, *Phil. Mag. Letters*, **83**, 1 (2003).
13. M. W. Chen, E. Ma, K. J. Hemker, H. W. Sheng, Y. M. Wang and X. M. Cheng, *Science*, **300**, 1275 (2003).
14. S. H. Oh, M. Legros, D. Kiener, P. Gruber and G. Dehm, *Acta Materialia*, **55**, 5558 (2007).

15. R. D. Nyilas and R. Spolenak, *Acta Materialia*, **56**, 5627 (2008).
16. P. A. Gruber, E. Arzt and R. Spolenak, *Journal of Materials Research*, **23**, 2406 (2008).
17. H. Gao, L. Zhang, W.D. Nix, C.V. Thompson and E. Arzt, *Acta Materialia*, **47**, 2865 (1999).
18. D. Weiss, H. Gao and E. Arzt, *Acta Materialia*, **49**, 2395 (2001).
19. P. A. Gruber, E. Arzt, and R. Spolenak, *Journal of Materials Research*, **24**, 1906 (2009).
20. N. Lu, Z. Suo and J. J. Vlassak, *Acta Materialia*, 2010. doi:10.1016/j.actamat.2009.11.010.
21. S. J. Horowitz and I. A. Blech, *Materials Science and Engineering*, **10**, 169 (1972).
22. I. A. Blech, *Journal of Applied Physics*, **48**, 473 (1977).
23. J. R. Black, Ieee Transactions on Electron Devices, **ED16**, 338 (1969).
24. I. A. Blech and E. S. Meieran, *Journal of Applied Physics*,. **40**, 485 (196).
25. C. L. Gan, C.V. Thompson, K.L. Pey, W.K. Choi, H.L. Tay, B. Yu and M.K. Radhakrishnan, *Applied Physics Letters*, **79**, 4592 (2001).
26. C. L. Gan, C.V. Thompson, K.L. Pey, and W.K. Choi, *Journal of Applied Physics*, **94**, 1222 (2003).
27. E. Arzt, O. Kraft, R. Spolenak and Y.C. Joo, *Zeitschrift Fur Metallkunde*, **87**, 934 (1996).
28. K. Fuchs, K., *Proc. Camb. Phil. Soc.* A, 1938. **34**: p. 100.
29. E. H. Sondheimer, *Advances in Physics*, **50**, 499 (2001).
30. A. F. Mayadas and M. Shatzkes, *Physical Review* **B 1**, 1382 (1970).
31. W. Steinhoegl, G. Schindler, G. Steinlesberger, M. Traving and M. Engelhardt,. *2003 Ieee International Conference on Simulation of Semiconductor Processes and Devices*, p. 27 (2003).
32. M. D. Uchic, D. M. Dimiduk, J. N. Florando and W. D. Nix,. *Science*, **305**, 986 (2004).
33. S. Buzzi, M. Dietiker, K. Kunze, R. Spolenak and J.F. Loffler, *Phil. Mag.*, **89**, 869 (2009).
34. H. Bei, S. Shim, E. P. George, M. K. Miller, E. G. Herbert and G.M. Pharr, *Scripta Materialia*, **57**, 397 (2007).
35. H. Bei, S. Shim, G.M. Pharr and E.P. George, *Acta Materialia*, **56**, 4762 (2008).
36. H. Gao, Y. Huang, W.D. Nix and J.W. Hutchinson, *Journal of the Mechanics and Physics of Solids*, **47**, 1239 (1999).
37. Y. Huang, Y., H. Gao, W.D. Nix, and J.W. Hutchinson, *Journal of the Mechanics and Physics of Solids*, **48**, 1 (2000).

CHAPTER 8

FORMATION OF HIGH-STRENGTH NANOCRYSTALLINE ALLOYS AND THEIR MECHANICAL PROPERTIES

Tohru Yamasaki

Division of Material Researches, University of Hyogo
2167 Shosha,Himeji, Hyogo 671-2280, Japan
E-mail: yamasaki@eng.u-hyogo.ac.jp

Mechanical properties of nanocrystalline alloys with grain size ranging from few to several tens nanometers have been examined. Sufficiently dense and structurally well characterized nanocrystalline materials have been prepared by dynamic compaction of mechanically alloyed nanocrystalline powders and also by electrodeposition techniques. Both dynamically consolidated (Fe, Ni)-TiN materials and electrodeposited Ni-W materials exhibited Hall-Petch hardening to grain sizes near about 10 nm whereas at grain size finer than about 10 nm, the hardness decreases in an apparent breakdown of the Hall-Petch relationship. High temperature hardness tests and indentation creep tests were performed between room temperature and up to 1000 K on (Fe, Ni)-TiN materials. With increasing temperature, softening occurs by two stages: hardness decreases slightly at first stage and decreases significantly at second stage from about 500 K. The grain size softening dependence suggests that softening in second stage occurs via grain boundary diffusion creep. Conversely, the tensile strength and ductility of nanocrystalline Ni-W electrodeposits have been strongly affected by the inclusion of impurities such as hydrogen and oxygen during the deposition process. Under optimal electrodeposition conditions, high-strength nanocrystalline Ni-W alloys with tensile strength of about 2300 MPa combined with good ductility were obtained.

273

1. Introduction

Mechanical properties of nanocrystalline materials have arouse strong interest over the past one or two decades because of the expectation that the grain size refinement of the materials into nanometer size range will enhance mechanical properties, such as strength and hardness, possibly without loss of ductility. The capability of superplastic forming at moderate temperatures and strain rates is also an attractive expectation for the nanocrystalline refinement. The remarkable present-day developments in this field were initiated by Gleiter and his collaborators.[1] They demonstrated that the plastic deformation of nanocrystalline ceramics at low temperature can be possible in the CaF_2 and TiO_2 nanocrystalline samples produced by using evaporation-condensation apparatus for the production of nanoscale powders. However, recent experimental observations reported that most of the nanocrystalline materials are generally limited for practical applications because of their severe brittleness.[2] It has not yet been well understood whether their brittle behavior is due to an intrinsic feature of nanocrystalline materials, or whether this is due to processing difficulties for the fine grain sizes such as an imperfect consolidation of the nanocrystalline powders.[3] Especially, the grain size dependence of the strength of nanocrystalline materials in the size range of a few to several tens nanometers is fundamentally important for understanding the mechanical behavior of nanocrystalline materials. The experimental results hitherto reported however are very scattered.

It is therefore required for studying the mechanical properties to make the experiment with sufficiently dense and structurally well characterized nanocrystalline materials. In this review, we show the following two methods for preparing well characterized high-strength nanocrystalline materials; (1) dynamic compaction of mechanically alloyed nanocrystalline powders by using a propellant gun, and (2) electrodeposition of nanocrystalline alloys. The dynamic compaction is a suitable tool for this purpose, since the powders are densely consolidated without grain growth. We prepared nanocrystalline Fe-TiN and Ni-TiN dual phase materials by dynamic compaction of Fe-TiN and Ni-TiN powders produced by mechanical alloying in a nitrogen gas[4-7], and investigated the grain size and the temperature dependences at a wide temperature range. On the other hand, electrodeposition is a superior technique for producing nanocrystalline materials having grain sizes anywhere from the essentially amorphous to nanoscaled materials for the grain sizes of about 5-50 nm in bulk form or as coatings with no post-processing requirements.[8] We also show that the nanocrystalline Ni-W alloys having both high hardness and high plasticity can be produced by electrodeposition if experimental parameters are carefully chosen.[9, 10]

2. Formation of Nanocrystalline Powders by Mechanical Alloying and Their Shock Consolidation

2.1. *Preparation of nanocrystalline Fe-TiN and Ni-TiN dual phase powders*

The Fe-TiN and Ni-TiN dual phase materials were selected because the high thermal stability of their nanocrystalline structure enables to investigate the grain size and the temperature dependences of hardness without the influences of irreversible structural changes. The mechanical properties of these nanocrystalline cermet materials will be interesting also from technological point of view.

Fe-63 vol. % TiN and Ni-17~64 vol. % TiN dual phase powders were prepared by ball milling $Fe_{50}Ti_{50}$, $Ni_{50}Ti_{50}$, $Ni_{70}Ti_{30}$ and $Ni_{90}Ti_{10}$ (in at. %) powder mixtures in a nitrogen gas for 360 to 540 ks, almost all the titanium atoms were nitrided to form cubic TiN, and nanocrystalline Fe-TiN and Ni-TiN dual-phase powder alloys were produced. The ball milling was performed using a vibrational ball mill in which the vial and balls were made of SUS316 and 304 stainless steel, respectively.

Figure 1 shows the schematic diagram of the vibrational ball mill for milling in a nitrogen atmosphere. The vial capacity was 1.3×10^{-3} m^3 and the ball diameter was 12.7 mm. The vial was connected to a nitrogen gas chamber with 3.4×10^{-2} m^3 capacity. The ball to powder weight ratio was 45:1. After evacuating the vial and the gas chamber, nitrogen gas was introduced to one atmospheric pressure. The decrease in the gas pressure was measured by a pressure gauge connected to the gas chamber. The atomic percent of nitrogen absorbed in the powders were calculated from the reduction in the gas pressure.

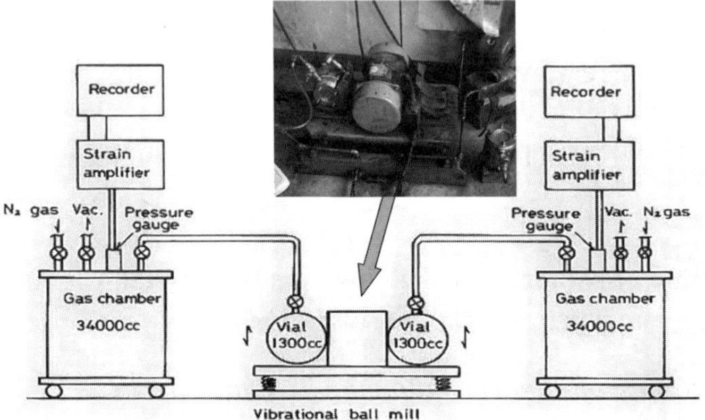

Fig. 1. The schematic diagram of the vibrational ball mill for milling in a nitrogen atmosphere. The vial capacity was 1.3×10^{-3} m^3 and the ball diameter was 12.7 mm. The vial was connected to a nitrogen gas chamber with 3.4×10^{-2} m^3 capacity.

Figure 2 shows the nitrogen contents of the $Fe_{75}Ti_{25}$, $Fe_{50}Ti_{50}$ and $Fe_{25}Ti_{75}$ (in at. %) milled powders as a function of milling time. After 50 h milling, the absorbed nitrogen was saturated. The amounts of saturated nitrogen increased with increasing Ti content, and almost all the titanium atoms in the Fe-Ti powder mixtures might be eventually nitrided to form cubic TiN.

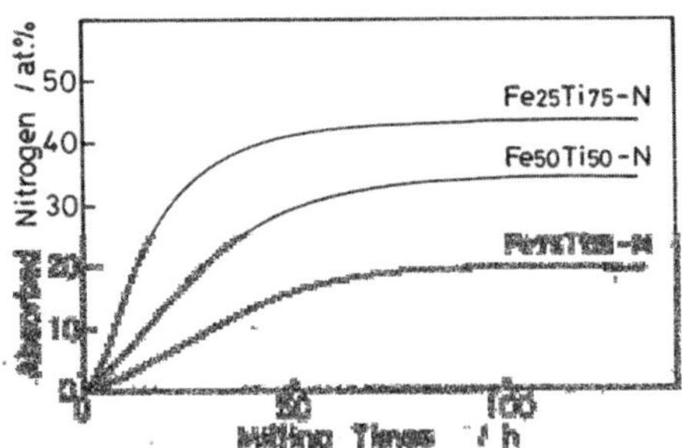

Fig. 2. Amounts of nitrogen absorbed by Fe-Ti powders during MA in N$_2$ atmosphere.

Figure 3 shows the X-ray diffraction patterns of the $Fe_{50}Ti_{50}$ powder mixture milled several times in a nitrogen atmosphere. Diffraction peaks representing crystalline Fe and Ti disappeared after 20 h of milling, and only broad peaks from the cubic TiN phase are observed.

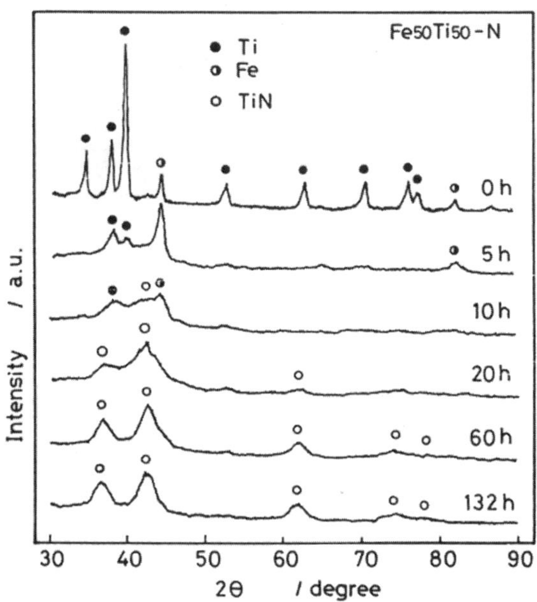

Fig. 3. The X-ray diffraction pattern of $Fe_{50}Ti_{50}$ powders milled in nitrogen atmosphere.

2.2. *Dynamic consolidation of the nanocrystalline powders*

The milled powders of about 0.2 μm in particle diameter were filled into steel container of 25 mm inner diameter and 10 mm depth by tapping to about 45 % of theoretical density, and dynamically compacted at room temperature and 850~900 K for the Fe-TiN and the Ni-TiN powders respectively, by impact of the flier launched by a propellant gun as illustrated in Fig. 4. The impact pressures were changed from 21.4 to 40.5 GPa by controlling the speed of the flier. Bulk materials of higher density were obtained by increasing the compacting pressure and temperature. The materials however structurally inhomogeneous; the density decreased at inner region. Therefore specimens of about 5 mm thick cut from the outer layer of higher density were used.

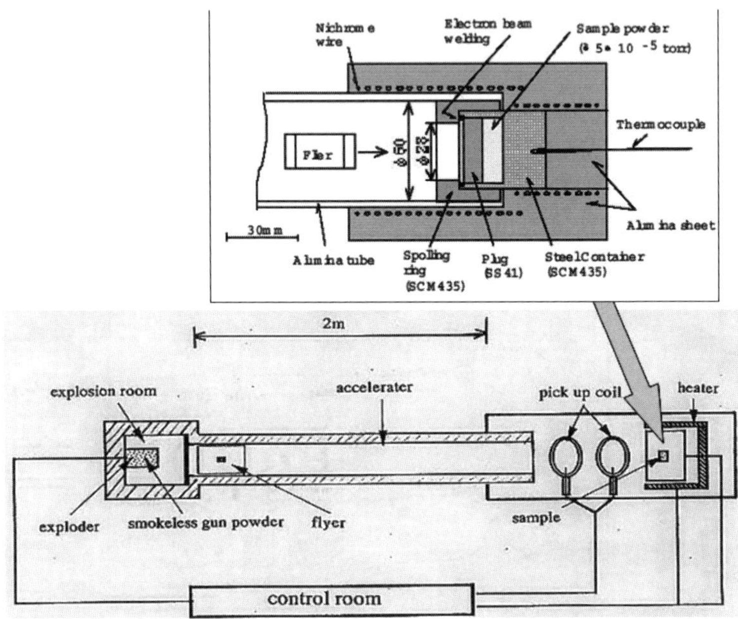

Fig. 4. Schematic diagram of a propellant gun.

Figure 5 shows the X-ray diffraction patterns of the Fe-63 vol. % TiN materials consolidated by the dynamic compaction at various pressures. With the materials consolidated by the dynamic compaction at pressures of 21.4 and 38.6 GPa, the diffraction patterns are similar to those of the as milled powders and only broad diffraction peaks from the cubic TiN are observed. The high-resolution transmission electron microscope (HR-TEM) observation of the consolidated material at the pressure of 38.6 GPa however revealed that the structure are composed of the TiN and α-Fe phases with the grain size of about 5 ~10 nm as shown in Fig. 6. With the materials consolidated at 40.5 GPa shown in Fig. 5, sharp peaks of α-Fe and Fe_3N phases appear superimposed on the TiN peaks, indicating that the grain growth occurred under the influence of an instantaneous heat evolution during dynamic compaction.

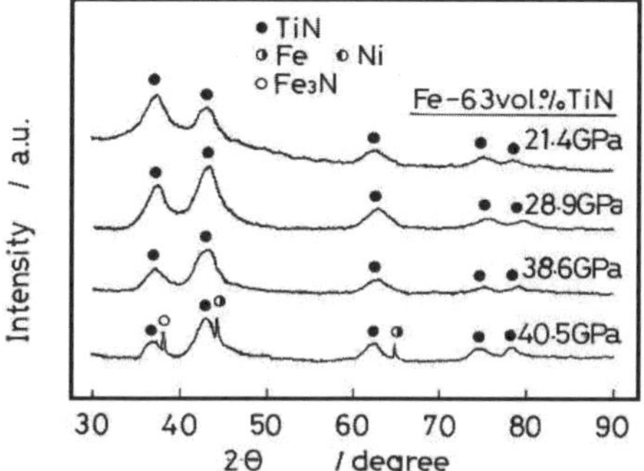

Fig. 5. X-ray diffraction patterns of dynamically compacted Fe-63 vol. % TiN materials by use of a propellant gun at various pressures.

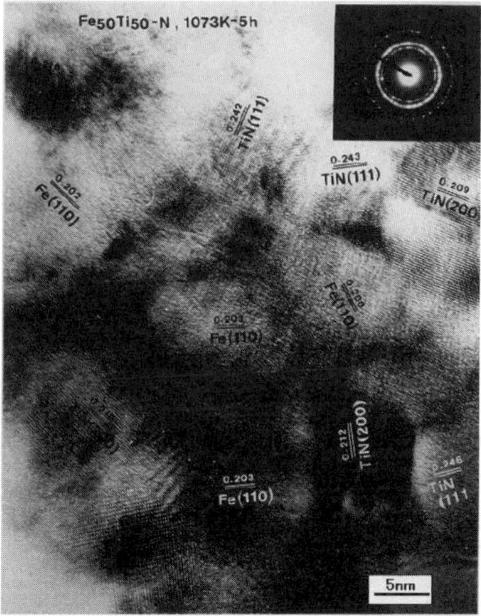

Fig. 6. The high-resolution transmission electron microscope (HR-TEM) observation of the shock consolidated Fe-63 vol. % TiN material at the pressure of 38.6 GPa. Structure was composed of the TiN and a-Fe phases with the grain size of about 5 ~10 nm.

Figure 7 shows the X-ray diffraction patterns of the Fe-TiN and the Ni-TiN materials consolidated at the pressure of about 39 GPa. The structure of the Ni-17, 44 and 64 vol. % TiN materials are also composed of the nanocrystalline TiN and *fcc*-Ni phases.

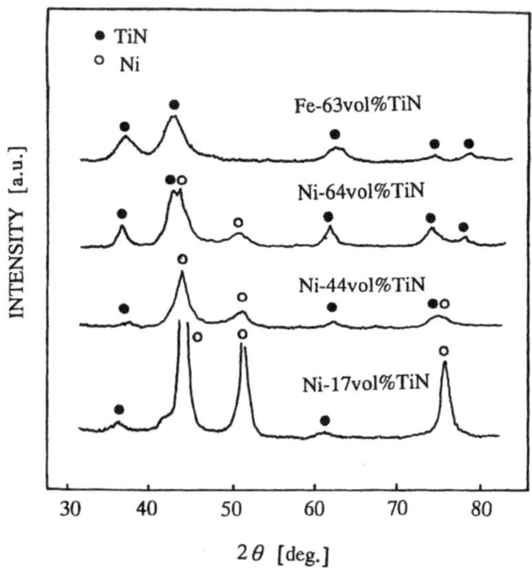

Fig. 7. X-ray diffraction patterns of dynamically consolidated Fe-TiN and Ni-TiN materials at the pressure of about 39 GPa.

Figure 8 shows grain sizes of the metal phases of the dynamically consolidated Fe-TiN and Ni-TiN materials as a function of annealing temperature. The nanocrystalline structure with grain sizes of 5 to 25 nm is stable up to about 1000 K, and then grain growth occurred at higher temperatures. The grain size of the TiN was approximately the same as that of metallic phases in Ni-44 & 64 vol. % TiM and Fe-63 vol. % TiN and smaller than that of the nickel phase in the Ni-17 vol. % TiN material. The average densities, as measured by the Archimedean method, were 92 % theoretical density for the Fe-TiN bulk material and 97 to 98 % theoretical density for the Ni-TiN bulk materials. Grain growth of these materials occurred at higher temperatures, and concomitantly density increased somewhat and approached the theoretical value.

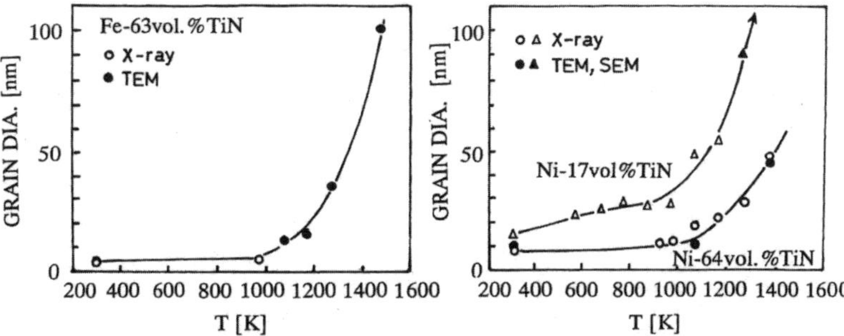

Fig. 8. Grain sizes in iron and nickel phases of dynamically consolidated Fe-63vol. % TiN and Ni-17 & 64 vol. % TiN materials at the pressure of about 38~39 GPa as a function of annealing temperature.

2.3. *Grain size dependence of hardness of dynamically consolidated Fe-TiN and Ni-TiN alloys*

Dynamically consolidated specimens with various grain sizes were prepared for hardness measurements by annealing the materials at 1000 K to 1500 K for 18 ks. The grain size was measured for each of the materials on TEM or SEM micrographs. Figure 9 shows the room temperature Vickers microhardness of the Fe-TiN and the Ni-TiN materials as a function of $d^{-0.5}$ (d is average grain diameter). For a comparison, the relationship for pure nickel extrapolated from the relationship in conventional grain sizes, which is calculated by a relation HV=3σ_y (σ_y is yield strength) with σ_y averaged for all the published data[11], is also shown. In the Fe-63 vol. % TiN and Ni-44, 64 vol. % TiN materials, the hardness increases with decreasing grain size to about 10 nm. At small grain size, it tends to decrease. In the Ni-17 vol. % TiN material, the softening might occur from a large grain size between 10 and 20 nm. The Hall-Petch slopes of 0.8 to 0.85 MPa-m$^{0.5}$ in the Ni-TiN materials are slightly higher than, but comparable to those of conventional nickel. Hall-Petch hardening with slopes similar to the present result has been observed for nanocrystalline iron[12] and nanocrystalline nickel[13, 14] prepared by ball milling and electrodeposition, respectively. In the less dense Fe-TiN materials that had

                    T. Yamasaki

been dynamically compacted at 28.9 GPa, the inverse grain size dependence was observed when the grain size was below 400 nm. This inverse grain size dependence should not be due to the size effect, but to densification during annealing.

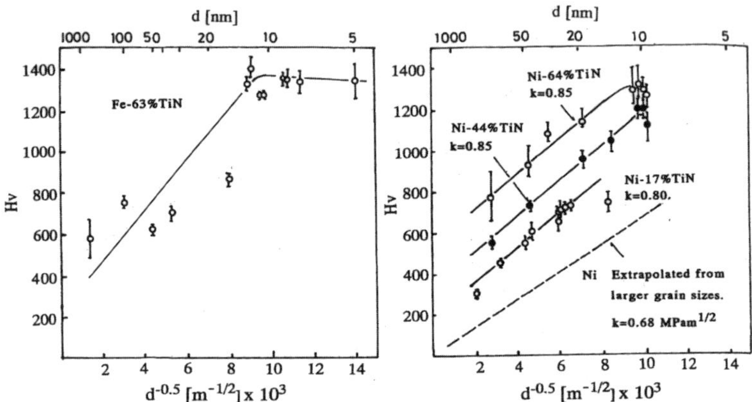

Fig. 9. Grain sizes in iron and nickel phases of dynamically consolidated Fe-TiN and Ni-TiN materials as a function of annealing temperature.

It has been noted in previous studies that the Hall-Petch slope of nanocrystalline materials depends on methods to vary grain size; it tends to decrease or even takes a negative value when grain size is varied by annealing.[15] This has been suggested to be due to an annealing effect, such as structural recovery of grain boundaries and densification of the materials.[15, 16] The present result indicates that the Hall-Petch hardening occurs significantly in nanoscale grain sizes even in annealing materials if they are sufficiently dense.

2.4. *Indentation creep in nanocrystalline materials*[6]

Figures 10 and 11 show results of the high-temperature Vickers microhardness measurements of Fe-63 vol. % TiN and Ni-64 vol. % TiN materials at the temperatures up to about 1000 K with a loading time of 10 seconds. The hardness measured on heating and cooling runs is reproducible even for materials with grain sizes below 10 nm. This is consistent with the high thermal stability of the nanocrystalline structure as mentioned in Fig. 8 and indicates that there are no annealing effects on the hardness. The softening occurs by two stages; the hardness slightly decreases at the first stage and significantly at the second stage with increasing temperature. The second stage softening tends to occur from lower temperatures; it occurs from about 500 K with grain sizes of 5 – 35 nm for both Fe-TiN and Ni-TiN materials. With the grain size increased up to about 100 nm, the second stage softening almost disappeared. The grain size dependence of the softening suggests that the second stage softening occurs via grain boundary diffusional creep.

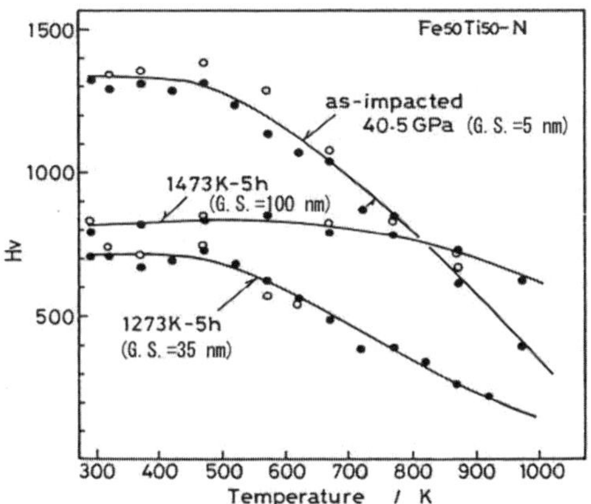

Fig. 10. High temperature Vickers microhardness of the dynamically compacted Fe-63 vol. % TiN powders at 40.5 GPa as a function of the temperature. Vickers hardness tester with a 1.96 N load up to 973 K and a loading time of 10 seconds.

So, indentation creep tests at constant temperatures between 523 K and 973 K were carried out on Fe-63 vol. % TiN and Ni-17, 64 vol. % TiN materials with the grain sizes of less than about 100 nm. In Figure 12, examples of the Ni-17 vol. % TiN materials are shown of hardness vs. loading time curves. The initial decrease in hardness is rapid, and then it slowly decreases. The analysis of indentation creep is complicated because applied stress and strain rate are not constant but a decreasing function of loading time.

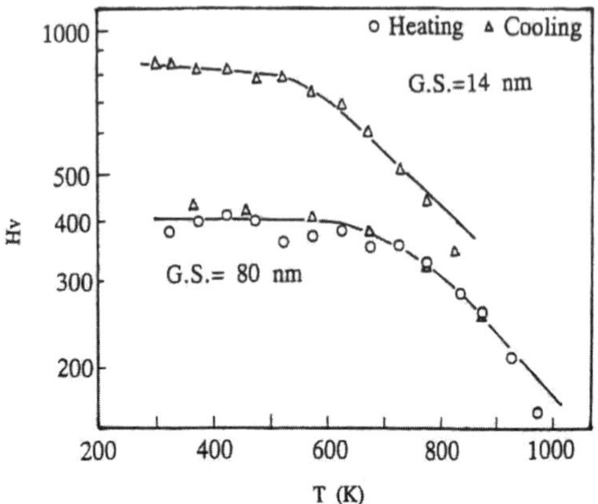

Fig. 11. High temperature Vickers microhardness of the dynamically compacted Ni-64 vol. % TiN powders at 38.5 GPa as a function of the temperature.

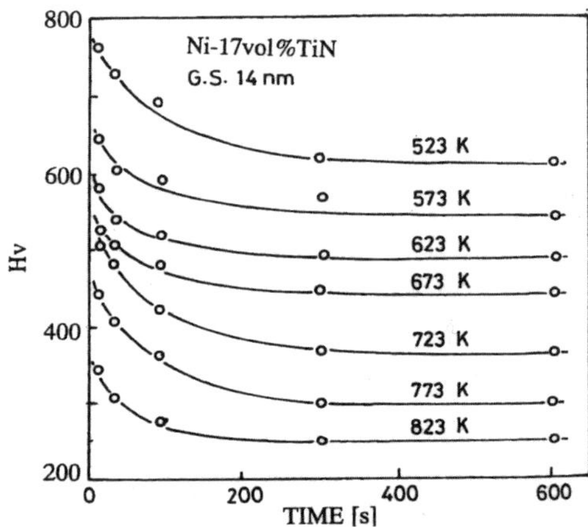

Fig. 12. Hardness vs. loading time curves for the Ni-17 vol. % TiN alloy measured at various temperatures.

Atkins *et al.*[17] have analyzed the indentation creep by using the following equation, which is consistent with the Andrade transient creep rate:

$$H^{-3} - H_0^{-3} = B\exp(-Q/3RT)(t^{1/3} - t_0^{1/3}) \tag{1}$$

where H_0 is the hardness at an initial loading time t_0, and B is a structure-dependence factor. It has been demonstrated that the Eq. (1) fits experimental relationships very well for a number of crystalline solids, and activation energies Q, very close to those for volume self-diffusion, and obtained consistently with a controlling mechanism based on dislocation climb when tested at temperatures higher than 0.5 T_m.[18, 19]

A linear relationship is predicted in Eq. (1) between $\ln(H^{-3} - H_0^{-3})$ and 1/T with slope of unity. Figure 13 and 14 show the $\ln(H^{-3} - H_0^{-3})$ and 1/T plots for the Fe-63 vol. % TiN, Ni-64 vol. % TiN and Ni-17 vol. % TiN materials with grain sizes to 10 to 80 nm and testing temperatures between 523 and 973 K under condition where the H_0 has been taken to the value at t_0=1s, which is determined by extrapolating from ln H and ln t linear relationship. It is seen in all the materials that a whole family of straight lines is obtained of constant slope. The slope is very close to unity, as predicted from the equation (1). Then the activation energy can be obtained by plotting $\ln(H^{-3} - H_0^{-3})$ against for 1/T for fixed loading times.

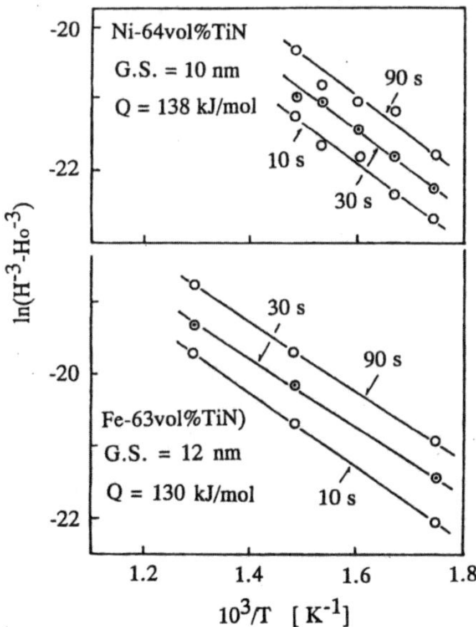

Fig. 13. Relationships between ln(H-3 – H0-3) and 1/T for Fe-63 vol. % TiN and Ni-64 vol. % TiN materials, plotted for several different loading times.

As shown in Figs. 13 and 14, straight-line relationships exist in all the materials and activation energies are obtained. The activation energies of about 130 kJ/mol for the Fe-TiN and Ni-TiN materials with grain sizes of 10 to 14 nm are about one-half of the activation energies for self-diffusion of nickel (280 kJ/mol) and α-iron (290 kJ/mol) and comparable to those for grain boundary self-diffusion, which have been measured to be 111 to 130 kJ/mol and 140 to 174 kJ/mol[19] for nickel and α-iron, respectively. As the grain size increases to 30 nm, the activation energy increases up to 200 kJ/mol, bujt is still lower than that for the self-diffusion.

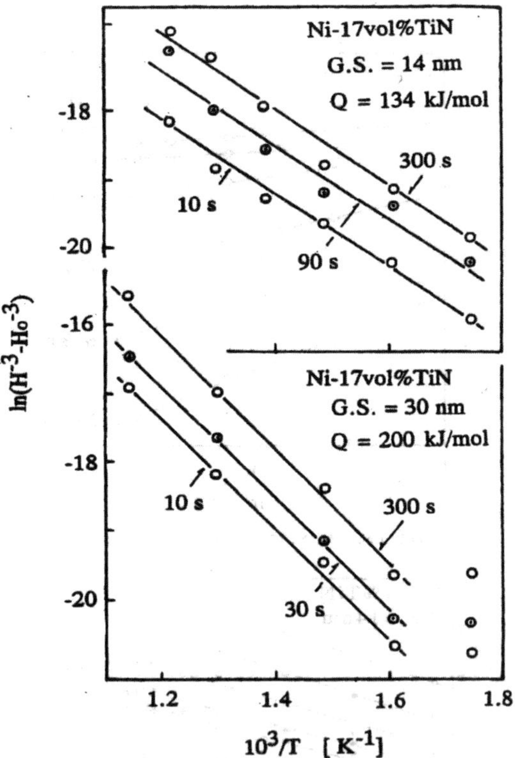

Fig. 14. Relationships between ln(H-3 − H0-3) and 1/T for Ni-17 vol. % TiN materials, plotted for several different loading times.

3. Formation of High-Strength Nanocrystalline Alloys by Electrodeposition

3.1. *Electrodeposition of nanocrystalline alloys*

The electrodeposition is a superior technique for producing nanocrystalline materials in bulk form or as coatings with no post-processing requirements. This is particularly important for applications to microfabrication technologies, especially to the molding process of LIGA technology that requires hard microstructures. Electrodeposited alloys having high hardness, however, are generally limited in these applications because of their severe brittleness.[20] For example, amorphous and nanocrystalline electrodeposited Ni-W alloys have some excellent properties such as high hardness, high corrosion resistance and high thermal stability. The process for electrodeposition of W-rich Ni-W alloys, however, has not yet been well developed and the electrodeposited alloys have been severely brittle.[20-24] In this section, we show an aqueous plating bath for the Ni-W electrodeposition that yields amorphous and nanocrystalline alloys of fairly high tungsten content, and also show that Ni-W alloys having both high hardness and high ductility can be produced by electrodeposition if experimental parameters are carefully chosen.[9, 10]

3.2. *Preparation of nanocrystalline Ni-W alloys and technique used for studies*

The plating bath compositions and conditions selected for this study are shown in Table 1. Trisodium citrate and ammonium chloride were introduced as complexing agents to form complexes with both Ni and W in the plating bath solution. Electrodeposition was done using Cu sheets as substrates prepared by electroplishing; a high-purity platinum sheet was used as the anode. The plating cell (600 mL beakers, each containing 500 mL of the bath) were controlled by a thermostat to maintain the desired bath temperatures.

Table 1. Plating Bath Compositions and Conditions for Ni-W electrodeposition.

Nickel Sulfate, NiSO4 -6H2O	0.06 mol/L
Sodium Tungstate, Na2WO4 -2H2O	0.14 mol/L
Trisodium Citrate, Na3C6H5O7 -2H2O	0.14 ~ 0.5 mol/L
Ammonium Chloride, NH4Cl	0.5 mol/L
(or Ammonium Sulfate, (NH4)2SO4)	(0.25 mol/L)
Bath Temperature	333 ~ 363 K
pH	7.5 ~ 8.5
Current Density	0.05 ~ 0.2 A/dm2

Figure 15 shows the schematic explanations showing the automatic controlling system for setting the desired pH-values and temperatures of the Ni-W electroplating bath solution.

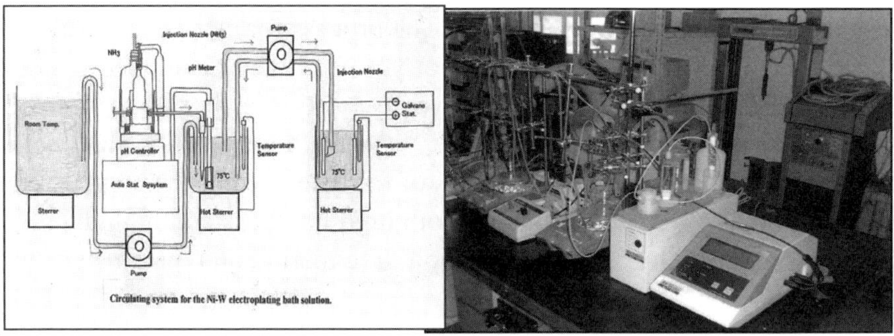

Fig. 15. Schematic explanations for the Ni-W electrodeposition showing the automatically controlling system for the pH-value and the temperature of the plating bath solution.

A fresh plating bath was made for each experiment, using analytical reagent grade chemicals and deionized water. The deposition rate of the Ni-W alloys was determined by weighting the substrate before and after the one-hour electrodeposition and calculating the additional mass per square centimeter. The electrodeposited Ni-W films were separated from the Cu substrates by immersing the samples in an aqueous solution containing CrO_3 (250 g/L) and H_2SO_4 (15 cc/L). The samples thus prepared were annealed at various temperatures in vacuum of about 10^{-3} Pa. The degree of ductility was determined by measuring the radius of curvature at which the fracture occurred in a simple bending test. The fracture strain on the outer surface of the specimen, ε_f, is estimated by the equation, $\varepsilon_f = t/(2r-t)$, where r is the radius of curvature on the outer surface of the bend sample at fracture and t is the thickness of specimen. Vickers microhardness was measured by using the as-electrodeposited and the annealed samples on Cu substrates with a 0.02 kg load and a loading time of 15 s in cross section.

3.3. *Formation of high-strength Ni-W electrodeposits*

3.3.1. *Plating bath temperature and applied current density.*[9]

The tungsten content of the electrodeposits is strongly influenced by the plating bath temperature and the applied current density. Figure 16 shows the X-ray diffraction patterns of the Ni-W electrodeposits and their tungsten contents for various plating bath temperatures between 333 K and 363 K with the trisodium citrate concentration of 0.5 mol/L at the applied current density of 0.2 A/cm^2.

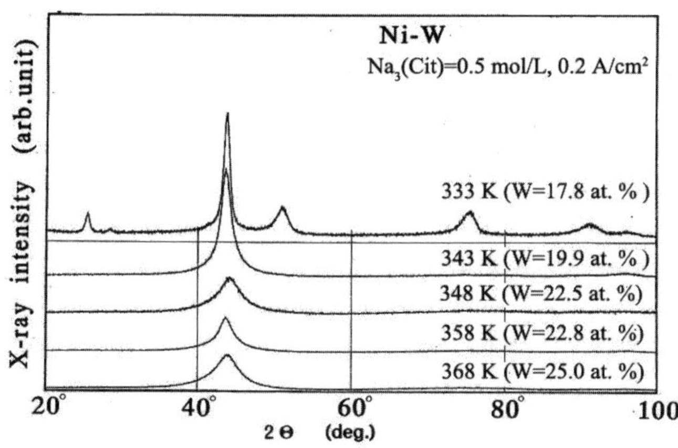

Fig. 16. X-ray diffraction patterns of the Ni-W electrodeposits and their tungsten contents for various plating bath temperatures between 333 K and 363 K at the applied current density of 0.2 A/cm^2.

The tungsten content of the electrodeposits increased with increasing plating bath temperature and the amorphous pattern appeared at a tungsten content of more than about 20 at. %. Figure 17 shows DTA measurements at a heating rate of 0.33 K/s of the as-electrodeposited Ni-W alloys for various plating bath temperatures.

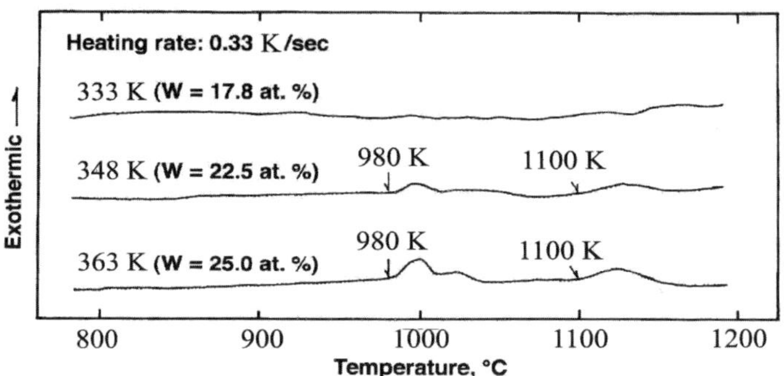

Fig. 17. DTA measurements at a heating rate of 0.33 K/s of the as-electrodeposited Ni-W alloys for various plating bath temperatures.

No distinct crystallization peaks were observed for the Ni-17.8 at. % W alloy electrodeposited at a plating bath temperature of 333 K. At a bath temperature of 348 K and above, the amorphous X-ray diffraction pattern appeared and the crystallization of these amorphous Ni-W alloys takes place in two steps. The first step crystallization at a temperature of about 980 K has been confirmed by X-ray analysis to be due to the crystallization of *fcc*-Ni-W solid solution. The second step of the DTA thermograms takes place at a temperature range between 1100 K and 1150 K. X-ray analysis has suggested that Ni_4W intermetallic compound precipitates during this step. Deposition rate and mechanical properties vs. plating bath temperature for the Ni-W electrodeposits are shown in Table 2. The Vickers microhardness of the electrodeposits increased continuously from 602 to 770 with increasing the plating bath temperature from 333 K to 363 K, respectively, while the deposition rate received its maximum value of 68.8 mg/cm^2 h at a plating bath temperature of 348 K and then decreased with increasing plating bath temperature.

Table 2. Deposition rate and mechanical properties vs. plating bath temperature for the Ni-W electrodeposits.

Bath Temp. K	W-content at. pct	Deposition rate Mg/cm2/h	Vickers Hardness HV	Fracture Strain ef	Structure judged by X-ray diffraction
333	17.8	49.2	602	0.0	Nanocrystalline
343	19.9	65.0	650	0.02	Nanocrystalline
348	22.5	68.8	685	1.00 Ductile	Amorphous
353	22.8	66.7	752	0.416	Amorphous
363	25.0	55.1	770	0.005	Amorphous

It may be noted that the as-electrodeposited Ni-22.5 at. % W alloy at the plating bath temperature of 348 K having high hardness of HV= 685 was ductile: it could be bent through an angle of 180° without breaking (ε_f=1.0). At the other plating bath temperatures, the electrodeposits were severely brittle. Figure 18 shows SEM micrographs of the Ni-22.5 at. % W alloys electrodeposited at the plating bath temperature of 348 K after bending through an angle of 180°.

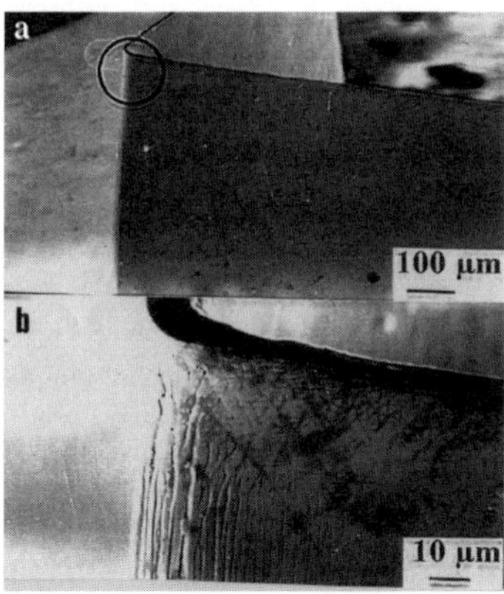

Fig. 18. SEM micrographs of the Ni-22.5 at. % W alloys electrodeposited at the plating bath temperature of 348 K after bending through an angle of 180°. Considerable spring back was observed. They deform plastically and extremely inhomogeneously: shear bands form on the bending edge showing the typical feature of the ductile metallic glasses.

They deform plastically and extremely inhomogeneously: shear bands form on the bending edge showing the typical feature of the ductile metallic glasses. The tungsten content of the Ni-W electrodeposits is also influenced by the applied current density.

Figure 19 shows X-ray diffraction patterns of as-electrodeposited Ni-W alloys for various current densities between 0.05 and 0.2 A/cm^2 at a plating bath temperature of 348 K. X-ray diffraction peaks of the deposited alloys broadened and the tungsten content increased with increasing applied current density. Under these plating conditions, the Ni-W electrodeposits having amorphous and nanocrystalline structures were ductile: they could be bent through an angle of 180° without breaking ($\varepsilon_f = 1.0$).

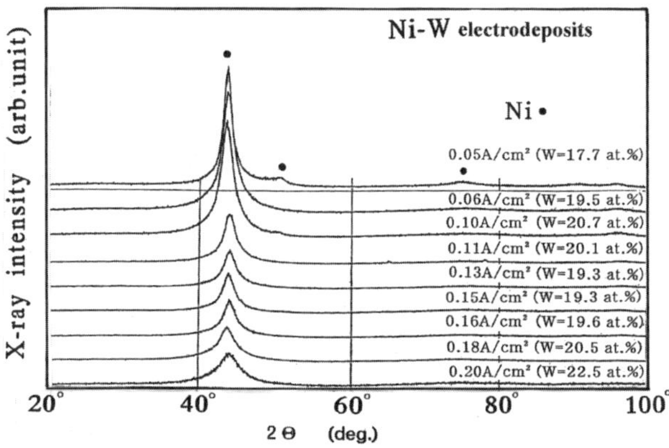

Fig. 19. X-ray diffraction patterns of as-electrodeposited Ni-W alloys for various current densities between 0.05 and 0.2 A/cm^2 at the plating bath temperature of 348 K. All of these Ni-W alloys are ductile; they can be bent through an angle of 180° without breaking.

Average grain sizes and Vickers microhardnesses of the as-electrodeposited Ni-W alloys for various current densities between 0.05 and 0.2 A/cm^2 are collected in Table 3. Average grain sizes in the Ni-W electrodeposits were obtained by applying the Scherrer formula to the diffraction lines of *fcc*-Ni (111) and the broad maximum of the amorphous phase. With increasing applied current density, the average grain size decreased and the Vickers microhardness increased continuously. On annealing these materials at 723 K for 24 h, grain growth occurred to 8.2 ~ 9.5 nm, and the hardness was largely increased to more than HV = 900.

Table 3. Average grain sizes and Vickers microhardness of as-deposited Ni-W alloys. Bath temperature: 348 K (723 K for annealing 24 h in vacuum).

Applied Current Density (A/cm2)	W-content at. pct	Grain Size, As-deposited nm	Vickers Hardness HV	Grain Size, 723 K-24 h nm	HV 723 K-24 h
0.05	17.7	6.8	558 Ductile	9.5	919 Brittle
0.10	20.7	4.7	635 Ductile	9.0	962 Brittle
0.15	19.3	4.7	678 Ductile	8.9	992 Brittle
0.20	22.5	2.5	685 Ductile	8.2	997 Brittle

3.3.2. *Co-deposited hydrogen and oxygen in the Ni-W electrodeposits*

As shown in Table 2, the as-electrodeposited Ni-W alloy at the plating bath temperature of 348 K was ductile. At the other plating bath temperatures, the electrodeposits were severely brittle. This temperature dependence of the ductility may be due to the effects of inclusion of impurities in the Ni-W electrodeposits such as co-deposited hydrogen and oxygen during the deposition process. Figure 20 shows the bath temperature dependence of amounts of the codeposited hydrogen and oxygen in the Ni-W electrodeposits with various concentrations of the trisodium citrate (Na_3(Cit.)) in the plating bath solution. In the case of the Na_3(Cit.) of 0.5 mol/L, amounts of the codeposited hydrogen and oxygen received their minimum values of 0.0006 at. % and 0.028 at. %, respectively, at the plating bath temperature of 348 K, and this temperature decreased with decreasing the Na_3(Cit.) concentration in the plating bath solution. Especially, highly ductile amorphous and nanocrystalline Ni-W alloys have been obtained at the plating bath temperatures between 323 and 348 K. As shown in Table 4, some of the

Ni-W electrodeposits having high hardness can be bent through an angle of 180° without breaking ($\varepsilon_f = 1.0$).

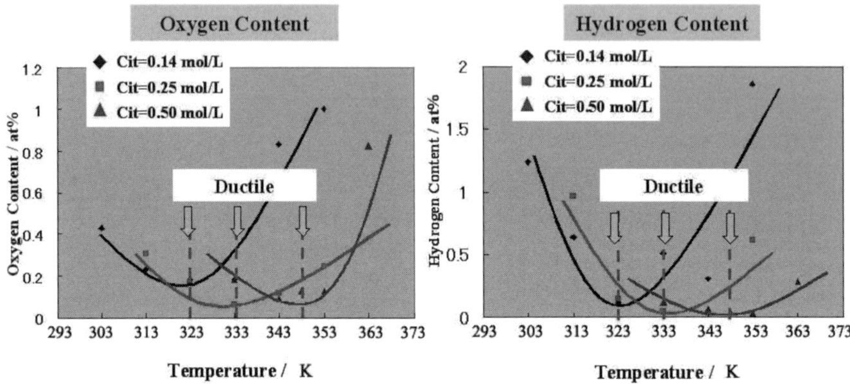

Fig. 20. Bath temperature dependence of amounts of the codeposited hydrogen and oxygen in the Ni-W electrodeposits with various concentrations of the trisodium citrate (Na3(Cit.)) in the plating bath solution.

Table 4. Plating conditions and various properties of the electrodeposited Ni-W alloys.

Composition (at. %)	Bath Temp. (K)	Na3(Cit.) mol/L	Current Density (A/cm²)	Hardness (HV)	Fracture Strain (ef)	Structure by X-ray Diff.
Ni - 25.0 % W	363	0.5	0.2	770	0.0002	Amorphous
Ni - 22.5 % W	348	0.5	0.2	685	1.0 Ductile	Amorphous
Ni - 20.7 % W	348	0.5	0.1	635	1.0 Ductile	3.0 nm
Ni – 17.7 % W	348	0.5	0.05	558	1.0 Ductile	6.8 nm
Ni - 12.3 % W	323	0.14	0.05	696	1.0 Ductile	5.2 nm
Ni - 10.6 % W	313	0.14	0.05	702	0.4	10.6 nm
Ni - 9.7 % W	303	0.14	0.05	695	0.02	5.4 nm + Oxide

3.4. Structural analysis of the Ni-W electrodeposits

3.4.1. *TEM and EXAFS analysis*

Figures 21 and 22 show low- and high-magnification TEM images and the corresponding selected area diffraction patterns of as-electrodeposited Ni-12.3 at. % W and Ni-20.7 at. % W alloys having both high hardness and high ductility, respectively. In the Ni-12.3 at. % W alloy shown in Fig. 21, grain size of about 3~5 nm are observed. The selected area diffraction patterns consist of the *fcc*-Ni-like Debye-rings.

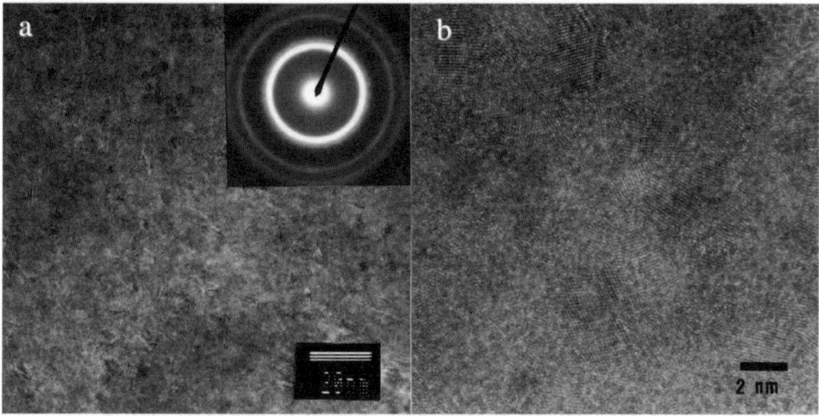

Fig. 21. TEM micrographs showing between about 3~5 nm particles in diameter and the selected area diffraction pattern in the as-electrodeposited Ni-12.3 at. % W alloy: a) low magnification, b) high magnification.

In the case of the Ni-20.7 at. % W alloy shown in Fig. 22, however, nanocrystalline grains with grain sizes between 2.5 and 3.5 nm which are roughly spherical and do not have clear interface in the intercrystalline regions are observed.

In the intercrystalline regions that are about 1 to 2 nm in widths, distorted lattice images (i.e., the amorphous-like images) are observed. The selected area diffraction pattern consists of well-broadened amorphous-like Debye-rings.

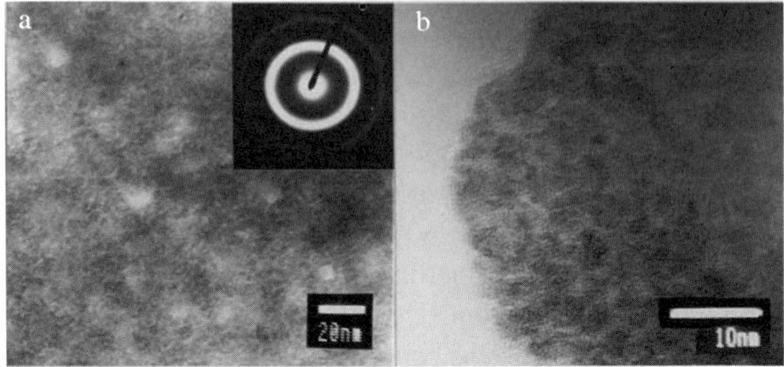

Fig. 22. TEM micrographs showing between about 3 nm particles in diameter and the selected area diffraction pattern in the as-electrodeposited Ni-20.7 at. % W alloy: a) low magnification, b) high magnification.

Nasu and his collaborators[25] have analyzed the local structure of the Ni-16.0 at. % W and the Ni-20.7 at. W electrodeposited Ni-W alloys having high hardness and high ductility by using the EXAFS measurements that were made at beam line 12C (2.5 GeV storage ring; maximum ring current of 200 mA of the Photon Factory in KEK, Japan) at 20K. The energy scanning was made near the Ni-K-edge and the W L_{III}-edge using a Si(111) monochromater in transmission mode. Figure 23 shows the radial distribution functions (RDFs) around Ni atoms for the electrodeposited Ni-W alloys obtained by Fourier transform of the

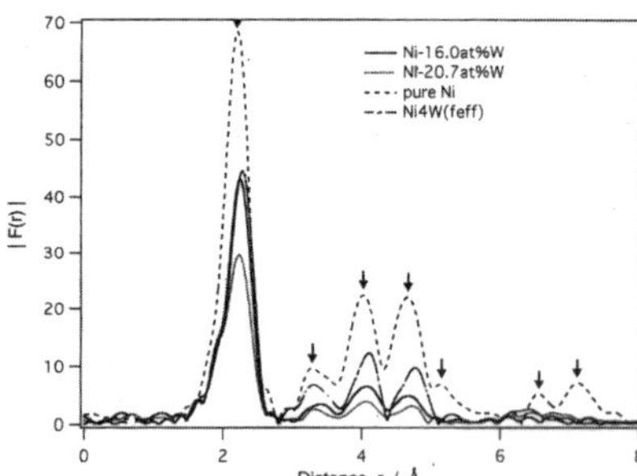

Fig. 23. Radial distribution functions (RDFs) around Ni atoms for the electrodeposited Ni-W alloys obtained by Fourier transform of the EXAFS spectra (0.36~1.75 nm-1 for the Ni K-edge).

EXAFS spectra (0.36~1.75 nm^{-1} for the Ni K-edge). The arrows indicate the locations of first, second, third and fourth nearest neighbors around the Ni atoms in crystalline nickel.

The RDFs for the electrodeposited Ni-W alloys were very similar to those for crystalline Ni and Ni$_4$W. The peak positions in the RDFs derived from the measured EXAFS spectra were closer to those of the Ni$_4$W crystal than to those for the *fcc*-Ni. The Ni$_4$W crystal structure is a combination of the structures of body-centered-tetrahedral tungsten (a=0.573 nm, c=0.355 nm) and face-centered-tetrahedral nickel (a=0.362 nm, c=0.355 nm) as schematically explained in Fig. 24.[26]

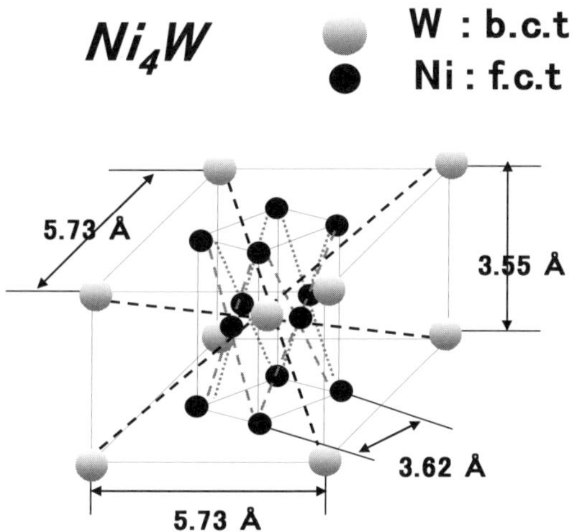

Fig. 24. Schematic diagram of the Ni$_4$W crystal structure which is a combination of the structures of body-centered-tetrahedral tungsten (a=0.573 nm, c=0.355 nm) and face-centered-tetrahedral nickel (a=0.362 nm, c=0.355 nm).

This is the reason why the RDFs around the Ni atom are very similar to that for the crystalline Ni and Ni$_4$W. The intensity of all peaks in the RDFs of the Ni-W alloys decreased, and the deviations of the distances of atomic pairs increased with increasing tungsten concentration. However, the alloy still had medium range order and maintained a crystal structure for a tungsten content of 20.7 at. %.

Figure 25 shows the RDFs around the W atom for the electrodeposited Ni-W alloys. There is no peak in the RDFs that corresponds to medium range order around the W atom, indicating considerable deviations in the atomic distances around the W atom.

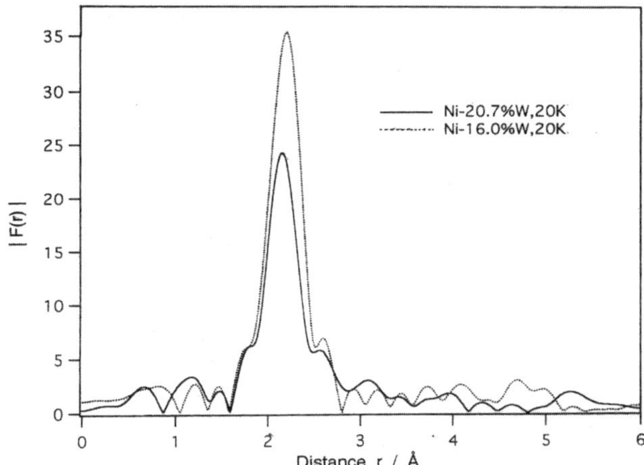

Fig. 25. RDFs around the W atom for the electrodeposited Ni-W alloys obtained by Fourier transform of the EXAFS spectra (0.36~1.60 nm-1 for the W LIII-edge).

3.5. *Mechanical properties of nanocrystalline Ni-W electrodeposits*

3.5.1. *Tensile and hardness test*

Figure 26 shows stress-strain curves obtained by tensile tests for the electrodeposited Ni-20.7 at. % W and Ni-12.3 at. % W alloys having grain sizes of about 3~5 nm. In order to clarify the effect of inclusion of codeposited hydrogen during deposition process, the curves for the degassed specimens at 353 K for 24 h in vacuum are also shown. The result varied considerably from specimen to specimen, being very sensitive to the residual hydrogen. In the case of the as-electrodeposited Ni-20.7 at. % W alloys, the tensile stress and the elongation at fracture were attained to 670 MPa and about ~0 %, respectively. After degassing at 353 K, the tensile stress and the elongation at fracture increased largely to a maximum value of 2333 MPa and about 0.5 %, respectively. In the case of the as-deposited Ni-12.3 at. % W alloy, the tensile strength and the elongation at fracture were attained to 1416 MPa and about 0.2 %, respectively. After degassing at 353 K, the tensile strength somewhat decreased to 800~1000 MPa.

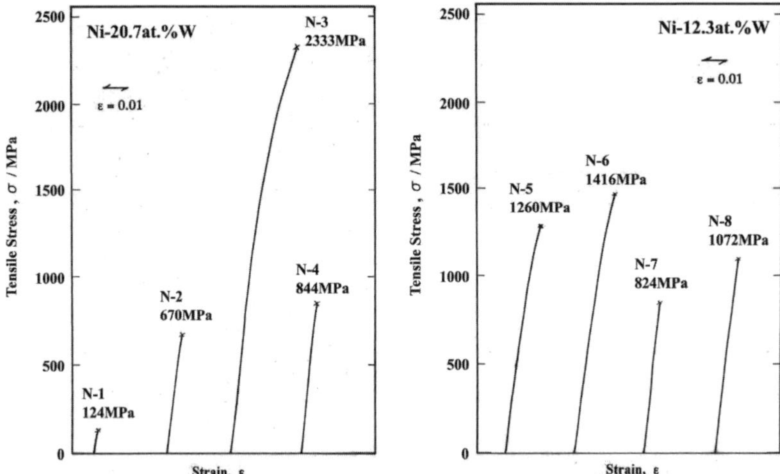

Fig. 26. Stress-strain curves obtained by tensile tests for Ni-20.7 at. % W and Ni- 12.3 at. % W alloys electrodeposited at the plating bath temperature of 348 K and 323 K, respectively (N-1) and (N-2): as-electrodeposited Ni-20.7 at. % W specimens, (N-3) & (N-4): degassed Ni-20.7 at. % W specimens at 80 °C for 24 h, (N-5) and (N-6): as-electrodeposited Ni-12.3 at. % W specimens, (N-7) and (N-8): degassed Ni-12.3 at. % W specimens at 80 °C for 24 h.

Figure 27 shows SEM images of fractured surfaces of tensile specimens of the electrodeposited Ni-20.7 at. % W and Ni-12.3 at. % W alloys after degassing at 353 K for 24 h. Their fracture surfaces are fairly typical of ductile amorphous alloys. Typical river or vein patterns on the fracture surfaces are observed and demonstrate local plastic deformation.

Figure 28 shows an example of micro-electroforming of the high-strength nanocrystalline Ni-20.7 at. % W alloy by using the UV-lithographic technologies. As shown in this figure, high precision forming in micrometer-size range can be possible during the Ni-W electrodeposittion, indicating this alloy system is particularly important for applications to microfabrication technologies.

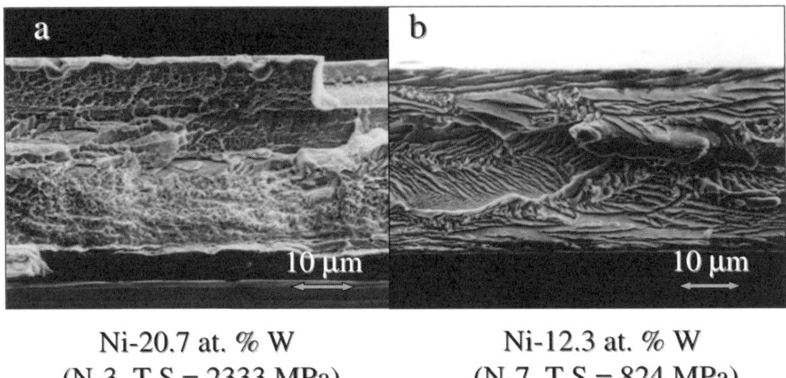

Ni-20.7 at. % W
(N-3, T.S.= 2333 MPa)

Ni-12.3 at. % W
(N-7, T.S.= 824 MPa)

Fig. 27. SEM images of fractured surfaces of tensile specimens of the degassed Ni-W electrodeposited alloys.

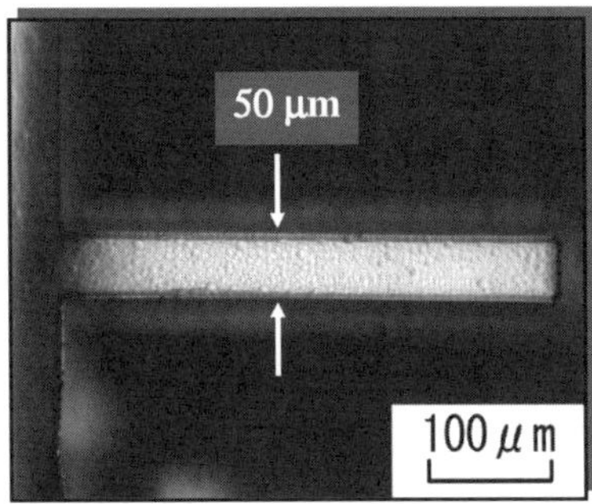

Fig. 28. Optical micrograph of an example of micro-electroforming of the high-strngth nanocrystalline Ni-20.7 at. % W alloys (Thickness : 50 mm).

3.5.2. *Critical grain size for high-strength nanocrystalline alloys*[10]

As shown in Table 4, nanocrystalline Ni-W alloys having grain sizes from the essentially amorphous to about 7 nm in diameter have exhibited the high hardness combined with the good ductility. Especially, the Ni-20.7

at. % W alloy having grain size of about 3 nm exhibited the high tensile strength of about 2300 MPa shown in Fig. 26. However, on annealing these materials, grain growth occurred, and they lost their ductility. Figure 29 shows the TEM micrographs and the corresponding selected area diffraction patterns of the Ni-20.7 at. % W and the Ni-12.3 at. % W alloys after annealing at 723 K for 2 h in vacuum. Grain growth occurred to about 10 nm in diameter for both Ni-W alloys, and they lost their ductility.

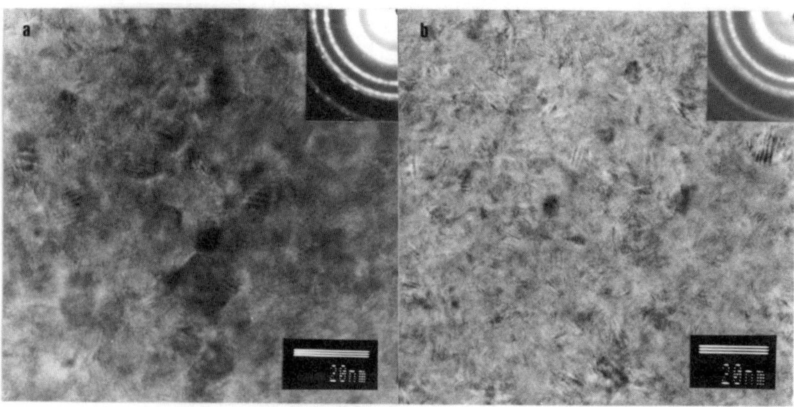

Fig. 29. TEM micrographs and the corresponding selected area diffraction patterns of the Ni-20.7 at. % W alloy a), and the Ni-12.3 at. % W alloy b) after annealing at 723 K for 2 h in vacuum.

In order to obtain the samples with various grain sizes, the samples thus prepared were annealed at various temperatures in vacuum of about 10^{-3} Pa. Figure 30 shows the fracture strain, ε_f, and the Vickers microhardness as a function of the grain size for the electrodeposited Ni-20.7 at. % W and Ni-12.3 at. % W alloys. Average grain sizes in the Ni-W electrodeposits were obtained by applying the Scherrer formula to the diffraction lines of *fcc*-like (111) peaks and also by direct observations from TEM micrographs. In both Ni-W alloys, when the grain size increased to about 9 nm, however, they exhibited the severe brittleness and the hardness attained to maximum values of about HV950 and HV850, respectively.

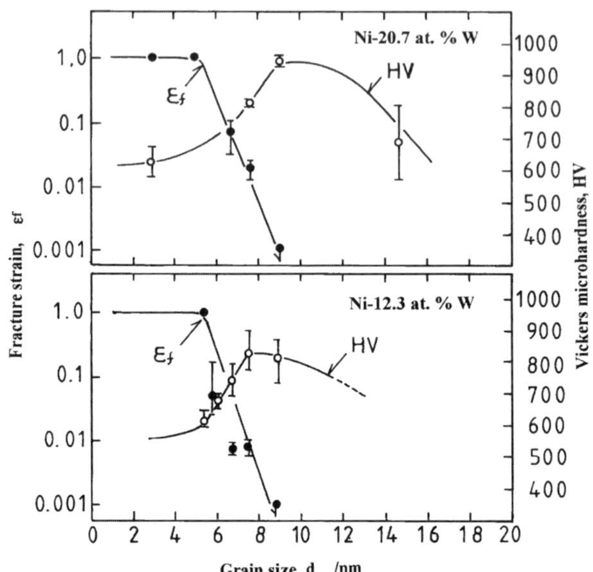

Fig. 30. The fracture strain, ef, obtained by simple bend tests and the Vickers microhardness as a function of grain size for the Ni-20.7 at. % W and Ni-12.3 at. % W alloys.

To estimate the critical grain size for keeping the good ductility in the nanocrystalline alloys, grain boundary sliding is considered to occur along suitable intercrystalline planes. Figure 31 shows schematic explanations of the modeling of deformation by grain boundary sliding with nanoscaled spherical grains being represented as a densely packed array of *fcc* crystal-type structure. As sliding occurs, close-packed planes of the nanoscaled grains do this easier than planes aligned in another direction. Thickness of these sliding planes (Δ) is derived from the following equation, as schematically explained in Fig. 32.

$$\Delta = (\sqrt{3}/2 - 1)d + D \qquad (2)$$

where d and D are the grain diameter and the grain boundary thickness, respectively. In this case, the grain diameter (d) is defined as the total length of the internal diameter of the grains and the grain boundary thickness. As shown in Fig. 31, thickness of the sliding planes (Δ)

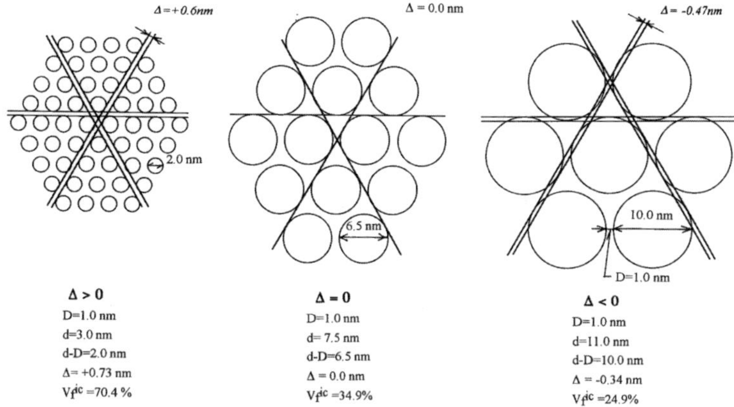

Fig. 31. Schematic explanations of the modeling of deformation by grain boundary sliding, with nanoscaled spherical hard grains being represented as a densely packed array of *fcc* crystal-type structure. As sliding occurs, close-packed planes of the nanoscaled grains do this easily than planes aligned in another direction.

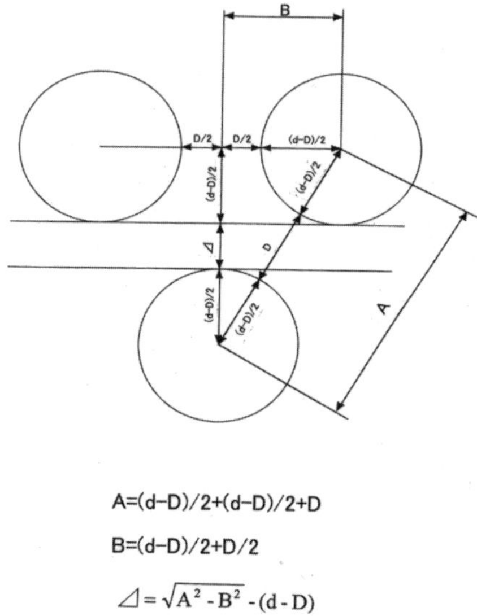

$$A=(d-D)/2+(d-D)/2+D$$

$$B=(d-D)/2+D/2$$

$$\triangle = \sqrt{A^2 - B^2} - (d-D)$$

Fig. 32. Schematic explanations for the thickness of sliding planes (D). This is derived from Eq. 2 where d and D are the grain diameter and the grain boundary thickness, respectively.

decreases with increasing the grain size (d) under the condition where the grain boundary thickness (D) is constant, and then Δ becomes zero and finally becomes negative. When the Δ is negative, grain boundary sliding becomes difficult and these materials should lose their ductility. We have defined the critical grain size (d_c) to keep the good ductility in nanocrystalline materials as the grain size under the condition where the Δ is zero.

Figure 33 shows the calculated Δ and intercrystalline volume fractions (V_f^{ic}) as a function of the internal diameter of the grains (d - D) where the D is changed over a range from 0.5 to 2.0 nm. The intercrystalline volume fraction is derived from the following equation,

$$V_f^{ic} = 1 - [\,(d - D)\,/\,d\,]^3. \tag{3}$$

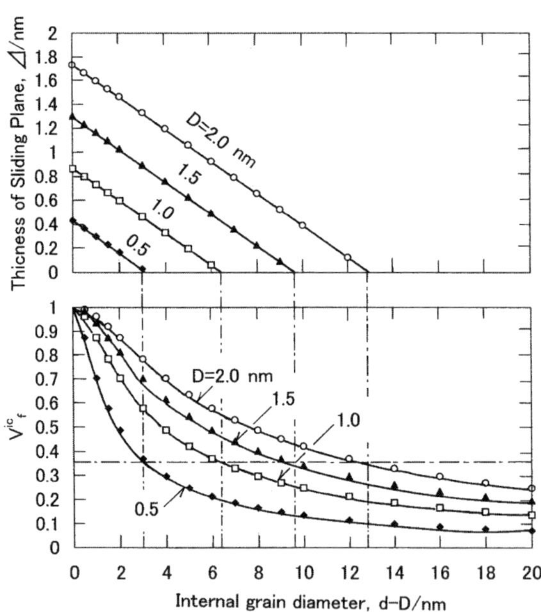

Fig. 33. Calculated thicknesses of the planes for the grain boundary sliding (D) and intercrystalline volume fractions (Vfic) as a function of the internal diameter of the grains (d - D) where the grain boundary thickness (D) is changed over a range from 0.5 to 2.0 nm.

As shown in this figure, the Δ decreases linearly with increasing the internal diameter of the grains (d - D). For example, when the D is taken as 1.5 nm, the (d - D) is about 9.7 nm under the condition $\Delta = 0$. Assuming the grain size obtained by the X-ray diffraction is equal to the internal grain diameter (d - D), the critical grain size (d_C) can be estimated to be about 10 ~ 11 nm in diameter for the grain boundary thickness of 1 to 2 nm as observed from the TEM micrographs. This data may be in good agreement with the critical value of d_C predicted from Eq. (2) for the grain boundary thickness of 1.5 nm.

In addition, suitable volume fraction of the crystallites to keep the high-strength in the nanocrystalline materials should be depending on the deformability of the intercrystalline regions. Under the condition of $\Delta = 0$, the values of V_f^{ic} attained to the same values of about 35 % where the D is changed over a range from 0.5 to 2.0 nm.

3.5.3. *Hall-Petch relationship in nanocrystalline Ni-W alloys*[26]

Figure 34 shows Vickers microhardness vs. annealing temperature for Ni-25.0 at. % W alloy after annealing at various temperatures. For comparison, same data for Ni-20 at. % P[27] and pure Ni[28] electrodeposits are also shown. For this temperature range, the hardness of the Ni-W alloy was sufficiently high as compared with those of the others.

The as-deposited Ni-25.0 at. % W alloy with an amorphous structure exhibited high hardness of about HV=770. On annealing this material at 673 K for 24 h in air or vacuum, the amorphous structure was maintained and the hardness was largely increased to about HV=1100. At a higher annealing temperature of 873 K for 24 h in vacuum, the.nanocrystalline structure with a grain size of about 12 nm was observed by X-ray diffraction and the hardness was further increased to a maximum value of Hv = 1450. On annealing at 973 K and above, the hardness was markedly decreased with increasing grain size.

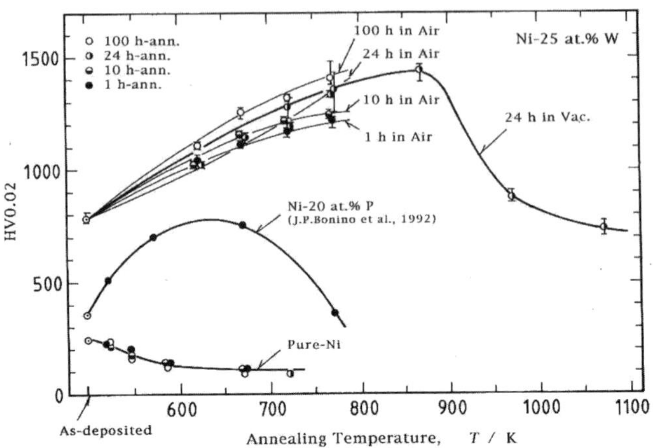

Fig. 34. Vickers microhardness vs. annealing temperature curves of Ni-25.0 at. % W, Ni-20 at. % P and pure-Ni electrodeposits for various annealing times.

Figure 35 shows the Vickers microhardness vs. $d^{-0.5}$ (d is mean grain dia.) relationship for the Ni-W alloy after annealing at various temperatures in air or vacuum. For comparison, the relationships for electrodeposited Ni[29] and conventional pure Ni extrapolated from the relationship at larger grain sizes[30] are also shown.

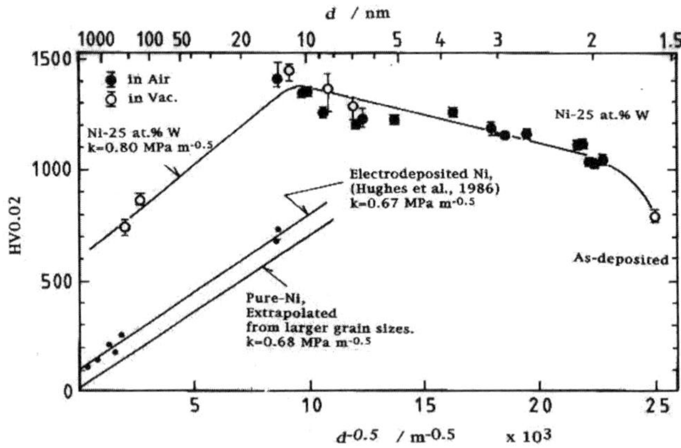

Fig. 35. Vickers microhardness as a function of d-0.5 (d is mean grain dia.) relationship for the Ni-25.0 at. % W alloy after annealing at various temperatures in air or vacuum.

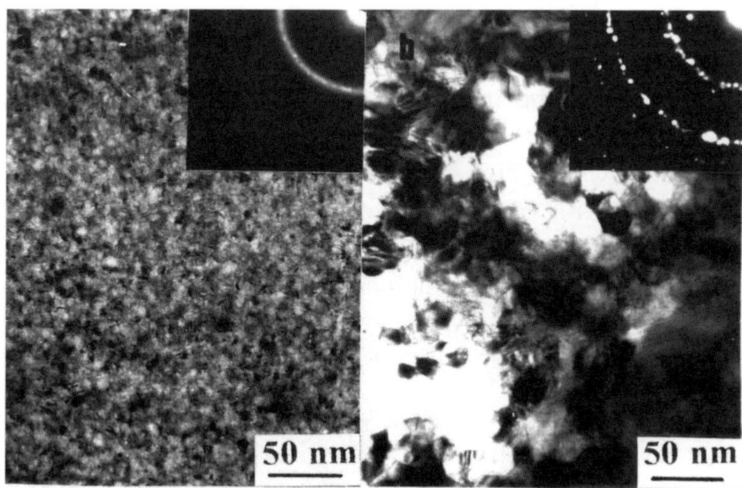

Fig. 36. TEM images and the selected area diffraction patterns of the Ni-25 at. % W alloys annealed at 723 K a), and 873 K b) for 24 h in vacuum.

In the Ni-W alloy, the hardness increases with decreasing the grain size to about 10 nm. The Hall-Petch slope of 0.8 MPa m$^{1/2}$ in the Ni-25 at. % W alloy is slightly higher, but comparable to those of the electrodeposited Ni and the conventional one. When the grain size is less than about 10 nm, the hardness decreases with decreasing grain size. Our recent results have also shown that the nanocrystalline pure-nickel produced by electrodeposition has exhibited the Hall-Petch strengthening to grain sizes near about 14 nm, and at a finer grain size of 12 nm, the hardness of nanocrystalline Ni decreases in an apparent breakdown of the Hall-Petch relationship.[31] So, such an apparent breakdown of the Hall-Petch relationship may be a general one when the grain size is less than about 10 nm.

Figure 36 shows the TEM images and the selected area diffraction patterns of the Ni-25.0 at. % W alloy annealed at 723 K and 873 K for 24 h in vacuum. On annealing at 723 K shown in Fig. 36 (a), nanocrystalline structure having grain sizes between 5 and 8 nm is observed. Noticeably, image contrast of the interface of individual grains is not clearly visible. The selected area diffraction pattern reveals the *fcc*-like lattice Debye rings indicating that the ultrafine grains become randomly oriented. On annealing at 873 K shown in Fig. 36 (b), grain size increased to about 15 nm and above, and the image contrast of the interface of individual grains is clearly visible.

Figure 37 shows the high resolution lattice image of the Ni-W alloy annealed at 723 K for 24 h in vacuum. Individual crystalline grains well randomly oriented with sizes between 5 and 8 nm can be observed. This image was obtained with a defocusing value of - 48 nm. In the intercrystalline regions that are about 1 to 2 nm in widths, distorted lattice images are observed.

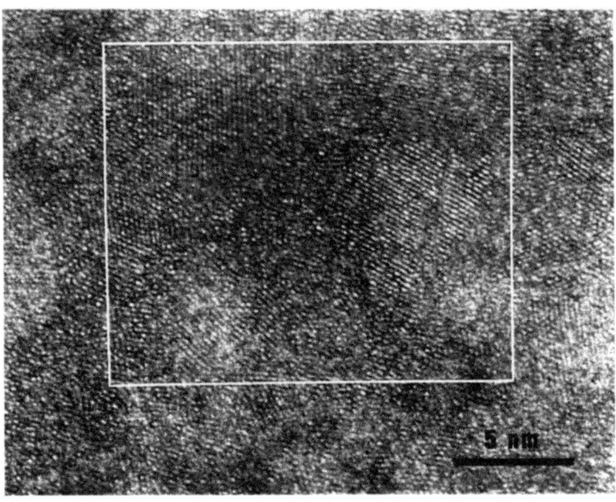

Fig. 37. High resolution TEM-lattice image of the Ni-W alloy annealed at 723 K for 24 h in vacuum.

As shown in Fig. 38, the lattice image tracing manually done of the framed part in Fig. 36 reveals the distorted lattice images in the intercrystalline regions having components of grain boundary and triple junction (i.e., intersection line of three or more adjoining grains). Especially, highly distorted lattice image is observed in the region of triple junction. When the defocusing value was changed from - 48 nm to + 16 nm, the straight line lattice image could not be observed in these intercrystalline regions. Similar features of HR-TEM observations have been reported in Ti-Mo alloys prepared by mechanical alloying and Ni_3Al by magnetron sputtering with grain sizes of less than 10 nm.[32, 33]

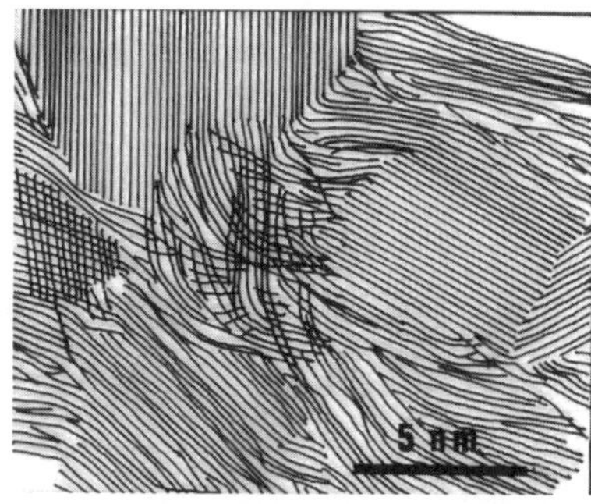

Fig. 38. The lattice image tracing manually done of the framed part in Fig. 3

3.5.4. *Hardness of nanocrystalline Ni-W alloys*

It is well known that the grain boundary structure of conventional coarse grained metals and alloys is well represented by the coincidence site lattice model (CSL-model) having periodic tilt boundary structures indexed by Σ values.[34] In the case of the Ni-W alloys shown in Fig. 35, the Hall-Petch strengthening was observed for the hardness extending to a finest grain size of about 10 nm. Therefore, when the grain size is above 10 nm, the grain boundary structures may be essentially the same as those of the coarse grained materials.

When the grain size is less than about 10 nm, decrease of the hardness was observed. Ovid'ko has proposed that quasi-periodic tilt grain boundaries having a structure disordered in a gas-like manner exist in nanocrystalline materials.[35] As shown in Figs. 37 and 38, grain boundary thickness of about 1 to 2 nm evaluated from the HR-TEM observation may be considerably thicker than that of the coarse grained materials (almost less than about 1 nm). As a result, the volume fraction of the grain boundary triple junction having the highly distorted lattice image should increase remarkably with decreasing the grain size as schematically explained in Fig. 39. Palumbo *et al.* have proposed to evaluate grain size dependence of volume fractions associated with both the grain boundary and the triple junction.[36, 37] The total intercrystalline fraction (V_f^{ic}) and the grain boundary fraction (V_f^{gb}) are derived from the following equations,

$$V_f^{ic} = 1 - [(d - D)/d]^3 \qquad (4)$$

$$V_f^{gb} = [3D(d - D)^2]/d^3 \qquad (5)$$

where d and D are the grain diameter and the grain boundary thickness, respectively. The fraction associated with the triple junction (V_f^{tj}) is given by,

$$V_f^{tj} = V_f^{ic} - V_f^{gb} \qquad (6)$$

Figure 40 shows the calculated volume fractions for V_f^{ic}, V_f^{gb} and V_f^{tj} as a function of $d^{-0.5}$ where D is changed over a range from 0.5 to 2.0 nm. The V_f^{gb} in Fig. 38 (b) increases to maximum values of about 45 % and then decreases with decreasing the grain size, while the V_f^{tj} in Fig. 39 (c) continues to increase. When the grain boundary thickness is 1.5 nm and above, the marked increase of the V_f^{tj} is observed in the range of grain sizes of less than about 10 nm.

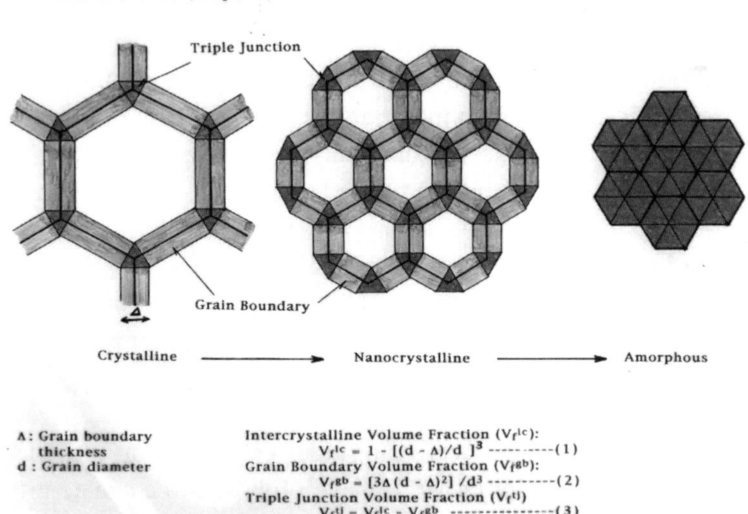

Fig. 39. Schematic explanations for grain size dependence of intercrystalline volume fraction associated with both the grain boundary and the triple junction. From Ref. 36.

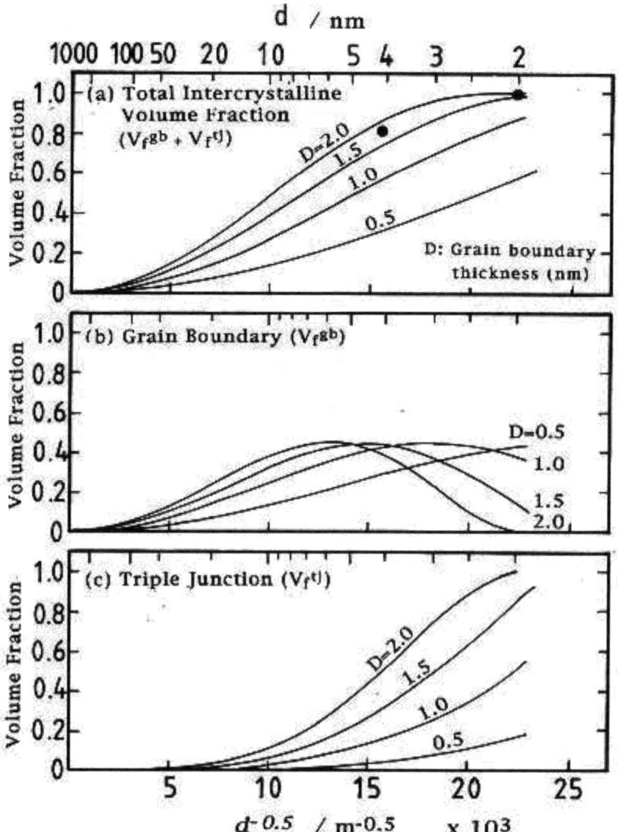

Fig. 40. Calculated volume fractions for Vfic, Vfgb and Vftj as a function of d-0.5 where D is changed over a range from 0.5 to 2.0 nm.

Figure 41 shows the Vickers microhardness vs. $d^{-0.5}$ relationship for the Ni- 25.0 at. % W alloy. In order to explain the decrease in hardness where the grain size is less than about 10 nm, we give a rough estimate of the grain size dependence of the hardness (H_{cal}) by following equation,

$$Hcal = Vf^{tj} Htj + (1 - Vf^{tj}) HHP \tag{7}$$

where V_f^{tj} and H_{tj} are the volume fraction and the hardness of the triple junction, respectively, and H_{HP} is a hardness obtained by the Hall-Petch relationship for the Ni-W alloy. On the basis of the calculated volume fractions shown in Fig. 40, we assume that the hardness of the triple junction (H_{tj}) is equal to that of the as-electrodeposited amorphous Ni-W alloy (Hv = 770). The H_{cal} vs. $d^{-0.5}$ relationships where the grain boundary thickness is changed in a range of between 0.5 and 2.0 nm are also shown in this figure. The decreasing tendency of the hardness under conditions where D = 1.5 and 2.0 nm can be well expressed by Eq. (6). Therefore, the significant increase of the triple junction fraction may be responsible for decreasing the hardness.

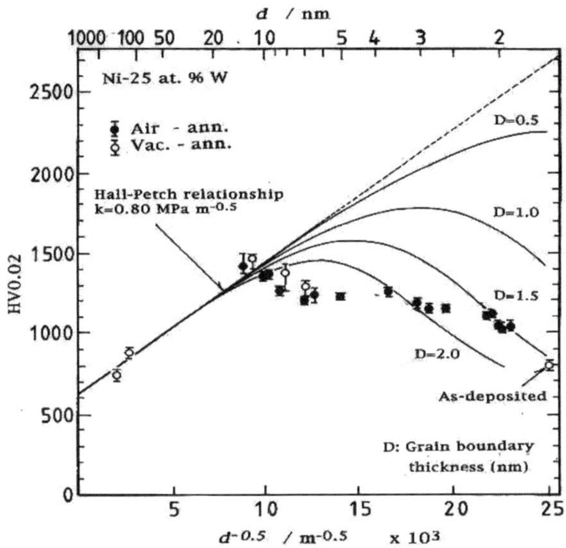

Fig. 41. Vickers microhardness vs. d-0.5 relationship for the Ni- 25.0 at. % W alloy. The Hcal vs. d-0.5 relationships where the grain boundary thickness is changed in a range between 0.5 and 2.0 nm are also shown.

As mentioned in above, when the grain size is less than about 10 nm, the grain boundary thickness evaluated from the HR-TEM observation may be considerably thicker than that of the coarse grained materials. Fujita has speculated atomic structure of ultrafine crystallites in a nanometer scale by using a volume free energy difference and a surface energy of an atom cluster.[38] Interface region of the nanocrystallites having a structure of non-periodic atomic array gradually expands into the center region when the size of crystallites decreases below a critical one. Swygenhoven *et al.*[39] have also calculated the influence of grain size of the mechanical properties of nanostructured Ni with the grain sizes between 3 and 10 nm by a molecular dynamics computer simulation, and have proposed terms of grain boundary viscosity giving rise to a viscoelastic behavior with the grain sizes below 5 nm.

These results support that the grain boundary thickness may increase with decreasing the grain size, and the hardness of the expanded grain boundary region may be lower than that of the inner grain region. Finally, the nanocrystalline structure can be transformed into the amorphous state under condition where the grain boundary thickness is equal to the grain size.

4. Conclusion

In order to clarify the mechanical properties of nanocrystalline alloys with the grain size range of a few to several tens nanometers, we have prepared the nanocrystalline materials with sufficiently dense and structurally well characterized materials as the following two methods; (1) dynamic compaction of mechanically alloyed Fe-TiN and Ni-TiN nanocrystalline powders by using a propellant gun, and (2) electrodeposition of nanocrystalline Ni-W alloys. Especially, electrodeposition is a superior technique for producing nanocrystalline materials and this is particularly important for applications to microfabrication technologies. Both of the dynamically consolidated (Fe, Ni)-TiN materials and the electrodeposited Ni-W materials have exhibited the Hall-Petch strengthening to grain sizes near about 10 nm, and at a finer grain size of about 10 nm, the hardness of these nanocrystalline materials decreases in an apparent breakdown of the Hall-Petch relationship. This decrease in hardness may be due to the significant increase of the intercrystalline volume fraction, especially the fraction associated with the triple junction of the grain boundaries.

The tensile strength and the ductility of nanocrystalline Ni-W electrodeposits may be strongly affected by the inclusion of impurities such as hydrogen and oxygen during the deposition process, and also by the grain size of the electrodeposits. Especially, the high-strength nanocrystalline Ni-20.7 at. % W alloy with average grain size of about 3 nm has been obtained; the tensile strength was attained up to 2333 MPa combined with good ductility. On annealing these materials at various temperatures, however, grain growth occurred and they lost their ductility. In order to estimate the critical grain size for keeping the good ductility in the nanocrystalline alloys, grain boundary sliding was considered to occur along the close-packed planes of the nanocrystalline grains under the condition where the nanoscaled spherical hard grains is represented as a densely packed array.

Some additional recent results concerning the mechanical properties of nanocrystalline Ni-W alloys are shown in references 40 to 44.

Acknowledgments

The author is deeply grateful to Prof. Y. Ogino of Himeji Institute of Technology, Japan for useful discussions and gratefully acknowledges financial supports of the Grand-in-Aid from the Minstry of Education, Culture, Sports, Science and Technology, Japan.

References

1. J. Karch, R. Birringer and H. Gleiter, *Nature*, **330,** 556 (1987).
2. C. C. Koch, D. G. Morris, K. Lu, A. Inoue, *Mater. Res. Soc. Bull.* **24,** 54 (1999).
3. D. G. Morris, *Mechanical Properties of Nanostructured Materials,* Materials Science Fundations, **Vol. 2** (Trans Tech Publications, Zurich, 1998) p. 61.
4. Y. Ogino, M. Miki, T. Yamasaki and T. Inuma, *Mater. Sci. Forum,* **88-90,** 795 (1992).
5. T. Yamasaki, Y. Ogino, K. Fukuoka, T. Atou and Y. Syono, *Materials Science Forum,* **179-181,** 647 (1995).
6. Y. Ogino, T. Yamasaki and Bao-Long Shen, *Metallurgical and Materials Transactions B,* **28B,** 299 (1997).

7. T. Yamasaki, Y. J. Zheng, Y. Ogino, M. Terasawa, T. Mitamura and T. Fukami, *Mater. Sci. & Eng. A, , **350**, 168 (2003).

8. A. M. El-Sherik and U. Erb, *Plating & Surface Finishing,* **82,** 85 (1995).

9. T. Yamasaki, R. Tomohira, Y. Ogino, P. Schlossmacher and K. Ehrlich, *Plating & Surface Finishing,* **87,** 148 (2000).

10. T. Yamasaki, *Scripta Materialia,* **44,** 1497 (2001).

11. A. Lasalmonic and J. L. Strudel, *J. Mater. Sci.,* **21,** 1837 (1986).

12. J. S. C. Jang and C. C. Koch, *Scripta Metall. Mater.,* **24,** 1599 (1990).

13. G. H. Hughes, S. D. Smith, C. S. Pande, H. R. Johnson and R. W. Armstrong, *Scripta Metall.,* **20,** 93 (1986).

14. A. M. El-Sherik, U. Erb, G. Palumbo and K. T. Aust, *Scripta Metall. Mater.,* **27,** 1185 (1992).

15. R. W. Siegel and G. E. Fougere, *Nanostructured Mater.,* **6,** 205 (1995).

16. H. Gleiter, *Prog. Mater. Sci.,* **33,** 223 (1989).

17. A. G. Atkins, A. Silverio and D. Tabor, *J. Inst. Met.,* **94,** 369 (1966).

18. R. M. Hooper and C. A. Brookers, *J. Mater. Sci.,* **94,** 4057 (1984).

19. G. Martin and B. Perraillon, *Grain Boundary Structure and Kinetics,* ASM, Metals park, OH, p. 239 (1979).

20. W. Ehrfeld, V. Hessel, H. Loewe, Ch. Schulz and L. Webwer, *Proc. 2nd Int. Conf. Micro Materials 97,* Berlin, 112 (1997).

21. L. E. Vaaler and M. L. Holt, *J. Electrochemical Soc.,* **90,** 43 (1946).

22. L. Domnikov, *Metal Finishing,* **62,** 68 (1964).

23. G. Rauscher, V. Rogoll, M. E. Baumgaertner and Ch. J. Raub, *Trans. Inst. Metal Fin.,* **71,** 95 (1993).

24. N. Atanassov, K. Cencheva and M. Bratoeva, *Plating & Surface Finishing,* **84,** 67 (1997).

25. T. Nasu, M.Sakura, T. Kamiyama, T. Usuki, O. Uemura and T. Yamasaki, *J. Non-crystalline Solids,* **312-314,** 319 (2002).

26. P. Villars and L. D. Calvert, *Pearson's Handbook: Crystallographic data for intermetallic phases,* ASM International;, Materials Park, OH (1991); T. Yamasaki, P. Schlossmacher, K. Ehrlich and Y. Ogino, *NanoStructuresd Materials,* **10,** 375 (1998).

27. J. P. Bonino, P. Pouderoux, C. Rossignol and A. Roussert, *Plating & Surface Finishing,* **79,** 62 (1992).

28. P. Graf, W. Schneider and H. Zimmermann (Forchungszentrum Karlsruhe), *private communication,* (1995).

29. G. D. Hughes, S. D. Smith, C. S. Pande, H. R. Johnson and R. W. Armstrong, *Scripta Metall.,* **20,** 93 (1986).

30. A. Lasalmonie and J. L. Strudel, *J. Mater. Sci.,* **21,** 1837 (1986).

31. C. A. Schul, T. G. Nieh and T. Yamasaki, *Scripta Materialia*, **46,** 735 (2002).

32. W.Y. Lim, E. Sukedai, M. Hida and K. Kaneko, *Material Science Forum,* **88-90** 105 (1992).

33. H.Van Swygenhoven, P. Boni, F. Paschoud, M. Victoria and M. Knauss, *Nanostructured Materials,* **6,** 739 (1995).

34. G. A. Chadwick and D. A. Smith, *"Grain Boundary Structure and Properties"* , (Academic Press, London, 1976).

35. I. A. Ovid'ko, *Nanostructured Materials,* **8,** 149 (1997).

36. G. Palumbo, S.J. Thorpe and K. T. Aust, *Scripta Metall.,* **24,** 1347 (1990).

37. G. Palumbo, U. Erb and K. T. Aust, *Scripta Metall.,* **24,** 2347 (1990).

38. H. Fujita, *Ultramicroscopy,* **39,** 369 (1991).

39. H.Van Swygenhoven and A. Caro, *Abstract of Int. Sympo. on Metastable, Mechanically Alloyed and Nanocrystalline Materials, (ISMANAM-97),* Barcelona,

40. H. Hosokawa, T. Yamasaki, N. Sugamoto, M. Tomizawa, K. Shimojima and M. Mabuchi, *Materials Transactions,* **45,** 1807 (2004).

41. H. Matsuoka, T. Yamasaki, Y. J. Zheng, T. Mitamura, M. Terasawa and T. Fukami, *Materials Sci. & Eng.* **A449-451,** 790, (2007).

42. T. Yamasaki, N. Oda, H. Matsuoka and T. Fukami, *Materials Sci. & Eng.,* **A449-451,** 833 (2007).

43. T. Yamasaki, M. Sonobe and H. Yokoyama, *Proc. of the 9th Int. Conf. On Technology Plasticity,* 1836 (2008).

44. T. Yamasaki, H. Yokoyama and T. Fukami, *Rev. Adv. Mater. Sci.,* **18,** 711 (2008).

CHAPTER 9

DISLOCATIONS AND PLASTICITY IN MINERALS WITH LARGE UNIT CELLS

Patrick Cordier and Philippe Carrez

Unité Matériaux et Transformations (UMET)
Université Lille 1, 59650 Villeneuve d'Ascq, France
Email: patrick.cordier@univ-lille1.fr

Minerals raise the question of the core structure and mobility of dislocations in materials with large unit cells and complex crystal chemistry. Completely unknown for a long time, the structure of dislocation cores in minerals has recently been studied by numerical modelling. The Peierls model and its recent numerical developments are first recalled since most of the studies have been performed within this framework. We discuss then some examples on perovskites, olivine, wadsleyite and garnets which illustrate some cases corresponding to Burgers vectors lengths ranging from 0.4 to 1.15 nm.

1. Introduction

Plastic deformation of solids cannot be understood without the introduction of the concept of crystal defects that carry plastic shear. Volterra[1] was the first to propose the concept of "distorsioni" created by introducing a step-like displacement discontinuity along a cut surface within a linear elastic continuum. The concept of dislocation was applied to plastic deformation of crystals in three independent studies performed by Taylor[2], Orowan[3] and Polanyi[4] in 1934. One of the main characteristics of dislocations is that they create a long-range stress field (originally studied by Love[5] who proposed the name "dislocation"). The centre of the dislocations is source of an elastic singularity which must

317

be artificially suppressed by a cut-off radius r_0. The elastic energy stored in the dislocation-containing crystal is large, precluding any equilibrium concentration of dislocations, even at high temperature. The elastic theory shows that the elastic energy is given by:

$$E_{el} = \frac{\mu b^2}{4\pi C} Ln\left(\frac{R}{r_0}\right) \tag{1}$$

where b is the length of the dislocation Burgers vector, μ is the shear modulus, R and r_0 are the outer and inner cut-off radii for the integration of the strain field, and C is a constant ranging from 1 to $1-v$ (v is the Poisson ratio). The fact that the energy scales with μb^2 per unit length has raised questions concerning the relevance of the concept of dislocations in case of materials with large unit cells. This question has been early examined by Frank[6] who proposed that when the magnitude of the Burgers vector exceeds a critical value, it would be energetically more favourable to remove the highly strained material at the centre of the dislocation and to create an additional free surface in the shape of a hollow core of radius r_h given by Frank's formula:

$$r_h = \frac{\mu b^2}{8\pi^2\gamma} \tag{2}$$

where γ is the surface energy. Srolovitz *et al.*[7] discussed the possible influence of anisotropic surface energies and local stress fields on the shape of a hollow core. Although there has been some report for hollow-core dislocations (also called micropipes, nanopipes, tunnel-like defects, etc) in the case of SiC, [8-11] $YBaCuO$[12] or GaN[13] their relevance in the context of mobile defects and plastic deformation has, to the best of our knowledge, not been established. The occurrence of dislocations in crystals with large unit cells has also been discussed by Nabarro[14] who highlighted the influence of bonding and suggested that the Peierls stress should be considered, and recognized that stacking faults and partial dislocations complicate the analysis.

Minerals represent a class of materials which shows numerous examples of unit cells that are large in 0, 1, 2 or 3 dimensions. From this point of view, some minerals represent situations that are intermediate between simple structures (either metallic or ceramics) and complex metallic alloys or quasicrystals (Fig. 1). In minerals, the elastic energy can be relaxed at the core and the singularity of the Volterra dislocation can be overcome by introducing a standard dislocation core with an increase of the discontinuity over a finite width either planar or non planar.

From a general point of view, this problem has been raised by Orowan and elaborated by Peierls[15] and Nabarro.[16] The formalism of the Peierls-Nabarro (PN) model is first recalled and a brief summary of some resolution methods of the PN equation is presented. Illustration of some dislocation core structures is presented on some phases that are important in the Earth's mantle: perovskite, wadsleyite and forsterite.

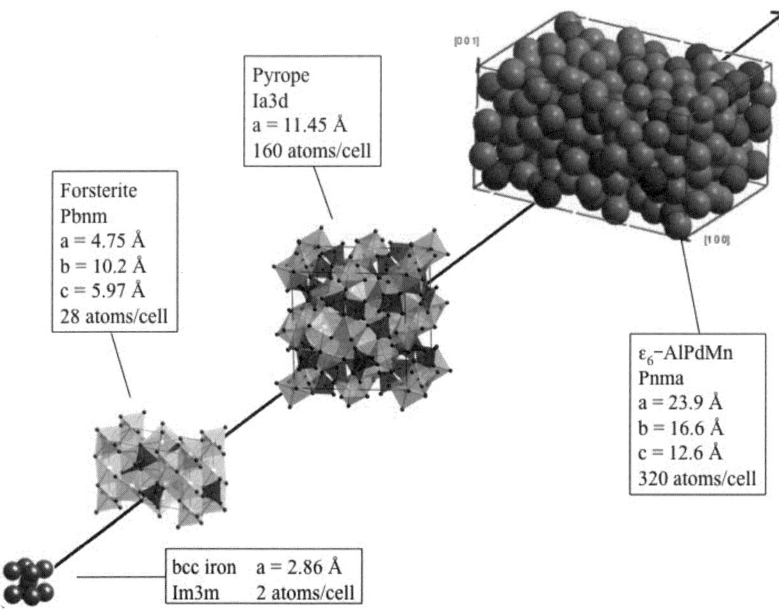

Fig. 1. Comparison of the unit cell dimensions of some minerals (Forsterite is the Mg-end member of the olivine group and Pyrope is a garnet) with simple metallic structures and complex metallic alloys.

2. The Peierls-Nabarro Model of Dislocations

2.1. *The volterra dislocation*

A Volterra dislocation can be introduced in a medium through the following process:

- a cut is made along a surface bounded by a contour *(C)*

- the two parts of the crystal separated by the cut are moved relative to each other by a translation vector of the lattice (further called Burgers vector).

- if necessary, some material can be filled or taken out to preserve the structure of the medium

- the surfaces of the cut are welded together and the applied forces are suppressed

The resulting defect results in a step-like discontinuity across the contour *(C)* (Fig. 2).

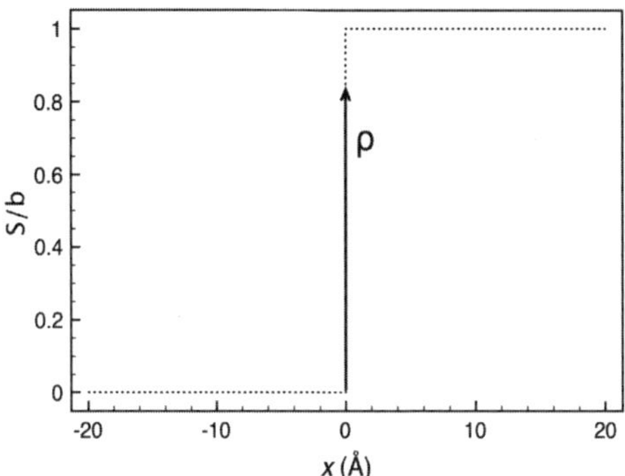

Fig. 2. Volterra dislocation, the shear profile across the dislocation core is a step function.

2.2. *The Peierls-Nabarro (PN) model*

The original PN model[15, 16] represents a useful and efficient approach to calculate the core properties of dislocations[17-19] based on the assumption of a planar core.[20] It has been shown to apply to a wide range of materials.[19, 21-25]

The crystal is divided into three regions (Fig. 3). The PN model assumes that the misfit region of inelastic displacement is restricted to the glide plane (between A and B), whereas linear elasticity applies far from it, above and below (regions a and b). The misfit region of inelastic displacement is characterized by a generalised stacking fault (GSF) energy (also called the γ-surface). A dislocation corresponds to a continuous distribution of shear $S(x)$ along the glide plane (x is the coordinate along the displacement direction of the dislocation in the glide plane) storing a misfit energy $\int \gamma(S(x))dx$.

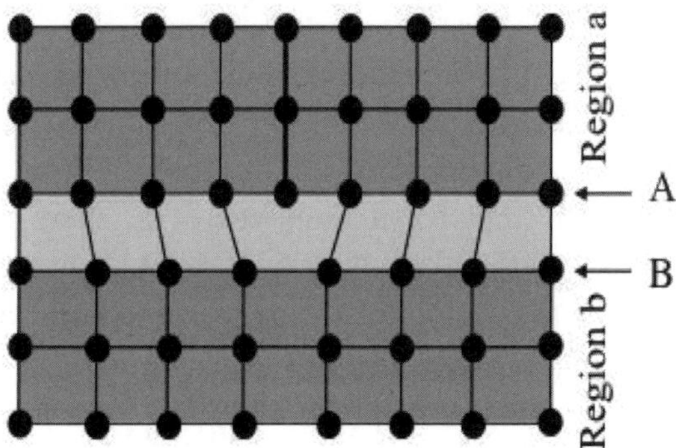

Fig. 3. The Peierls-Nabarro model localizes inelastic displacements in the grey layer between A and B where the core will be allowed to spread. Linear elasticity will apply to regions "a" and "b".

$S(x)$ represents the disregistry across the glide plane and the stress generated by such a displacement can be represented by a continuous distribution of infinitesimal dislocations with density $\rho(x)$ for which the total summation is equal to the Burgers vector b. The expression of the total dislocation energy functional taking into account the misfit energy in region A-B, and the elastic energy in regions (a) and (b) is thus

$$U_{tot} = \int \gamma(S(x))dx - \frac{K}{2} \iint \rho(x)\rho(x')\ln|x - x'|dxdx' \qquad (3)$$

where $\rho(x) = dS(x)/dx$ and K is a combination of elastic constant, reflecting the energy coefficient of the dislocation character. This coefficient can be calculated within the frame of the Stroh theory[26, 27] to take anisotropic elasticity into account.

The equilibrium structure of a dislocation is thus determined by minimizing the above energy functional. A variational derivative of (3) with respect to ρ leads to an integro-differential equation (4) nowadays referred to as the Peierls Nabarro equation.

$$\frac{K}{2\pi} \int_{-\infty}^{+\infty} \frac{1}{x-x'} \left[\frac{dS(x')}{dx'} \right] dx' = \frac{K}{2\pi} \int_{-\infty}^{+\infty} \frac{\rho(x')}{x-x'} dx' = F(S(x)) \qquad (4)$$

Originally the PN equation was introduced considering that the restoring force F acting between atoms on either sides of the interface is balanced by the resultant stress of the distribution. The significance of the GSF energy is that for a fault vector S, there exists an interfacial restoring stress derivating from γ which has the same formal interpretation as the initial restoring force F introduced originally by Peierls.[28]

$$\vec{F}(S) = -\overrightarrow{grad}\,\gamma(S) \qquad (5)$$

2.3. *The analytical solution*

As largely developed by Joos & Duesbery (1997), an analytical solution of the PN equation can be found by introducing a sinusoidal restoring force:

$$F(S(x)) = \tau^{max} \sin(\frac{2\pi S(x)}{b}) \qquad (6)$$

where b is the dislocation Burgers vector and τ^{max} is the ideal shear strength (ISS) which is defined as the "maximum resolved shear stress that an ideal, perfect crystal can suffer without plastically deforming" (Paxton *et al.* 1991). Incorporating this expression in the PN equation leads to a classic solution for the disregistry function:

$$S(x) = \frac{b}{\pi} \tan^{-1} \frac{x}{\zeta} + \frac{b}{2} \qquad (7)$$

where $\zeta = (Kb)/(4\pi\tau^{max})$ represents the half-width of the $\rho(x)$ dislocation distribution. One can easily check that the disregistry across the glide plane $S(x)$ verifies the boundary conditions, $S(x) = 0$ when $x \to -\infty$ (far from the dislocation on one side), $S(x) = b$ when $x \to +\infty$ (on the other side) and the summation of $S(x)$ over the whole space is equal to the Burgers vector b of the dislocation.

As pointed out by numerous authors (e.g. Ref. 24), one of the main achievements of the PN model is that it provides reasonable estimate of the dislocation core size. However, equally important is the possibility to evaluate the Peierls stress, i.e. the stress needed to move the dislocation without help of temperature. However, the original PN equation or the expression of the energy functional (3) is invariant with respect to an arbitrary translation of the dislocation density. Indeed, both hold for an elastic continuous medium which does not take into account the lattice discreteness whereas $S(x)$ can actually only be defined where an atomic plane is present (e.g. refs. 27 and 29). The simplest way to obtain the misfit energy corresponding to the Peierls dislocation and to determine the Peierls stress, is to perform a summation of the local misfit energy at the positions of atoms rows parallel to the dislocation line. The misfit energy can be thus considered as the sum of misfit energies between pairs of atomic planes (e.g. refs. 18 and 30) and can be written as

$$W(u) = \sum_{m=-\infty}^{+\infty} \gamma(S(ma'-u)) \cdot a' \qquad (8)$$

where a' is the periodicity of W, taken as the shortest distance between two equivalent atomic rows in the direction of the dislocation's displacement. The Peierls stress is then given by:

$$\sigma_P = \max\left\{ \frac{1}{b} \frac{dW(u)}{du} \right\} \qquad (9)$$

In order to solve the misfit energy function analytically, Joos & Duesbery[18] introduced a dimensionless parameter $\Gamma = \zeta/a'$. Then, some simple formulas can be derived for the extreme cases of very narrow ($\Gamma << 1$) and widely spread ($\Gamma >> 1$) dislocations:

$$\sigma_p(\Gamma << 1) = \frac{3\sqrt{3}}{8} \tau^{max} \frac{a'}{\pi \zeta} \qquad (10)$$

and

$$\sigma_p(\Gamma >> 1) = \frac{Kb}{a'} \exp(\frac{-2\pi\zeta}{a'}) \qquad (11)$$

The range of the limits was recently extended to the case $0.2 < \Gamma < 0.5$ where equation (10) has to be modified to take into account an exponential correction term.[31]

The relevance of this analytical model to predict the easiest slip systems in a given material has been assessed by Wang.[19] Although some trends are observed, this approach has shown its limits and cannot be regarded as fully quantitative.

2.4. *Numerical solutions based on generalized stacking faults*

The reason for this limitation is that the analytical solution presented above can only be applied in the case of a sinusoidal restoring force and with the knowledge of the ISS value. The range of application is thus very limited. However, a larger range of possibilities is offered considering that the restoring force introduced in the PN model is the gradient of the γ-surface.

Nowadays, the GSF energy is relatively easy to calculate using atomistic calculations (either *ab initio* or using semi-empirical potentials). To calculate a GSF for a given slip plane, one has to consider a perfect crystal cut across the slip plane into two parts which are subjected to differential displacement through an arbitrary fault vector S. The faulted lattice exhibits an extra energy per unit area $\gamma(S)$. Considering all possible S vectors in the slip plane generates the γ

surface. Practically, the calculation is made using supercell methods and minimisation of the excess energy by relaxation of atomic positions perpendicularly to the glide plane. Depending on the atomistic calculations techniques used, there are several possibilities to build supercells which have been recently reviewed by Bulatov *et al.*[32]

Introducing real restoring forces, calculated at the atomic scale, in the PN equation renders analytical solutions of S or ρ more complicated to evaluate. Among the various ansatz proposed for the expression of $S(x)$[33,34, 35], one may retain the formulation of Joos *et al.*[30] based on a set of variational constant. Restricting the treatment to one dimension of the γ-surface along the Burgers vector direction, the disregistry distribution in the dislocation core $S(x)$ can be obtained by searching for a solution in the form:

$$S(x) = \frac{b}{2} + \frac{b}{\pi} \sum_{i=1}^{N} \alpha_i \cdot \arctan \frac{x - x_i}{c_i} \qquad (12)$$

where α_i, x_i and c_i are variational constants. The idea is to search for a solution which can be written as a sum of N fractional dislocation density. Using the previous form of $S(x)$, the infinitesimal dislocation density $\rho(x)$ is given by

$$\rho(x) = \frac{dS(x)}{dx} = \frac{b}{\pi} \sum_{i=1}^{N} \alpha_i \frac{c_i}{(x - x_i)^2 + c_i^2} \qquad (13)$$

As the disregistry $S(x)$ and the density $\rho(x)$ must be solution of the normalisation condition, $\int \rho(x)dx = b$ the α_i are constrained by $\sum_{i=1}^{N} \alpha_i = 1$.

Substituting equation (13) into the left-hand side of the PN equation (4), gives the restoring force

$$F^{PN}(x) = \frac{Kb}{2\pi} \sum_{i=1}^{N} \alpha_i \cdot \frac{x - x_i}{(x - x_i)^2 + c_i^2} \qquad (14)$$

The variational constants α_i, x_i and c_i are obtained from a least square minimisation of the difference between F^{PN} and the restoring force F

derivated from the γ-surface. A typical value of *N=3* was used by Joos *et al.*[30] for dislocation in Si. Recent calculations in mineral structures[36, 37] used *N=6*, especially for calculations along a GSF characterized by a stable stacking fault. Indeed, stable stacking fault reflects the tendency of dislocation to dissociate into two partial dislocations of collinear Burgers vector. As a consequence, this treatment is suitable to study dislocation core structure for a planar dislocation.

2.5. *The Peierls-Nabarro galerkin method*

During the last ten years, several improvements have been proposed. Among then we find the Semi-discrete Variational PN (SVPN) model which appears to be able to deal with three-dimensional displacement fields as encountered in case of non-collinear dissociation.[21, 24] SVPN is based on a nodal resolution of the PN equation. In spirit, the concept of nodal description is close to the one of a Finite Element method. At the same time, some coupling between FE methods and discrete representations of dislocation have also been proposed (see for instance the work of Lemarchand *et al.*[38]). However, based on a dislocation dynamics approach, the constraint of modelling dislocation with no core structure is a strong limitation to study the Peierls stress. The same kind of argument applies to the Phase Field methods which are able to reproduce dissociated dislocations core structure but are unable to predict dislocation core width.[39] Very recently, Denoual[40, 41] rewrote in a very promising way the PN model in the element free Galerkin method (comparable to a FE method). Such a treatment should account for complex boundary conditions or material structures and allows for the calculation of core structure as presented in the following. The Peierls-Nabarro Galerkin (PNG) method is a generalization of the PN method to allow for multiple glide planes and complex (possibly three-dimensional) cores. As in the initial PN model, the dislocation core structure naturally emerges from minimizing an elastic energy (through an approximation of a continuous field representation) and an interplanar potential. For the sake of clarity, let us consider one unique slip plane Σ. In the PNG model, two distinct fields are used: $u(r)$ a three-dimensional displacement field of the volume V and a two dimensional $S(r)$ field which is expressed in the normal basis of the Σ plane. Thus, $u(r)$ represents a homogenous strain whereas the displacement jump when crossing Σ is measured by $\tilde{S}$. The problem consists therefore in minimizing the following Hamiltonian H with respect to u and $\tilde{S}$

$$H = \int_V \{ E^e[u,S] + \frac{1}{2}\Omega u\acute{Y}^2 \}dV + \int_\Sigma E^{isf}[S]d \qquad (15)$$

In equation (15), Ω corresponds to the material density. E^e corresponds to the elastic strain energy whereas E^{isf} is the inelastic stacking fault energy which depends on the material and controls the spreading of dislocation. E^{isf} is a function of the GSF energies γ from which the linear elastic part has been subtracted.[40-42] In doing so, we strictly comply with the overall linear elastic behaviour except for small applied strains. In practice, minimization with respect to S is achieved by means of a time dependent Ginzburg Landau equation whereas the Galerkin method is used to compute the evolution of $u(r)$.

A simulation can be divided into three phases. A discrete dislocation is first introduced in a finite volume, the boundary of which is given by imposing a displacement equal to the result of a convolution of an elementary elastic solution with the dislocation density (see Pillon *et al.*[43] for implementations details). The imposed boundary conditions are therefore consistent with a dislocation in an infinite media, without any influence on the core structure. The equilibrium between the displacement jump field S and the density of dislocation ρ, as defined as the derivative of S by the position coordinates, is obtained through a viscous relaxation scheme. During a second step, a homogeneous strain is progressively superimposed at a rate that allows quasi-static equilibrium, even with noticeable evolution of the core structure. During the last step, a rapid evolution of the dislocation core structure followed by a displacement of one (or more) Burgers vector leads to a relaxation of the measured stress. The ultimate macroscopic stress is the Peierls stress.

3. Some Examples

3.1. *Cubic perovskites*

A large number of important materials in both mineralogy and materials science exhibit a perovskite structure, or a structure derived from it. Perovskite is the name of an orthorhombic mineral of chemical composition $CaTiO_3$ which was originally described by Rose,[44] a collaborator of von Humboldt, from metasomatic rocks in the Urals. It is also the name of a structure of cubic symmetry (space group *Pm3m*) from which $CaTiO_3$ deviates slightly. Strontium titanate (also named Tausonite when it is from natural origin) is commonly regarded as the archetypal cubic perovskite. $SrTiO_3$ has a very large dielectric constant. It is also widely used as a substrate for epitaxial growth of high-temperature superconducting (and many oxides) thin film. As generally observed in perovskites[45], the most common slip systems observed in deformed $SrTiO_3$ are $\langle 110 \rangle \{1\bar{1}0\}$ and $\langle 100 \rangle \{010\}$.[46-49] It is however possible, following growth processes, to find other dislocations like $\langle 1\bar{1}0 \rangle \{001\}$[50] or $\langle 001 \rangle \{1\bar{1}0\}$.[51] Plasticity of $SrTiO_3$ exhibits a very surprising evolution with temperature with a very strong flow stress anomaly eventually leading to an inverse brittle-ductile transition.[48, 49] Such a behaviour suggests that dislocations may exhibit several core structures.

3.1.1. *Core spreading within the PN model*

Four linear GSF have been calculated[52] that correspond to the potential slip systems in $SrTiO_3$: $\langle 100 \rangle \{010\}$, $\langle 100 \rangle \{011\}$, $\langle 110 \rangle \{001\}$ and $\langle 110 \rangle \{1\bar{1}0\}$ (see the supercells used on Fig. 4 and the GSF on Fig. 5). All the GSF exhibit a single peak shape that can be significantly different from a sinusoidal profile, in particular for the case of $\langle 110 \rangle \{1\bar{1}0\}$ which is characterized by a plateau around γ^{max}. This justifies to go beyond the analytical approach which is clearly poorly adapted.

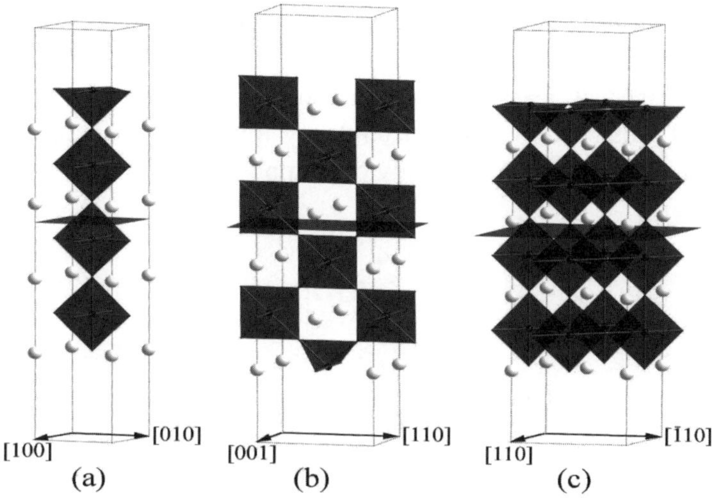

Fig. 4. Supercells used to calculate the GSFs for $SrTiO_3$. Light gray spheres correspond to Sr atoms; oxygen octahedra are displayed in gray with a small black sphere in the centre corresponding to the Ti atom. The shear plane located in the middle of the supercell is represented in gray. For all the supercells, a vacuum buffer is added in the direction normal to the slip plane to avoid interaction between repeated stacking faults resulting from the use of periodic boundary conditions. Atoms present on outer surface layers are kept fixed during the GSF calculation to mimic the action of a surrounding bulk material: a) supercell used for [100](001), b) supercell used for [001](110) and [110](110), c) supercell used for [110](001). (After ref. 52)

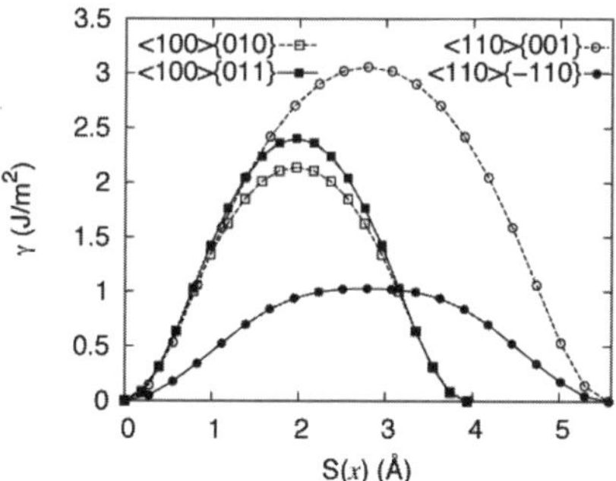

Fig. 5. Generalised stacking faults (GSF) calculated for $SrTiO_3$. The GSF are plotted against the shear displacement vector S(x). (After Ref. 52)

The dislocation densities ρ determined in this study are presented on Fig. 6 for edge characters. All the cores are relatively wide with an extension over at least one lattice repeat along the shear direction. $\langle 110 \rangle \{1\bar{1}0\}$ dislocations show a tendency toward dissociation although the partial dislocations are strongly overlapping.

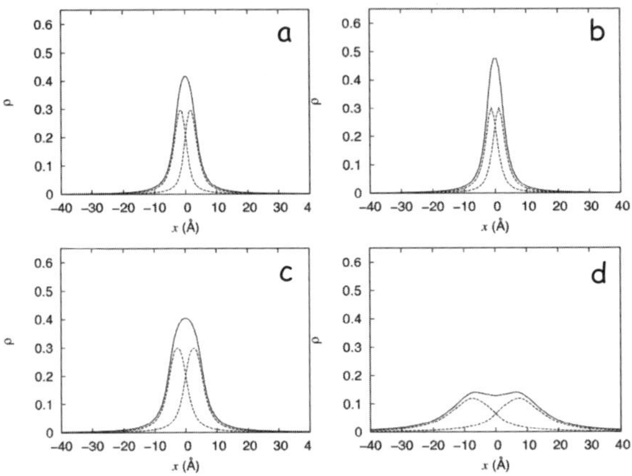

Fig. 6. Dislocation density (ρ) for edge dislocations in $SrTiO_3$ perovskite calculated for the following slip systems: a) <100>{010}, b) <100>{011}, c) <110>{001} and d) <110>{1010}.

The core structure is illustrated on Fig. 7 which shows that a stable stacking fault appears in the core where TiO_6 octahedra are joining along their edges in a pattern very similar to the so-called post-perovskite structure which is found in some structures at pressures above the stability field of perovskite. This core spreading has tremendous implications on the dislocation mobility since, as shown in Table 1, $\langle 110 \rangle \{1\bar{1}0\}$ is the easiest slip system with very little lattice friction (compared to the other slip systems). It is worth noticing that this particular core structure and related mobility favours a 5.571 Å long Burgers vector over a 3.939 Å long Burgers vector (<100> dislocations). This dislocation core structure is likely to account for the easy glide of dislocations in $SrTiO_3$ at low temperature.

Fig. 7. atomistic model of a $\langle 110\rangle\{110\}$ dislocation from the PN modelling (relaxation is only achieved along the shear direction).

Table 1. Peierls-Nabarro stresses for screw and edge dislocations in $SrTiO_3$.

Slip system	σ_p (GPa)	
	Screw	Edge
<100>{010}	0.7	0.6
<100>{011}	9.9	0.6
<110>{001}	0.9	1.2
<110>{110}	0.006	0.004

3.1.2. *Climb dissociation*

The high-temperature inverse brittle-ductile transition observed in $SrTiO_3$ must involve a different core structure with a reduced mobility as suggested by Brunner *et al.*[48], Gumbsch *et al.*[49] and Sigle *et al.*[53] Let us examine here the possibility of another dissociation scheme of these dislocations, involving local climb, as suggested by these authors.

The PN model described above considers the possibility of dislocation splitting into partial dislocations separated by a stacking fault (Fig. 8b) in the glide plane (called the "glide" fault). At high temperature, another configuration called *climb dissociation* can be obtained by redistributing matter between the two half planes "A" and "B" (Fig. 8c). This process involves atomic diffusion although it remains onservative. It must be noted that the two planes "A" and "B" must exhibit the same chemical composition.

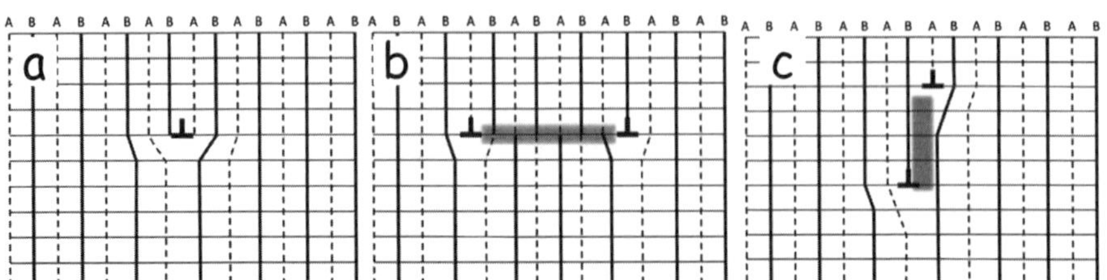

Fig. 8. Glide vs climb dissociation: a) unrelaxed core of a dislocation in a material with two different lattice planes "A" and "B"; b) splitting into two partial dislocations separated by a (glide) stacking fault (shaded); c) climb dissociation into two partial dislocations separated by a (climb) stacking fault.

Combining elastic energy of interactions between parallel dislocations (as proposed in ref. 27) and stacking fault energy, Mitchell & Heuer[54] proposed that the energy of a climb dissociation can be written as the sum of the interaction energy W_{12} between the two partials (1 and 2) and the (climb) stacking fault energy per unit length:

$$\frac{W_{12}}{L} + \frac{W_{SF}}{L} = \frac{\mu b_1 b_2}{2\pi(1-\nu)}\left[-\ln\frac{d}{r_0} - \sin^2\theta\right] + \gamma(\theta).d \qquad (16)$$

where b_1 and b_2 are the Burgers vectors of the two partials (assumed to be collinear), r_0 is the core radius, $\gamma(\theta)$ is the stacking fault energy, d is the spacing of the partials and θ is the angle between the fault plane and the slip plane ($\theta=0$ for a glide dissociation and $\theta=90$ for a climb dissociation). For an isotropic elastic medium, such as $SrTiO_3$, and without applied stress, the equilibrium distance between partial dislocations is given by:

$$d_{eq} = \frac{\mu b_1 b_2}{2\pi(1-\nu)\gamma(\theta)} \qquad (17)$$

The energy difference between the climb and glide configurations is:

$$\frac{W_{climb}}{L} - \frac{W_{glide}}{L} = \frac{\mu b_1 b_2}{2\pi(1-\nu)}\left[\ln\frac{\gamma_{climb}(90)}{\gamma_{glide}(0)} - 1\right] \qquad (18)$$

Figure 9 shows how the "A" and "B" plane are arranged in the $SrTiO_3$ structure. Along [001], the layer "A'" is made of strontium atoms whereas "B'" is made of titanium and oxygen. Since these layers are chemically distinct, it is not possible to redistribute matter from one to the other. We conclude that climb dissociation of <100> dislocations is not possible. On the contrary, <110> dislocations exhibit two strictly identical layers perpendicular to the Burgers vector which makes climb dissociation chemically possible. Moreover, the climb fault made by removing a "A" or a "B" layer (see Fig. 9) has exactly the same structure as the $\frac{1}{2}\langle 110\rangle\{1\bar{1}0\}$ glide fault.

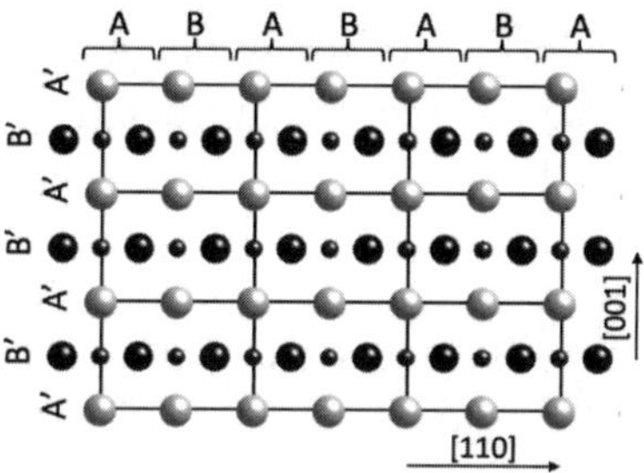

Fig. 9. Structure of $SrTiO_3$ view down [110]. Layers "A" and "B", perpendicular to [110] are indicated. They help discussing the climb faults of <110> dislocations (whatever their actual glide plane). Similarly, climb dissociation of <100> dislocations can be appraised from the layering "A'" and "B'" perpendicular to the corresponding shear direction.

Thus, the $\gamma_{climb}(90)$ stacking fault energy is the same as the $\gamma_{glide}(0)$ energy obtained for the $\langle 110\rangle\{1\bar{1}0\}$ slip system. The climb dissociation mode of $\langle 110\rangle\{1\bar{1}0\}$ edge dislocations is thus always more favourable at high temperature. This sessile core configuration is probably at the origin of the inverse brittle-ductile transition observed in $SrTiO_3$.

3.2. The [001] dislocation in olivine

Olivine form a solid-solution between two end-member phases: forsterite Mg_2SiO_4 and fayalite Fe_2SiO_4. The structure of olivine is based on a distorted hexagonal close-packed oxygen sublattice. The Bravais lattice is orthorhombic with, for forsterite, a = 0.475 nm, b = 1.019 nm and c = 0.597 nm (*Pbnm* space group). The unit cell contains four formula units. The most common dislocations correspond to the shortest lattice repeats: [100] and [001]. [010] dislocations are not activated. The common slip systems at high temperature involve [100] slip on several planes: (010), {031}, {021}, {011} and (001) whereas at low temperature and high stresses, slip occurs along [001] in (100), {110} and (010). [001] screw dislocations are always found to exhibit very marked crystallographic characters and are less mobile than edge dislocations. Hence, they control plastic deformation of olivine at low temperature. The structure of the core of [001] screw dislocations has been modelled by Durinck *et al.*[55] using the PN model. [001] screw dislocations tend to dissociate in the (100) and {110} planes, but not in the (010) where they spread as a single dislocation (Fig. 10a).

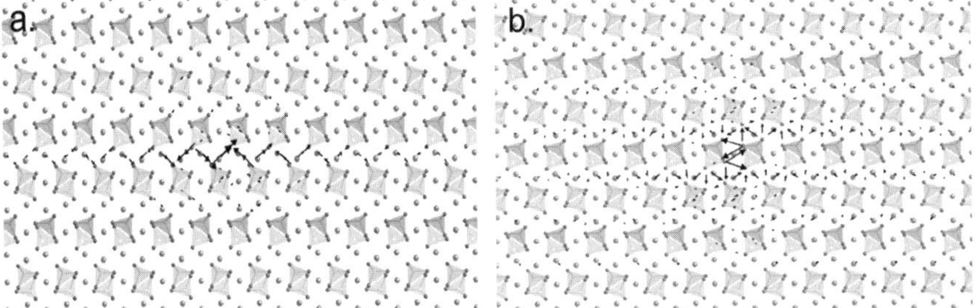

Fig. 10. Differential displacement maps for the [001] screw dislocation in forsterite. The dislocation is viewed end on (i.e. along the [001] direction). The (010) plane is horizontal: a) Peierls-Nabarro dislocation, b) dislocation calculated by direct simulation by Walker *et al.*[56]

Planar core structures derived from the PN model fail to account for the low mobility of screw dislocations, at least in the (010) plane. Another approach is to perform a direct atomistic simulation. This has been done by Walker *et al.*[56] and discussed by Carrez *et al.*[57] Such calculations show (Fig. 10b) that [001] screw dislocations can exhibit a complex 3D core structure with core spreading in two adjacent (010) planes connected by a highly sheared region in between (i.e. in the (100) plane). The energy of this 3D configuration is 2.7 eVÅ^{-1} less than that of the PN configuration. The [001] screw dislocation is found to exhibit (at least) two possible core structures:

- a high-energy configuration which is planar and hence mobile
- a low-energy configuration which is 3D and hence sessile

The high lattice friction observed on [001] screw dislocations is thus probably due to the difficulty to bring the dislocation core from the low-energy configuration to the mobile configuration. Finally, it can be anticipated that a 3D core will respond to the whole stress field and not only to the resolved shear stress in the shear plane (as predicted by the Schmid law).

3.3. *The [010] dislocation in wadsleyite*

Wadsleyite is the high-pressure polymorph of olivine that is stable at depths between 410 and 520 km in the mantle. We will focus on the Mg end-member Mg_2SiO_4. Wadsleyite has a spinelloid structure made of MgO_6 octahedra and SiO_4 tetrahedra. The Bravais lattice is orthorhombic (space group *Imma*) with eight formula units per unit cell. The lattice parameters at 18.5 GPa are a = 0.56983 nm, b = 1.1438 nm and c = 0.82566 nm.[58] [010] (and to a less extend [001]) dislocations have large Burgers vectors and their core structure is of interest in the present context. We will focus here on the [010] dislocations. Calculations have been made using the PNG model. Figure 11 shows how the mesh is constructed to model a [010] dislocation. A large number of (100), (001) and {101} planes are introduced in the calculation around the initial dislocation core position. They are all assigned a 2D-GSF calculated using the GULP code[59] and the "THB1" potential parameterization[60, 61] developed for oxides. Figure 12 shows the case of a [010] edge dislocation which was initially introduced in the (100) plane. After relaxation, the core is shown to widely spread into this plane and give rise to four well-defined partial dislocations which extend over more that 35 Å. When a [010] dislocation with a screw character is introduced in the same plane, it starts spreading into the (100) plane, but very rapidly,

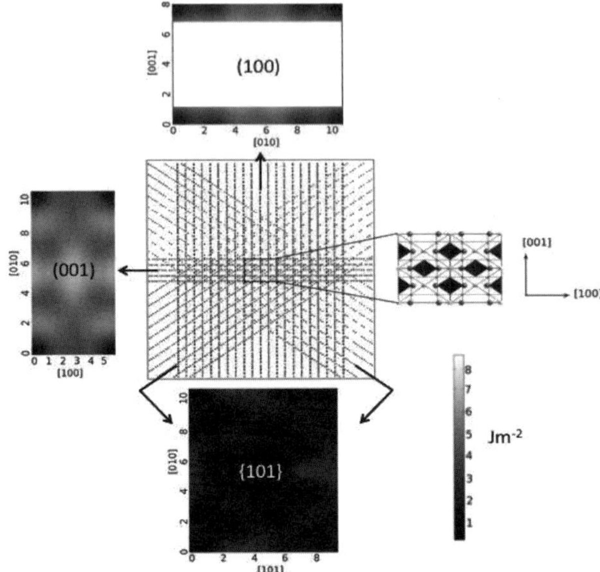

Fig. 11. Description of the mesh used to model the [010] screw dislocation. The dislocation line is viewed edge-on. In between the glide planes, the material has a purely elastic behaviour (the corresponding interpolation points are not presented). The geometry of the glide planes (in grey), introduced when non-linear elastic behaviour is suspected to occur, corresponds to the crystal symmetry. The location and number of glide planes around the dislocation line is shown. The central portion enclosing the dislocation line is enlarged on the right to show the influence of the crystal chemistry on the choice of the planes.[71]

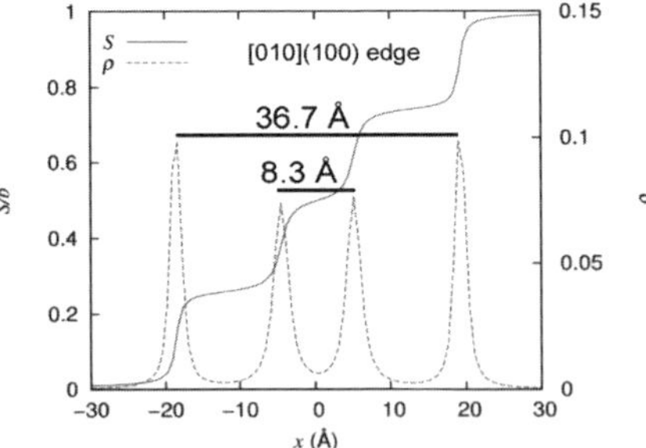

Fig. 12. [010](100) edge dislocations. Shear profile S and dislocation density ρ are given along [010].[71]

screw partial dislocations cross-slip into a weaker (see the GSF on Fig. 11) {101} plane. The final core structure exhibits four partial dislocations distributed into three different planes (Fig. 13). This core cannot glide as it is. Under stress, partial dislocations must cross-slip back into the primary plane (under a resolved shear stress of 300 MPa) before the planar dislocation can glide (under 510 MPa). Once again, non-Schmid effects can be anticipated.

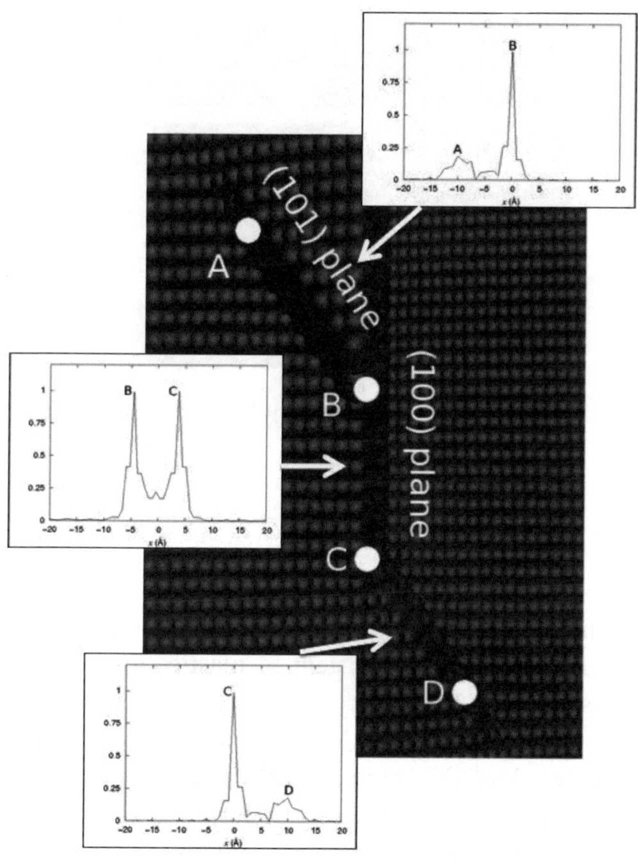

Fig. 13. Structure of the [010] screw dislocation introduced initially in the (100) plane which spreads after relaxation into several planes (the density profile in each plane is represented).[71]

Finally, it is worth noticing that these complex core structures have been actually observed by Thurel and Cordier[63] and Thurel *et al.*[64] in synthetic wadsleyite samples deformed at high-pressure and high-temperature in a multianvil apparatus (Fig. 14).

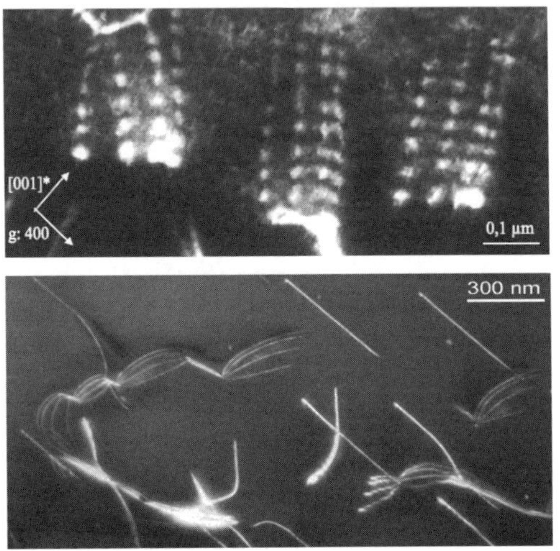

Fig. 14. Complex dissociation of [010] dislocations in Mg_2SiO_4 wadsleyite deformed at 17 GPa, 1500°C.

3.4. *Dislocations in garnets*

Garnets represent an important mineral family for the rheological behaviour of the upper mantle, and the mantle transition zone. They are often regarded as being very resistant to plastic flow. The structure of garnets is based on a bcc lattice (space group *Ia3d*). The unit cell contains eight structural units $X_3Y_2Z_3O_{12}$. It can be viewed as a three-dimensional network of ZO_4 tetrahedra (SiO_4 in case of silicate garnets) and YO_6 octahedra sharing corners and forming dodecahedral cavities filled with X cations. Therefore garnets represent a family of complex silicates with widely varying chemical composition (pyrope represented on Fig. 1 has a chemical composition $Mg_3Al_2Si_3O_{12}$). The lattice parameters of garnets vary usually in the range 1.15-1.20 nm.

In cubic garnets, the shortest lattice repeat is 1/2<111> and, as in bcc metals, one can expect plastic deformation to be dominated by 1/2<111> dislocations. One must note however that, given the large unit cell of the garnet structure, such dislocation Burgers vector exhibit an unusually large magnitude (close to 1 nm). Most studies have shown that plastic deformation of garnets is mostly carried by 1/2<111> dislocations, however, occurrence of <100> glide has been demonstrated experimentally (Fig. 15a). Experiments as well as studies of natural materials suggest that garnets undergo a brittle-ductile transition which, in nature, occurs at about 600-800°C.[65, 66] In the low-temperature, brittle, regime, crystal plasticity, if it occurs, is restricted to the vicinity of

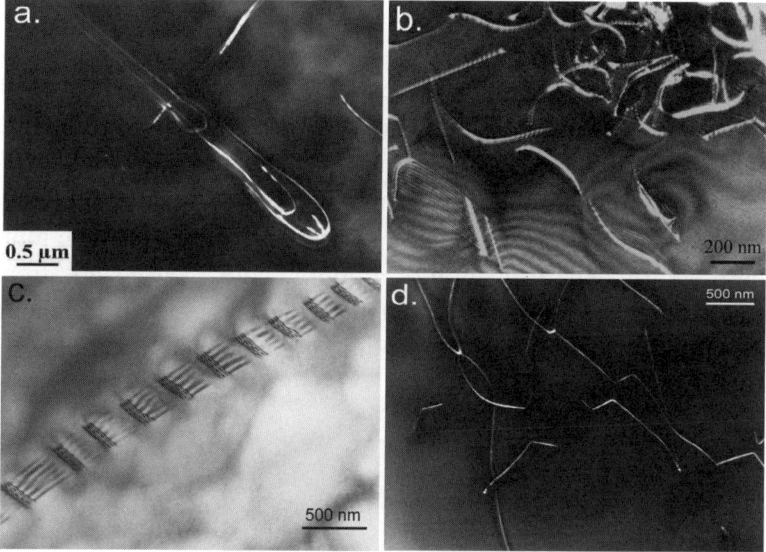

Fig. 15. Dislocations in pyrope-almandine garnets: a) [100] dislocations introduced in samples deformed experimentally in a multianvil apparatus at 900°C, 6 GPa. TEM weak-beam dark-field micrograph. g: 400.; b) Eclogites from the Alps. High density of dissociated dislocations near a crack. TEM weak-beam dark-field micrograph. g: 400.; c)Eclogites from the Alps. Dissociated dislocations associated with stacking faults. TEM bright-field micrograph. g: 400.; d) garnet from Koidu pipe. Tangled perfect dislocations. TEM weak-beam dark-field micrograph. g: 400.

cracks (Figs. 15b and c). Dislocations are nucleated due to the stress concentration at the crack tips, but they don't expand very far from the tip. These dislocations often exhibit complex dissociation patterns (Figs. 15b and c). Above the brittle-ductile transition, garnets deform by motion of perfect dislocations (Fig. 15d). 1/2<111> dislocations glide in three kinds of planes: {110}, {112} and {123}.

No modelling of dislocation cores in garnets has been performed so far and it is difficult to understand how such a large core can glide. Observations suggest that very high stresses are necessary to generate a planar core (see complex dissociations of Figs. 15b and c). The crystals are then close to their brittle regime. At high temperature, no evidence for core dissociation can be observed, and we have suggested that the core needs local diffusion to drag an extended core. A suggestion that has not been verified nor contradicted yet.

Some evidences for hollow cores have been found in garnets from natural origins.[65] Figure 16 shows the case of a dislocation from a garnet eclogite which is connected to a fluid inclusion (microscopic water-rich fluid inclusions are frequent in natural rocks due to the circulation of fluids in the crust and to some finite solubility of water in some minerals at high-pressure). The formation of a hollow core clearly starts from the fluid inclusion and is probably related to the preferential dissolution of garnets within the dislocation core. Such preferential etching is probably

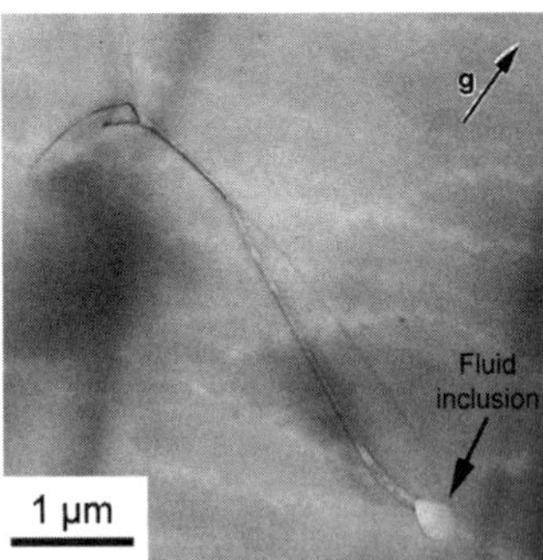

Fig. 16. Eclogite garnet from from Yakutia pipe. This dislocation connected to a fluid inclusion shows a hollow core. TEM Bright-field, *g*: 400.

related to the formation of tunnel-like defects during growth of magnetic garnets[67] and has been used in the past to create etch channels and observe dislocation substructures in naturally deformed garnets.[68, 69] Apart from these cases where etching obviously play a role, we are not aware of any occurrence of hollow cores of dislocations in deformed garnets in the context of plastic deformation.

4. Concluding Remarks

Many minerals have crystal chemistry that challenge the Frank criterion for the stability of the dislocation core. Hollow cores seem to represent an option in static situations only (when the dislocation doesn't have to move) and during etching or re-dissolution. A large body of experimental evidence demonstrate that dislocations with large Burgers vectors (slightly above 1 nm) do exist and can even participate to plastic deformation under some circumstances. TEM studies (in particular using weak-beam techniques) have shown in many cases that core spreading and eventually dislocation dissociation is the mechanism which allows to stabilize a dislocation core. Modelling based on the PN model, has highlighted the importance of the GSF's in constraining the ability of a dislocation core to spread in one or more planes. Depending on the relative ability of the crystallographic planes to be sheared, the dislocation will spread in one plane (the glide plane, and its mobility will be increased, see $SrTiO_3$ at low temperature), in several planes (and its mobility will be hampered, see forsterite or wadsleyite) or in a volume around the core (is it the case of garnets?). At high temperature, another relaxation mechanism can be made possible by diffusion and the core can be, in some cases, able to reach a climb configuration (hampering glide and possibly promoting climb). It is thus obvious that these relaxation mechanisms have paramount implications on the dislocation mobilities and hence on the mechanical properties of minerals. They are also likely to have consequences on the possibility of diffusing species along the dislocation cores (the so-called pipe diffusion, see for instance

Gaboriaud 2009) [70] more rapidly. A continuing effort in modelling the fine core structure of important mineral phases is clearly necessary to understand their mechanical properties.

References

1. V. Volterra, *Ann. Ec. Norm.*, **24**, 401 (1907).
2. G. Taylor, *Proc. Roy. Soc. A*, **145**, 363 (1934).
3. E. Orowan, *Z. Phys.*, **89**, 605 (1934).
4. M. Polanyi, *Z. Phys.*, **89**, 660 (1934)
5. A. E. H. Love, *The mathematical theory of elasticity*. (Cambridge University Press, Cambridge, 1920) p. 221;
6. F. C. Frank, *Acta Cryst.*, **4**, 497 (1951).
7. D. J. Srolovitz, N. Sridhar, J. P. Hirth and J. W. Cahn, *Scripta Mater.*, **39**, 379 (1998).
8. I. Sunagawa and P. Bennema, *J. Cryst. Growth*, **53**, 490 (1981).
9. W. M. Si, M. Dudley, R. Glass, V. Tsvetkov and C. Carter, *J. Elect. Mater.*, **26**, 128 (1997).
10. J. Heindl, W. Dorsch, P. P. Strunk, S. G.Muller, R. Eckstein, D. Hofmann and A. Winnacker, *Phys. Rev. Lett.*, **80**, 740 (1998).
11. M. Grodzicki, P. Mazur, S. Zuber, G. Urbanik and A. Ciszewski, *Thin Solid Films*, **516**, 7530 (2008).
12. M. F. Chisholm and D. A. Smith, *Phil. Mag. A*, **59**, 181 (1989).
13. W. Qian, M. Skowronski, K. Doverspike,L. B. Rowland and D. K. Gaskill, *J. Cryst. Growth*, **151**, 396 (1995).
14. F. R. N. Nabarro, in *"Dislocations 2004"*. Eds. P. Veyssière, L. Kubin and J. Castaing, (CNRS, Paris, 1984) p.19.
15. R. E. Peierls, *Proc. Phys. Soc. Lond.*, **52**, 34 (1940).
16. F. R. N. Nabarro, *Proc. Phys. Soc. Lond.*, **59**, 256 (1947).
17. Q. Ren, B. Joos and M. S. Duesbery, *Phys. Rev.* **B52**, 13223 (1995).
18. B. Joos and M. S. Duesbery, *Phys. Rev. Lett.*, **78**, 266 (1997).
19. J. N. Wang, *Mater. Sci. Eng. A*, **206**, 259 (1996).
20. G. Schoeck, *Mater. Sci. Eng. A*, **400-401**, 7 (2005).
21. V. V. Bulatov and E. Kaxiras, *Phys. Rev. Lett.*, **78**, 4221 (1997).
22. B. von Sydow, J. Hartford and G. Wahnström, *Comp. Mat. Sci.*, **15**, 367 (1999).
23. G. Lu, N. Kioussis, V. V. Bulatov and E. Kaxiras, *Philos. Mag. Lett.*, **80**, 675 (2000).
24. G. Lu, G. in *"Handbook of Materials Modeling. Volume 1 Methods and Models"*. Ed; S. Yip (Springer, Berlin, 2005), 1.

25. C. R. Miranda and S. Scandolo, *Comp. Phys. Comm.*, **169**, 24 (2005).

26. J. W. Steeds, *Introduction to Anisotropic Elasticity of Dislocations* (Oxford University Press, 1973) .

27. J. P. Hirth and J. Lothe. *Theory of dislocations.* (John Wiley and Sons, Inc, New York, 1982).

28. V. Vítek, *Phil. Mag.*, **18**, 773 (1968)

29. G. Schoeck, *Philos. Mag. A*, **79**, 2629 (1999).

30. B. Joos, Q. Ren and M. S. Duesbery, *Phys. Rev. B*, **50**, 5890 (1994).

31. B. Joos and J. Zhou, *Phil. Mag. A*, **81**, 1329 (2001).

32. V. V. Bulatov, W. Cai, R. Baran and K. Kang, *Phil. Mag.*, **86**, 3847 (2006).

33. F. Kroupa and L. Lejcek, *Czech. J. Phys. B*, **22**, 813 (1972).

34. J. Hartford, B. von Sydow, G. Wahnström and B. I. Lundqvist, *Phys. Rev.* **B58**, 2487 (1998).

35. J. A. Yan, C. Y. Wang and S. Y. Wang, *Phys. Rev.* **B70**, 174105 (2004).

36. P. Carrez, P. Cordier, D. Mainprice and A. Tommasi, *Eur. J. Mineral.*, **18**, 149 (2006) .

37. P. Carrez, D. Ferré and P. Cordier, *Philos. Mag. A*, **87**, 3229 (2007).

38. C. Lemarchand, B. Devincre and L. P. Kubin, *J. Mech. Phys. Solids*, **49**, 1969 (2001).

39. C. Shen and Y. Wang, *Acta Mat.*, **52**, 683 (2004).

40. C. Denoual, *Phys. Rev.* **B70**, 024106 (2004).

41. C. Denoual, *Comp. Methods Appl. Mech. Engrg.*, **196**, 1915 (2007).

42. Y. P Pellegrini, C. Denoual and L. Truskinovsky, (2008) in *IUTAM Symposium on variational concepts with applications to the mechanics of materials.* Ed. K. Hackl (Bochum, Germany) in press.

43. L. Pillon and C. Denoual, *Phil. Mag.*, **89**, 127 (2009).

44. G. Rose, *Poggendorff Ann. Phys. Chem.*, **48**, 551 (1839).

45. J. P. Poirier, S. Beauchesne and F. Guyot, in "*Perovskite a structure of great interest to geophysics and materials science*". Eds. A. Navrotsky and D. Weidner (AGU, Washington DC, 1989) p.119.

46. J. Nishigaki, K. Kuroda and H. Saka, *Phys. Stat. Sol. (a)*, **128**, 319 (1991).

47. T. Matsunaga and H. Saka, *Phil Mag Letters*, **80**, 597 (2000).

48. D. Brunner, S. Taeri-Baghbadrani, W. Sigle and M. Rühle, *J. Am. Ceram. Soc.*, **84**, 1161 (2001).

49. P. Gumbsch, S. Taeri-Baghbadrani, D. Brunner, W. Sigle and M.. Rühle, *Phys. Rev. Lett.*, **87**, (2001).

50. Z. Mao. and K. M. Knowles, *Phil Mag A*, **73**, 699 (1996).

51. C. L. Jia, A. Thust and K. Urban, *Phys. Rev. Lett.*, **95**, 225506 (2005).

52. D. Ferré, P. Carrez and P. Cordier, *Phys. Rev.* **B77**, 014106 (2008).

53. W. Sigle, C. Sarbu, D. Brunner and M. Rühle, *Phil. Mag. A* **86**, 4809 (2006).

54. T. E. Mitchell and A. H. Heuer, in *"Dislocations in Solids, Vol 12"* (Elsevier Science, Amsterdam, 2004), p.339.

55. J. Durinck, P. Carrez and P. Cordier, *Eur. J. Mineral.*, **19**, 631 (2007).

56. A. M. Walker, J. D. Gale, B. Slater and K. Wright, *Phys. Chem. Chem. Phys.*, **7**, 3235 (2005).

57. P. Carrez, A. M. Walker, A. Metsue and P. Cordier, *Philos. Mag. A*, **88**, 2477 (2008).

58. H. Horiuchi and H. Sawamoto, *Am. Mineral.*, **66**, 568 (1981).

59. J. D. Gale and A. L. Rohl, *Molecular Simulation*, **29**, 291 (2003).

60. M. J. Sanders, M. Leslie and C. R. A. Catlow, *J. Chem. Soc.*, 1271 (1984).

61. G. V. Lewis and C. R. A. Catlow, *J. Phys. C*, **18**, 1149 (1985).

62. A. T. Paxton, P. Gumbsch and M. Methfessel, *Phil. Mag. Lett.*, **63**, 267 (1991).

63. E. Thurel and P. Cordier, *Phys. Chem. Minerals*, **30**, 256 (2003).

64. E. Thurel, P. Cordier, D. Frost and S. Karato, *Phys. Chem. Minerals*, **30**, 267 (2003).

65. V. Voegele, P. Cordier, V. Sautter, T. G. Sharp, J. M. Lardeaux, and F. O. Marques, *Phys. Earth Planet. Inter.*, **108**, 319 (1998).

66. Z. C. Wang and S. C. Ji, *Canadian Mineralogist*, **37**(2), 525 (1999).

67. S. Takasu and S. Shimanuki, *J. Crystal Growth*, **24-25**, 641 (1974).

68. H. Carstens, *Contr. Mineral. Petrol.*, **24**, 348 (1969).

69. H. Carstens, *Contr. Mineral. Petrol.*, **32**, 289 (1971).

70. R. J. Gaboriaud, *J. Phys. D: Appl. Phys.* **42**, 135410 (2009).

71. A. Metsue, PhD Thesis, University of Lille (2010).

CHAPTER 10

INORGANIC NANOTUBES BASED ON TRANSITION METAL DICHALCOGENIDES: SYNTHESIS AND MECHANICAL PROPERTIES

Maja Remškar

Solid State Physics Department
Jozef Stefan Institute, Ljubljana, Slovenia
Email:maja.remskar@ijs.si

Inorganic and carbon nanotubes distinguish by important peculiarities, from growth mechanisms to various physical and chemical properties and possible applications. A list of inorganic compounds synthesized in a cylindrical shape using a variety of preparation techniques is presented and the most promising directions of applications are indicated. In particular, structural and morphological properties of MoS_2 and WS_2 nanotubes determining their chemical stability and mechanical properties are explained in details. Recently discovered hybrid inorganic nanotube structures, i.e. WS_2 nanobuds and MoS_2 "mama" tubes are shown in a view of their morphological hybrid structure.

1. Introduction

In the last two decades, nanotubes especially carbon ones together with C_{60} represent symbols of developing nanotechnology. Their discovery attracted an interest of scientific and technological community followed by strong financial support and broadened with the production of a variety of nanosized materials and devices. Since 1992, inorganic nanotubes have also played an increasingly important role in nanotube research with a variety of physical possibilities and promising chemistry. Cylindrical geometry of a tubular crystal with curved and therefore

345

strained lattice is a counterpart of relaxed nanowire or nanoparticle morphologies and brings new technological challenges.

2. Synthesis of Inorganic Nanotubes

The first reports on carbon nanotubes in 1991 by Iijima[1] and on inorganic (WS_2) nanotubes by Tenne in 1992[2], have been followed by intense experimental and theoretical research on hollow cylindrical structures. This chapter is confined to inorganic nanotubes (NTs) only. The chronological list of the first reports of different kinds of nanotubes or rolled-up structures shows the degree of activity in this field in the first ten years: *1992*: WS_2[2]; 1993: MoS_2[3]; *1995*: BN[4], SiO_2[5]; *1998*: TiO_2[6], VO_x[7], $NiCl_2$[8]; 2000: $NbSe_2$[9], Au[10], Co and Fe[11]; *2001*: CdS[12], $CdSe$[13], ZnS[14], NiS[15], $Cu_{5.5}FeS_{6.5}$[16], Al_2O_3[17], In_2O_3 and Ga_2O_3[18], GaN[19]; *2002*: ZrS_2 and HfS_2[20], NbS_2 and TaS_2[21], (Er, Tm, Yb, Lu) oxide[22], ZnO[23], $BaTiO_3$ and $PbTiO_3$[24], Cu and Ni[25], Te[26], ReS_2[27], silicon nanotubes[28], etc.

The structural, optical and electrochemical properties of inorganic nanotubes, particularly MoS_2, WS_2, and of fullerene-like particles have been reviewed by Tenne[29,30]. Methods for synthesizing inorganic nanotubes, together with simulations of their structures and predictions of their properties, have been reviewed by Ivanovski.[31]

The objective of the present paper is to review the status of experimental research of inorganic nanotubes since 2004[32] with an emphasis on their structural properties. The number of compounds synthesized in cylindrical shapes was continuously growing as well as variety of preparation techniques.

Several families of inorganic nanotubes and fullerene-like particles have been synthesized up to now. The current list of some representative compounds for each of these families is as follows:

a) Transition metal chalcogenide NTs: MoS_2[3], $MoSe_2$[33], WS_2[2], WSe_2[33], NbS_2[21], $NbSe_2$[9], TaS_2[21], ZrS_2[20], HfS_2[20], TiS_2[34] ZnS[14], NiS[15], $CdSe$[13], CdS[12], VS_2[35];

b) Oxide NTs:

- transition metal oxides: TiO_2^6, ZnO^{23}, GaO/ZnO^{36}, VO_x^{37}, $W_{18}O_{49}^{38}$, $V_2O_5^{39}$, $Al_2O_3^{17}$, $AlOOH^{40}$, $In_2O_3^{18}$, $InVO_4^{41}$, $Ga_2O_3^{18}$, $BaTiO_3^{42}$, $PbZr_{0.52}Ti_{0.48}O_3^{43}$, $PbTiO_3^{42}$, SnO_2^{44}, $MgAl_2O_4^{45}$; MoO_3^{46}; RuO_2^{46}

- silicon oxide: $SiO_2^{5,\,39}$; Ag_2SiO_3/SiO_2^{47}, Fe-doped chrysotile[48];

- rare earth oxides: (Er, Tm, Yb, Lu) oxide [22]; Y_2O_3:(Eu,Tb,Dy)[49], (Y, Gd, Tb, Dy, Ho, Er, Tm, Yb, Lu) oxide [50]

c) Transition metal halogenous NTs: $NiCl_2^8$

d) Mixed phase and metal doped NTs: $PbNb_nS_{2n+1}{}^{51}$, $Mo_{1-x}WS_2{}^{52}$, $W_xMo_yC_zS_z{}^{53}$; $Nb\text{-}WS_2{}^{54}$, WS_2-carbon NTs[55], NbS_2-carbon NTs[56]; $Au\text{-}MoS_2{}^{57}$, $Ag\text{-}WS_2{}^{57}$, $Ag\text{-}MoS_2{}^{57}$, $Cu_{5.5}FeS_{6.5}{}^{58}$; $BaWO_4{}^{59}$, $SbPS_{4-x}Se_x{}^{60}$, $InGaAs/GaAs^{61}$

e) Boron and silicon based NTs: BN^{39}, BCN^{62}, Si^{63};

f) Metal and metal oxide nanotubes: Au [64], Co [65], Fe^{65}, Cu [66], Ni^{66}, Te^{67}, Bi^{68}, $Ni(OH)_2{}^{69}$, $LaNiO_3{}^{70}$

g) Rare earth fluorides : $NaHoF_4{}^{71}$, $NaSmF_4{}^{71}$

h) Organic/inorganic nanotubes: $PVP\text{-}TiO_2{}^{72}$, $K_4Nb_6O_{17}\text{-}(C_nH_{2n+1}NH_3)^{+}$ [73], $(C_4H_{12}N)_{14}[(UO_2)_{10}(SeO_4)_{17}(H_2O)]^{74}$, Eu_2O_3.carbon NTs[75]

i) Transition metal nitrides and carbides: AlN^{76}, WC^{77}, GaN^{78}

The most important methods for growing inorganic nanotubes can be divided broadly into: a) sulphurization, b) decomposition of precursor crystals, c) template growth, d) precursor assisted pyrolysis, e) misfit rolling, f) direct synthesis from the vapour phase, g) electrospinning, h) precipitation, i) self-assembly, etc.. Some nanotubes grow only by the combination of several processes.

3. Applications of Inorganic Nanotubes

Superior physical properties of some inorganic NTs have been predicted and some of them have already been confirmed by experiment. These properties include the shock-wave resistance of WS_2 NTs,[79] use of the WS_2 NTs as ultra sharp tips in scanning probe microscopy,[80, 81] superconductivity in $NbSe_2$ nanotubes and nanorods[82], superconductivity in mixed-phase $W_xMo_yC_zS_2$ NTs,[53] enhanced magnetic coercivity in Ni NTs in comparison with bulk nickel,[66] and the unusual magnetic state in lithium-doped $MoS_{2-x}I_y$ NTs.[83] The $MoS_{2-x}I_y$ sub-nm diameter nanotubes can be used as storage material for reversible lithium batteries[84] or as electron field emitters.[85] The catalytic activity of WS_2 for the hydrodesulphurization of thiophene has been enhanced by formation of unique shapes of the panicles, such as tubes and/or fibres[86] Quantum effects have been demonstrated as, for example, a blue-shift of 100 nm in the CdSe nanotubes relative to the bulk polycrystalline material. The explanation of this effect in terms of quantum confinement in two dimensions has been confirmed by photoluminescence spectroscopy.[12] Nanotubes with important physical and chemical properties have been prepared, like ferroelectric $PbZr_{0.52}Ti_{0.48}O_3$ NTs[43], wide-band $K_4Nb_6O_{17}$[73] semiconducting nanotubes, and the first polymer-amorphous TiO_2 composite materials in tubular structure and of unlimited length were synthesized by electrospinning[72]. The first uranyl selenate[74] nanotubes important for radioactive waste management demonstrate the possibility of tubular nanostructures for actinides in higher oxidation states. The synthetic Fe-doped chrysolite[48] nanotubes represent an alternative to carbon nanotubes for innovative technological applications. The $LaNiO_3$[70] NTs with walls composed of 3-5 nm crystals were prepared as the first step to establish perovskite nanotubes with metal conductity and Pauli paramagnetism. Tungsten carbide (WC) NTs were prepared as particularly important nanomaterial for applications due to a high melting point, superior hardness, low friction coefficient, high oxidation resistance and superior electrical conductivity of bulk compound. Arrays of $InVO_4$ nanotubes[41] revealed twice as a high charge capacity in lithium-ion intercalation than films of the same composition. Reversible electrochemical copper-ion intercalation in the VS_2-alkylamine nanotubes was demonstrated.[35] Functionalization of inner surface of gold nanotubes by peptide-nucleic-acid are used for DNA binding and by such a manner for label-free quantification of complementary DNA sequences.[87] The In_2O_3 NTs have been demonstrated as efficient room-temperature NH_3 gas sensors.[88]

Several further applications of inorganic NTs can be foreseen. Due to their cylindrical geometry, these nano-materials have a low mass density, a high porosity and an extremely large surface to weight ratio. Their potential applications range from high porous catalytic and ultralight anticorrosive materials to non-toxic strengthening fibres. This may lead to more efficient use and increased durability of materials. Doping these semiconducting nano-structured materials may also allow further miniaturization of electronic systems leading to new optoelectronic materials. The helical structure of undoped tubes, with their semiconductor behaviour and optical activity, open up possible applications in non-linear optics and in solar cell technology. Theoretical predictions of dependence of the self-inductance and resistance of chiral conducting nanotubes on the square of the frequency[89] can be tested experimentally in millimetre long inorganic nanotubes. Silicon tubular structures were predicted to have semiconducting features independent on the tube diameter and chirality,[90] which is important for nano-devices in already established silicon nanotechnology.

4. Theoretical Description of Cylindrical Geometry

The basis of the growth mechanism of inorganic nanotubes is the lack of resistance to bending of thin quasi two-dimensional crystal flakes.[91] This bending can be spontaneous, as in transition metal dichalcogenide NTs[92] and in misfit layer compounds[93] grown from the vapour phase, or geometrically conducted by a template growth in channels of alumina membrane[94] and as coatings of nanofibres.[95]

Theoretical considerations suggest that, due to the strain energy coupled with bending, the narrow tubular structure might be less energetically favoured than a finite strip. But when the diameter passes a critical value, the strain in the tubes becomes smaller than the energy associated with the edges (dangling bonds) in the layered strips and the cylindrical geometry, with self-closed layers, becomes clearly the most stable structure.[96] As an example, MoS_2 tubes, with a diameter larger than 6.2 nm, have been shown to be more stable than stripes $\pi \times 6.2$ nm wide. In accordance with these calculations, the smallest inner diameter for single-wall NT should not be smaller than 6 or 7 nm. The smallest stable multi-wall NT has an inner diameter about 5 nm and a corresponding outer diameter of about 12 nm or larger.[96] Nanotubes with radial dependent interlayer distances, which are found experimentally, have not yet been described theoretically.

The simple rolling of crystal flake does not usually lead to nanotube longitudinal growth if the stacking order and orientation relationship between two adjacent turns is not satisfied. It is true that the geometry of the folded flake is cylindrical, but such a formation can only grow in the radial direction until the strain energy no longer exceeds the energy gained by van der Waals interactions between layers. The interaction between the layers is weak and interlayer distances are arbitrary.

On the contrary, if the conditions for stacking order and orientation relationship are satisfied, the interaction between the adjacent turns is stronger, but two limit cases still exist from the macroscopic point of view. In the case where van der Waals interaction energies between the planes are much smaller than those associated with covalent bonds, dislocations will be generated on bending. In other cases, dislocations will not form on bending and the system will remain coherent.[97] Microscopically, the matching of adjacent layers at the atomic level, with consequent extension and/or contraction effects, contributes to the elastic energy caused by the enlarged circumference and by the bending of the molecular layers. The total elastic energy per primitive cell is larger in nanotubes than in microtubes, where the elastic energy is distributed over a larger number of atoms.

The symmetry of inorganic NTs is explained in terms of line groups[98], which describe the quasi one-dimensional systems holding the translational periodicity of elementary structural units along one axis. These structural units have their own point group symmetry and are arranged by operations of the generalized translation group. The theory is formulated for nanotubes with walls one, two or three molecular layers thick.

5. MoS$_2$ and WS$_2$ Nanotubes

The most accurately studied lattice structures of inorganic nanotubes were those of MoS$_2$ and WS$_2$ tubes. Several growth mechanisms have been observed[91] but not all can be scaled up for producing industrially relevant amount of nanotubes. The largest reported quantities were obtained using a fluidized bed reactor, wherein synthesis of multi grams quantities of very long WS$_2$ nanotube phases, was accomplished.[99] The nanotubes obtained in this reaction are open ended, 15-20 nm in diameter and 4-8 layers thick. Some are as long as a few hundred microns, very uniform in shape, many of them have quite perfect crystallinity. Contrary to the self-sacrificial growth mode of different molybdenum or tungsten compounds, where the cylindrical geometry is caused by minimization of free energy by self-termination of molecular layers growing in the sulpurization process and hollow geometry is due to difference in mass density of metal oxides and metal sulfides, thin folded flakes of layered crystals like MoS$_2$ can also directly roll up and adopt the cylindrical shape. The folds of strongly undulated layered crystals at a micro level size can serve as an origin of turbulent gas flow of transported molecules, which promotes a tube-like growth mode. Instability of weakly bonded molecular layers at plate-like crystal edges and at surface growth steps against bending causes the growth of very narrow nanotubes. Several tubes can nucleate by such a way at the same crystal edge, where they combine and continue in growth like a rope.

Recently discovered MoS$_2$ nanopods called "mama"- tubes[100] belong to the most fascinating hybrid inorganic nanomaterials, where nanotubes serve as nanoreactors and afterwards as nanocontainers of MoS$_2$ fullerene-like particles, which have in-situ grown in a confined geometry of nanotubes.

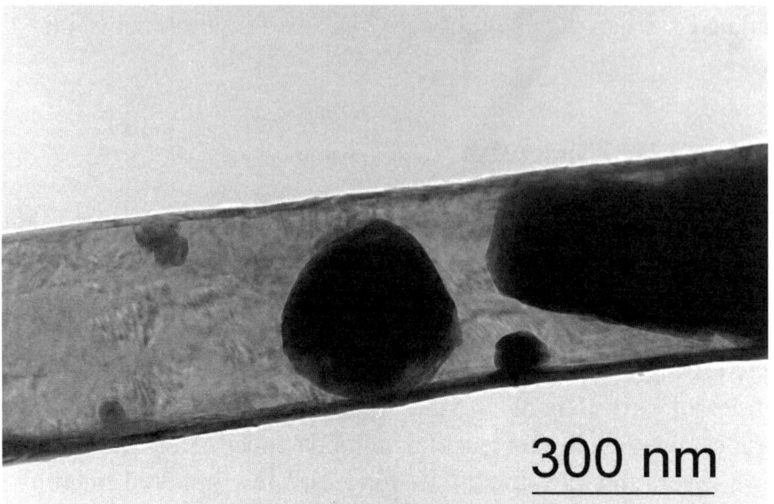

Fig. 1. A transmission electron micrograph of a MoS_2 "mama"-tube with encapsulated MoS_2 fullerene-like particles (Figure originally published in Advanced Materials[100]).

Encapsulated nanoparticles are protected against influence of an environment and any uncontrolled release. On stimulation, thin nanotubes walls break and MoS_2 fullerene-like particles can be released in a controlled way for an application. "Mama"- tubes bear the first inorganic analogy to carbon peapods, where encapsulated C_{60} molecules are situated inside single wall carbon nanotubes.

The temperature controlled transformation of $Mo_6S_2I_8$ precursor nanowires to MoS_2 results in a very rich system of MoS_2 nanotube-hybrid structures[101]. Besides "mama"-tubes, the morphologies like nanotubes with split walls, MoS_2 nanobuds or completely released MoS_2 nano-onions can be grown in a controlled way. The discovered synthetic route based on Mo_6 clusters may apply to other transition metal clusters in combination with different chalcogenides to give rise to a chalcogenide-nanotube technology.

6. Conclusion

Numerous growth techniques are used for synthesis of transition metal dichalcogenide NTs. After the initial enthusiasm at successful synthesis of cylindrical crystals from new inorganic compounds, additional demands on their properties are being made. Control of their dimensions is desirable, and crucial for some applications. The second important demand is structural perfection, which is especially important for their mechanical and electric properties in the construction of nanodevices. As in carbon nanotubes, these demands are rarely satisfied simultaneously in inorganic nanotubes. A further problem is a small quantity of inorganic nanotubes and/or the difficulty with their purification. Many physical experiments and most of the potential applications are restricted by the micro-gram quantities of some nanotubes currently available. There is still plenty of room for technological as well as for basic phenomenological research on inorganic nanotubes.

References

1. S. Iijima, *Nature*, **354**, 56 (1991).
2. R. Tenne, L. Margulis, M. Genut, and G. Hodes, *Nature*, **360**, 444 (1992).
3. Y. Feldman, E. Wasserman, D. J. Srolovitz and R. Tenne, Science 1995, **267**, 222.
4. N.G. Chopra, R.G. Luyken, K. Cherrey, V.H. Crespi, M.L. Cohen, S.G. Louie, and A. Zettl, Science, **269**, 966 (1995).
5. H. Nakamura and Y. Matsui, J.Am.Chem.Soc., **117**, 2651 (1995).
6. P. Hoyer, *Langmuir*, **12**, 1411 (1996).
7. M.E. Spahr, P. Bitterli, R. Nesper, M. Müller, F. Krumeich and H.U. Nissen, *Angew.Chem., Int.ed.*, **37**, 1263 (1998) (*Angew.Chem.*, **110**, 1339(1998))
8. Y.R. Hacohen, E. Grunbaum, R. Tenne, J. Sloan and J.L. Hutchison, *Nature*, **395**, 337 (1998).
9. D.H. Galvan, J.H. Kim, M.B. Maple, M. Avalos-Berja and E. Adem, *Fullerene Sci Technol.*, **8**, 143 (2000).
10. J.C. Hutleen, K.B. Jirage and C.R. Martin, *J.Am.Chem.Soc.*, **120**, 6603 (2000).
11. G. Tourillon, L. Pontonnier, J.P. Levy and V. Langlais, *Electrochem.Solid-State Lett.*, **3**, 20 (2000).

12. C.N.R.Rao, A.G. Govindaraj, F.L. Deepak, N.A. Gunari and M.Nath, *Appl.Phys.Lett.*, **78**, 1853 (2001).

13. A. Govindaraj, F.L. Deepak, N.A. Gunari and C.N.R. Rao, *Israel J. Chem.*, **41**, 23 (2001).

14. L. Dloczik, R. Engelhardt, K. Ernst, S. Fiechter, I. Seiber and R. Könenkamp, *Appl.Phys.Lett.*, **78**, 3687 (2001).

15. X. Ziang, Y. Xie, L. Zhu, W. He and Y. Qian, *Adv.Mater.*, **13**, 1278 (2001).

16. Y. Peng, Z. Meng, C. Zhong, J. Lu, L. Xu, S. Zhang and Y. Qian, *New J. Chem.*, **25**, 1359(2001).

17. L. Pu, X. Bao, J. Zou and D. Feng, *Angew.Chem., Int. Ed.*, **40**, 1490(2001).

18. B. Cheng and E.T. Samulski, *J.Mater.Chem.*, **11**, 2901(2001).

19. J.Y. Li, X.L. Chen, Z.Y. Qiao, Y.G. Cao, H. Li, *J.Mater.Sci.Lett.*, **20**, 1987 (2001).

20. M. Nath and C.N.R. Rao, *Angew.Chem., Int.Ed.*, **41**, 3451 (2002).

21. M. Nath and C.N.R. Rao, *J.Am.Chem.Soc.*, **123**, 4841 (2001).

22. M. Yada, M. Mihara, S. Mouri, M. Kuroki and T. Kijima, *Adv. Mater.*, **14**, 309 (2002).

23. J.Wu, S. Liu, C. Wu, K. Chen and L. Chen, *Appl.Phys.Lett.*, **81**, 1312 (2002).

24. B. A. Hernandez, K.S. Chang, E.R. Fisher and P.K. Dorhout, *Chem. Mater.*, **14**, 480 (2002).

25. C.C. Han, M.Y. Bai and J.T. Lee, *Chem. Mater.*, **13**, 4260 (2001).

26. B. Mayers and Y. Xia, *Adv. Mater.*, **14**, 279 (2002).

27. M. Brorson, T.W. Hansen and C.J.H. Jacobsen, *J.Am.Chem.Soc.*, **124**, 11582 (2002).

28. J. Sha, J. Niu, X. Ma, J. Xu, X. Zhang, Q. Yang and D. Yang, *Adv.Mater.*, **14,** 1219 (2002).

29. R. Tenne, Colloids and Surfaces A: Physicochem. Eng. Aspects, **208**, 83 (2002).

30. R. Tenne, *Nature Nanotechnology,***1**, 103 (2006).

31. A.L. Ivanovskii, *Russian Chemical Reviews*, **71**, 175 (2002).

32. M. Remskar, *Adv.Mater.*, **16**,1497 (2004).

33. M. Nath and C.N.R. Rao, *Chem.Commun.* **21**, 2236 (2001).

34. A. Margolin, R. Popovitz-Biro, A. Albu-Yaron, A. Moshkovich, L. Rapoport and R. Tenne,, *Current Nanoscience*, **1**, 253 (2005).

35. H.A. Therese, et al. , *Angew. Chem. Int. Edn*, **44,** 262 (2005).

36. J. Hu, Y. Bando, and Z. Liu, *Adv.Mater.*, **15**, 1000 (2003).

37. M. Niederberger, H.J. Muhr, F. Krumeich, F. Bieri, D. Günther and R. Nesper, *Chem.Mater.*, **12**, 1995 (2000).

38. W.B. Hu, Y.Q. Zhu, W.K. Hsu, B.H. Chang, M. Terrones, N. Grobert, H. Terrones, J.P. Hare, H.W. Kroto and D.R.M. Walton, *Appl.Phys. A*, **70**, 231 (2000).

39. B.C. Satshkumar, A. Govindaraj, M. Erasmus Vogl, Lipika Basumallick and C.N.R. Rao, *J.Mater.Res.*, **12**, 604 (1997).

40. J.G. Lu, J. Zhang, W.P. Ding, B. Shen and X.F. Guo, *Chinese J.Inorg.Chem.*, **23**, 897 (2007).

41. Y. Wang and G.Z. Cao, *J.Mater.Chem.*, **17**, 894 (2007).

42. B. A. Hernandez, K.S. Chang, E.R. Fisher and P.K. Dorhout, *Chem. Mater.*, **14**, 480 (2002).

43. Y. Luo, I. Szafraniak, N.D. Zakharov, V. Nagarajan, M. Steinhart, R.B. Wehrspohn, J.H. Wendorff, R. Ramesh and M. Alexe, *Appl.Phys.Lett.*, **83**, 440 (2003).

44. Y. Wang, J.Y. Lee and H.C. Zeng, *Chem.Mater.*, **17**, 3899 (2005).

45. H. Fan, M. Knez, R. Scholz, K. Nielsch, E. Pippel, D. Hesse, U. Gësele and M. Zacharias, *Nanotechnology* **17** 5157 (2006).

46. B.C. Satishkumar, A. Govindaraj, M. Nath and C.N.R. Rao, *J.Mater.Chem.*, **10**, 2115 (2000).

47. J. Sha, J. Niu, X. Ma, J. Xu, X. Zhang, Q. Yang and D. Yang, *Adv.Mater.*, **14**, 1219 (2002).

48. E. Foresti, M.F. Hochella, H. Kornishi, I.G. Lesci, A.S. Madden, N. Roveri and H.F. Xu, *Adv.Func.Mater.*, **15**, 1009 (2005).

49. G.X. Liu and G.Y. Hong, *J.Nanosci.Nanotech.*, **6**, 120 (2006).

50. M. Yada, C. Taniguchi, T. Torikai, T. Watari, S. Furuta and H. Katsuki, *Adv. Mater.*, **16**, 1448 (2004).

51. D. Bernaerts, S. Amelincx, G. Van Tendeloo and J. Van Landuyt, *J. Cryst. Growth*, **172**, 433 (1997).

52. M. Nath, K. Mukhopadhyay and C.N.R. Rao, *Chem.Phys.Lett.*, **352**, 163 (2002).

53. W.K. Hsu, Y.Q. Zhu, C.B. Boothroyd, I. Kinloch, S. Trasobares, H. Terrones, N. Grobert, M. Terrones, R. Escudero, G.Z. Chen, C. Colliex, A.H. Windle, D.J. Fray, H.W. Kroto, and D.R.M. Walton, *Chem.Mater.*, **12**, 3541 (2000).

54. Y.Q. Zhu, W.K. Hsu, S. Firth, M. Terrones, R.J.H. Clark, H.W. Kroto and D.R.M. Walton, *Chem.Phys.Lett.*, **342**, 15 (2001).

55. R.L.D. Whitby, W.K.Hsu, C.B. Boothroyd, H.W. Kroto and D.R.M. Walton, *Chem.Phys.Lett.*, **359**, 121 (2002).

56. Y.Q. Zhu, W.K. Hsu, H.W. Kroto and D.R.M. Walton, *J.Phys.Chem. B*, **106**, 7623 (2002).

57. M. Remskar, Z. Skraba, P. Stadelmann and F. Levy. *Adv.Mater.*, **12**, 814 (2000).

58. Y. Peng, Z. Meng, C. Zhong, J. Lu, L. Xu, S. Zhang and Y. Qian, *New J. Chem.*, **25**, 1359 (2001).

59. D. Li, H. Wu and Z. Li, X.F. Cong, J. Sun, Z.H. Ren, L.P. Liu, Y. Li, D.W. Fan and J.C. Hao, *Colloids Surf. A*, , **274**, 18 (2006).

60. C.D. Malliakas and M.G. Kanatzidis, *J. Am. Chem. Soc.*, **128,** 6538 (2006).

61. A.V. Prinz and V. Ya. Prinz, *Surf.Sci.*, **532-535**, 911 (2003).

62. O.Stephan, P.M. Ajayan, C. Colliex, Ph. Redlich, J.M. Lambert, P. Bernier and P. Lefin, *Science*, **266**, 1683 (1994).

63. J. Sha, J. Niu, X. Ma, J. Xu, X. Zhang, Q. Yang and D. Yang, *Adv.Mater.*, **14**, 1219 (2002).

64. J.C. Hutleen, K.B. Jirage and C.R. Martin, *J.Am.Chem.Soc.*, **120,** 6603 (2000).

65. G. Tourillon, L. Pontonnier, J.P. Levy and V. Langlais, *Electrochem.Solid-State Lett.*, **3**, 20 (2000).

66. C.C. Han, M.Y. Bai, and J.T. Lee, *Chem. Mater.*, **13**, 4260 (2001).

67. B. Mayers and Y. Xia, *Adv. Mater.*, **14**, 279 (2002).

68. Y.D. Li, J.W. Wang, Z.X. Deng, Y.Y. Wu, X.M. Sun, D.P. Yu and P.D. Yang, *J.Am.Chem.Soc.*, **123**, 9904 (2001).

69. S.L. Chou, F.Y. Cheng and J. Chen, *Eur.J.Inorg.Chem.*, **20**, 4035 (2005).

70. M. Tagliazucchi, R.D. Sanches, H.E. Troiani and E.J. Calvo, *Solid State Comm.*, **137**, 212 (2006).

71. L.F. Liang, H.F. Xu, Q. Su, H. Konishi, Y.B. Jiang, M.M. Wu, Y.F. Wang and D.Y. Xia, *Inorg. Chem.*, **43**, 2004 (2004).

72. D. Li and Y. Yia, *Nano Lett.*, **4**, 933 (2004).

73. G. Du, Y. Yu and L.M. Peng, *Chem.Phys.Lett.*, **400**, 536 (2004).

74. S.V. Krivovichev, V. Kahlenberg, I.G. Tananaev, R. Kaindl, E. Mersdorf and B.F. Myasoedov, *JACS*, **127**, 1072 (2005).

75. L. Fu, Z. Liu, Y. Liu, B. Han, J. Wang, P. Hu, L. Cao and D. Zhu, *Adv. Mater.*, **16**, 350 (2004).

76. Q. Wu, Z. Hu, X.Z. Wang, Y.N. Lu, X. Chen, H. Xu and Y.J. Chen, *Am.Chem.Soc.*, **125**, 10176 (2003).

77. S.V. Pol, V.G. Pol and A. Gedanken, *Adv.Mater.*, **18,** 2023 (2006).

78. J. Goldberger, R. He, Y. Zhang, S. Lee, H. Yan, H.J. Choi and P. Yang, *Nature*, **422**, 599 (2003).

79. Y.Q.Zhu, T. Sekine, K.S. Brigatti, S. Firth, R. Tenne, R. Rosentsveig, H.W. Kroto and D.R.M. Walton, *J.Am.Chem.Soc.*, **125**, 1329 (2003).

80. A. Rothschild, S.R. Cohen and R. Tenne, *Appl.Phys.Lett.*, **75**, 4025 (1999).

81. H. Dai, J.H. Hafner, A.G. Rinzler, D.T. Colbert and R.E. Smalley, *Nature*, **384**, 147 (1996).

82. M. Nath, S. Kar, a.K. Raychaudhuri and C.N.R. Rao, *Chem.Phys.Lett.*, **368**, 690 (2003).

83. D. Mihailovic, Z. Jaglicic, D. Arcon, A. Mrzel, a. Zorko, M. Remskar, V.V. Kabanov, R. Dominko, M. Gaberscek, C.J. Gómez-García, J.M. Martínez-Agudo and E. Coronado, *Phys.Rev.Lett.*, **90**, 146401 (2003).

84. R. Dominko, D. Arcon, A. Mrzel, A. Zorko, P. Cevc, P. Venturini, M. Gaberscek, M. Remskar and D. Mihailovic, *Adv.Mater.*, **14**, 1531 (2002).

85. V. Nemanic, M. Zumer, B. Zajec, J. Pahor, M. Remskar, A. Mrzel, P. Panjan and D. Mihailovic, *Appl.Phys.Lett.*, **82,** 4573 (2003).

86. E. Furimsky, *Appl.Catalysis A*, **208**, 251 (2001).

87. G. Jágerszki, R.E. Gyarcsányi, L. Höfler and E. Pretsch, *Nano Letters*, **7**, 1609 (2007).

88. N. Du, H. Zhang, B. Chen, X. Ma, Z. Liu, J. Wu and D. Yang, Adv.Mater., **19**, 1641 (2007).

89. Y. Miyamoto, A. Rubio, S.G. Louie and M.L. Cohen, *Phys.Rev. B*, , **60**, 13885 (1999).

90. S.Y. Jeong, J.Y. Kim, H.D. Yang, B.N. Yoon, S.H. Choi, H.K. Kang, C.W. Yang and Y.H. Lee, *Adv.Mater.*, **15**, 1172 (2003).

91. M. Remškar, Z. Skraba, F. Cleton, R. Sanjines and F Levy, *Appl. Phys. Lett.*, **69**, 351 (1996).

92. M. Remškar, Z. Skraba, R. Sanjinés and F. Lévy, *Appl. Phys. Lett.*, , **74**, 633 (1999).

93. D. Bernaerts, S. Amelincx, G. Van Tendeloo and J. Van Landuyt, *J. Cryst. Growth*, **172,** 433 (1997).

94. B. Cheng and E.T. Samulski, *J.Mater.Chem.*, **11,** 2901 (2001).

95. R.L.D. Whitby, W.K.Hsu, C.B. Boothroyd, H.W. Kroto and D.R.M. Walton, *Chem.Phys.Lett.*, **359,** 121 (2002).

96. G. Seifert, T. Köhler and R.Tenne, *J. Phys. Chem. B*, **106**, 2497 (2002).

97. D.J. Srolovitz, S.A. Safran, and R. Tenne, *Phys.Rev. E*, **49**, 5260 (1994).

98. I. Milosević, T. Vuković, M. Damjanović and B. Nikolić, *Eur.Phys.J.*, **17**, 707 (2000).

99. R. Rosentsveig, A. Margolin, Y. Feldman, R. Popovitz-Biro and R. Tenne, *Chem. Mater.* **14**, 471 (2002).

100. M. Remškar, A. Mrzel, M. Virsek and A. Jesih, *Adv. Mater.* **19**, 4276 (2007).

101. M. Remškar, M. Viršek and A. Mrzel, *Apply. Phys. Lett.* **95**, 133122 (2009).

CHAPTER 11

AN INTRODUCTION TO SPIN ELECTRONICS

J. M. D. Coey

CRAN and School of Physics, Trinity College
Dublin 2, Ireland
E-mail: jcoey@tcd.ie

Spin electronics exploits the spin angular momentum and magnetic moment of the electron to add new functionality to electronic devices. A first generation of applications was based on magnetoresistive sensors and magnetic memory. The sensors have numerous applications, not least in digital recording. Read heads are usually spin-valves exhibiting giant magnetoresistance (GMR) or tunnelling magnetoresistance (TMR). Magnetic random-access memory (MRAM) is based on switchable TMR spin valve cells, similar in structure to the read head. New devices are under development where the angular momentum of a spin-polarized current is used to exert spin transfer torque, or the flow of spin-polarized electrons is controlled via a third-electrode in a transistor-like structure.

1. Introduction

A hugely successful electronics industry has been built around the manipulation of electronic charge in semiconductor microcircuits. The operations needed for computation are conducted using complementary metal-oxide semiconductor (CMOS) logic. The semiconductors can be doped n or p-type so that the charge carriers may be electrons or holes. Binary data are stored as charge on the gates of field-effect transistors (FETs). An important feature of CMOS logic is that it consumes power only when the transistors are switching between the *on* and *off* states. It is scalable technology, which has been repeatedly miniaturized since its introduction in 1982. According to the semiconductor industry roadmap, the minimum feature size in silicon circuits was 45 nm in 2008, and it is projected to decrease to 22 nm in 2011. It is unclear what will take over at the end of the roadmap, when feature sizes of less than 10 nm will make CMOS unsustainable. Some form of spin-based electronics may be an option. Ultimately, we might imagine binary information encoded in the spin of a singe electron, or quantum coherence of electronic states being exploited in q-bits which offer a new paradigm for computation.

Semiconductor random-access memory can be dynamic (DRAM) or static (SRAM). Both are volatile. Information stored on the gate of an FET is lost, when the power is shut down. DRAM requires only one transistor per memory cell, but it demands periodic refreshment every few milliseconds as the charge leaks away. SRAM requires less power and is faster, but the memory cell uses six transistors. Error rates are about one per month per gigabit of memory, but these are bound to increase as the cells become ever smaller and cosmic rays and other background radiation disturb their stored charge. Permanent information storage demands nonvolatile memory, and this is where magnetic recording comes in. Ferromagnetic information storage has been the partner of semiconductor electronics in the information revolution; it owes its continuing success to the scalability of the magnetic dipole field, which varies as m/r^3, where $\mathsf{m} \approx Md^3$ is the magnetic moment of a magnetized bit of material with magnetization M and dimension d. Hard discs and magnetic tapes are the repository for mind-boggling amounts of archived data in computers and servers. The recorded information is erasable and the media can be reused. The amount of data stored on servers and other hard discs around the world is astonishing, of order 1 exobit (10^{21} bits). The text of this article is a mere 10^7 bits, and the figures account for a further 10^9 bits. Each and every one is stored in an individually-addressable magnet on a hard disc. It is sobering to realize that the numbers of magnets and transistors manufactured in fabs exceeds the number of grains of rice and corn grown in fields!

Flash is nonvolatile computer memory that can be electrically erased and reprogrammed. It is based on FET cells that have two gates instead of one. A fully-insulated floating gate stores the charge. Flash is economical, but rather slow, with limited re-writeability. Portable memory sticks based on flash are challenging magnetic hard disc storage for low-end applications.

Conventional electronics pays no attention to electron spin. It treats the electrons simply as a mobile charged particles. Charge currents surge around silicon chips in ever-smaller and faster semiconductor circuits, where the number of transistors per chip has been doubling every two years in accordance with Moore's law. Named after the founder of INTEL, this is not a physical principle but an empirical observation based on over 30 years of industrial development. Like all exponential growth, Moore's law growth must eventually come to an end. By 2020 the extrapolated dimensions of a transistor approach those of the atom! But for the present, Moore's law retains the status of a self-fulfilling prophesy. Nonvolatile storage of vast quantities of digital data on hard discs and tapes has remained competitive because magnetic storage densities have followed their own Moore's law, with densities doubling even faster than for semiconductors (Fig. 1).

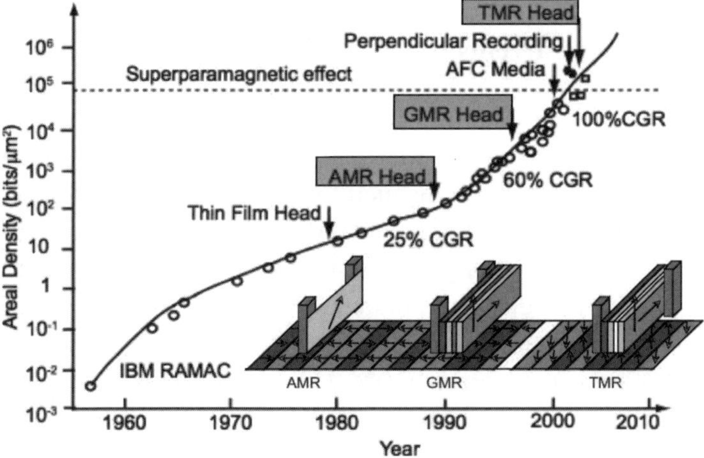

Fig. 1. Progress of hard disc magnetic storage. Three generations of magnetoresistive readers are illustrated.

This progress has been facilitated by three innovations in the stray field sensor, which is the read head. First was the introduction of a thin film reader based on anisotropic magnetoresistance (AMR), an intrinsic property of ferromagnetic metals discovered in 1857. Next, following the discovery of giant magnetoresistance (GMR) by Albert Fert and Peter Grünberg in 1988 for which they received the 2007 Nobel Prize in Physics, was the introduction of the GMR spin valve in 1998. More recently, it is being replaced by a tunnel magnetoresistance (TMR) spin valve, and a changeover from longitudinal to perpendicular recording.

The relentless trend towards extreme miniaturization of electronic logic and memory has led to the consideration of schemes where the electron spin plays a key role in the operation of new thin-film devices. A highly-successful first generation of spin electronics is based on the magnetoresistive sensors and related bistable memory elements. These are two-terminal devices. Future three-terminal devices, including various types of spin transistor, may offer spin gain. The vision for spin electronics is that it will integrate magnetism with electronics at chip level, a marriage already achieved with optics in various optoelectronic devices.

2. Spin-Polarized Currents

Together with its charge e and mass m_e, the electron carries quantized angular momentum $\hbar m_s$ where $m_s = \pm\frac{1}{2}$ for the $\uparrow$ and $\downarrow$ states. The arrow indicates the direction of the magnetic moment of the electron, which is opposite to the direction of angular momentum because of the negative sign of the electronic charge. The $\uparrow$ electrons are the majority-spin electrons in a ferromagnet, or a paramagnet in an applied field. The spin moment of the electron is proportional to its spin angular momentum, $m = \gamma\hbar m_s = \pm 1$ Bohr magneton, where $\gamma = e/m_e$ is the gyromagnetic ratio. Spin angular momentum of the electron is the basis of both solid-state magnetism and spin electronics. An electric current is always a flow of charge; but it can also be a flow of angular momentum. The important difference is that angular momentum, unlike charge, is not conserved in an electric circuit. Electrons can be flipped from ↑ to ↓ states, or vice versa, by scattering processes which are relatively uncommon compared with the normal scattering events that modify the momentum and occasionally the energy of the electron. The comparative rarity of spin-flip scattering means that conduction can be thought of as taking place in two independent, parallel channels for ↑ and ↓ electrons — the two current model proposed by Mott in 1936 for conduction in metals. A brief review of spin polarized electronic conduction is appropriate, before embarking on the applications.

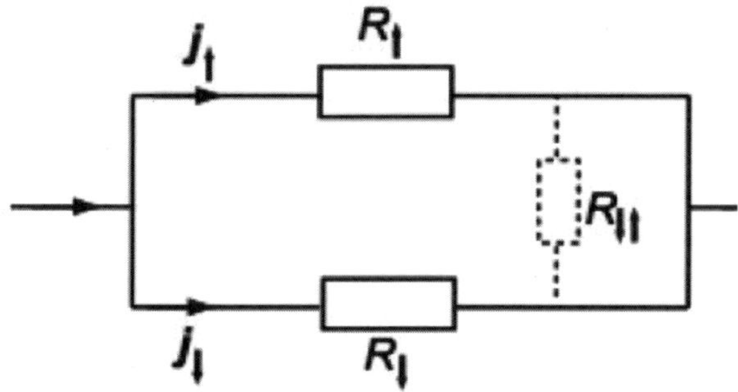

Fig. 2. The two-current model.

2.1. *Conduction mechanisms*

Conduction in metals is normally a diffusive process. The electrons are continually being scattered, and the mean free path between scattering events may be different for ↑ and ↓ electrons in a ferromagnetic metals. The conductivity for each channel is then proportional to its mean free path $\lambda_\uparrow$ or $\lambda_\downarrow$. Consider first conduction in copper, the nonmagnetic metal having a filled *3d* band with ten electrons per atom, and a half-filled *4s* band with one electron. Conduction is due to the 4s electrons which acquire a drift velocity v_d in the direction of an applied electric field. The current density is $j = nev_d$, where n is the electron density. In copper, n is 8.45×10^{28} m^{-3}. Typical current densities in electronic circuits are $\sim 10^7$ A m^{-2}, making v_d of order 1 mm s^{-1}.

The mean free path λ is the average distance traveled by an electron in the time τ between collisions, which is known as the momentum relaxation time. Then $\lambda = v_F \tau$ where v_F is the Fermi velocity. In the free-electron model $v_F = (\hbar/m_e)(3\pi^2 n)^{1/3}$, which in the case of copper is 1.6×10^6 ms^{-1}. The Fermi wavelength of the electron in copper, $\lambda_F = h/m_e v_F$, is therefore 0.5 nm or about two interatomic spacings. The relation between conductivity σ and mean free path λ is

$$\sigma = ne^2\lambda/m_e v_F \tag{1}$$

Since the conductivity of pure copper is $\approx 10^8$ S m^{-1}, the mean free path there is about 40 nm, and the momentum relaxation time is about 25 fs. Impurities reduce σ, λ and τ. A rough rule of thumb for monovalent metals is $\lambda \approx 10^{-15}\sigma$ meters. An electron, like a billiard ball, will undergo several elastic collisions with momentum transfer before it experiences an inelastic collision, which entails energy loss (or gain) as well. The electron undergoes many collisions, of order $\nu = 100$ or more, before it experiences a spin-flip scattering event that involves an exchange of angular momentum with the lattice. The distance traveled in time t $\gg \tau$ in a diffusive process is $l = \sqrt{(D_e t)}$, where D_e is the diffusion constant (units m^{2}s^{-1}). Since electrical conduction is essentially diffusion of electrons in the direction of the applied electric field, the spin diffusion length is given by the three-dimensional random-walk expression

$$l_s = \sqrt{(D_e \tau_s)} \tag{2}$$

where τ_s is the time between spin-flip scattering events, and the diffusion constant for electrons, which is proportional to the conductivity, is $D_e = (1/3)v_F^2\tau = (1/3)v_F\lambda$. Hence $l_s = \sqrt{[(1/3)v_F\lambda^2]}$. The distance traveled by the electron along its path between spin flip events is $\lambda_s = v_F\tau_s$. Conductivity is related to the density of states at the Femi level $N(E_F)$ and the diffusivity by the Einstein relation

$$\sigma = e^2 N(E_F)D_e \tag{3}$$

For Cu, $D_e = 21\ 10^{-3}\ m^2\ s^{-1}$, so the estimate $\tau_s = 100\tau$ gives $l_s = 240$ nm. The value for highly pure copper is larger than this, but copper in thin film devices contains defects and diffused impurity atoms which decrease the conductivity and reduce the mean free path and spin diffusion length below these estimates.

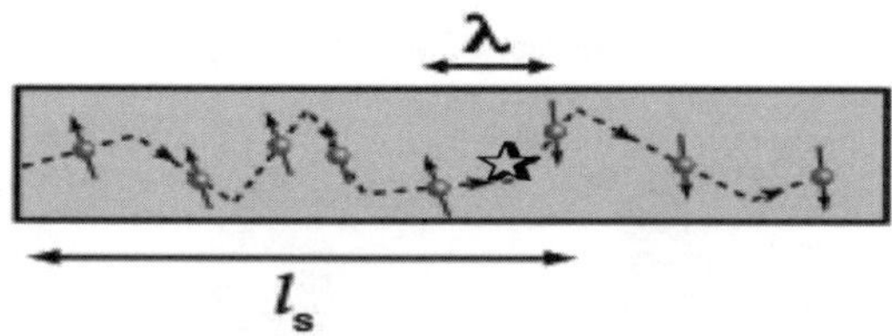

Fig. 3. The diffusive process characteristic of electrical conduction in metals. Since spin-flip scattering is much less common than normal momentum scattering events, the spin diffusion length l_s is much longer than the mean free path λ. Generally $\lambda_s \gg l_s > \lambda$.

Passing now from copper to nickel, a strong ferromagnetic metal with one less electron per atom, the $3d$ band is spin split, as shown in Fig 4. The electronic configuration of Ni is $3d^{9.4}4s^{0.6}$. Conduction is still mainly by $4s$ electrons, which have a much greater mobility than their $3d$ counterparts, but the 4s band is partially spin polarized, and the $\uparrow$ and $\downarrow$ carriers are now scattered differently. The $\uparrow$ electrons behave much as in copper since the $3d^{\uparrow}$ band is full, but the $\downarrow$ electrons can be scattered into empty $3d^{\downarrow}$ states at the Fermi level. The mean free path of $\downarrow$ electrons in nickel is about five times shorter than that of their $\uparrow$ counterparts. The spin diffusion length in nickel is only about 10 nm. Cobalt is more effective as a spin polarizer. Indicative values of the scattering lengths for some metals and alloys commonly featured in thin film devices are listed in Table 1.

Table 1. Estimates of mean free paths and spin diffusion lengths, in nm.

	Al	Fe	Co	Ni	$Fe_{20}Ni_{80}$	Cu
$\lambda_{\uparrow}$	12	8	20	5	15	20
$\lambda_{\downarrow}$	12	8	1	1	1	20
l_s	350	50	40	10	3	200

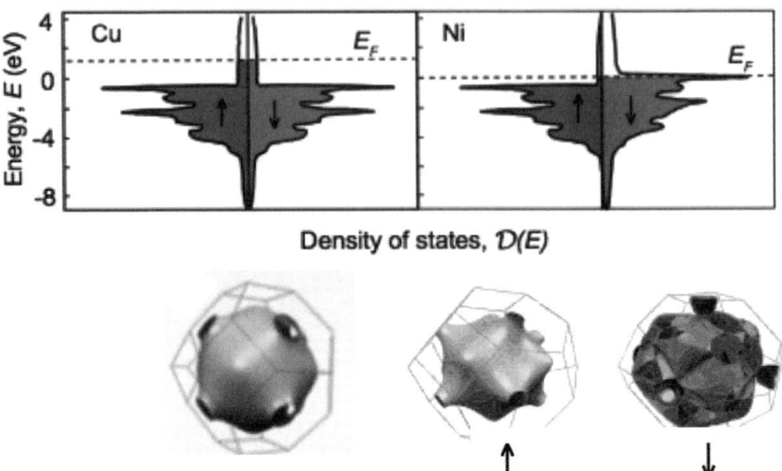

Fig. 4. Comparison of the densities of states and Fermi surfaces of copper and nickel.

Besides diffusion, three other modes of electron transport are encountered in solids. One is *ballistic transport*, where the mean free path of the electrons exceeds the dimensions of the conductor, so the electrons traverse the conductor in a single shot, without scattering. This arises in small point contacts, and in highly-perfect conductors with a low electron density, such as semiconductors or carbon nanotubes. There is no distinction between ballistic transport of ↑ or ↓ carriers. In diffusive transport, the electrons move ballistically between collisions.

The resistance of electrical circuits with a constriction, and therefore their magnetoresistance, is dominated by the resistance of the constriction. The resistance for a large contact, where the transport is diffusive, is simply $R = \rho t/\pi a^2$. Here ρ is the resistivity of the material in the contact, a is its radius and t is the thickness. When the diffusive contact is thin, t $\ll$ 2a, the resistance $R_M = \rho/2a$ is known as the Maxwell resistance. On the other hand, if the transport is ballistic, the resistance of a small contact is given by the Sharvin formula

$$R_S = \{2h/e^2\}/(a^2 k_F^2) \tag{4}$$

The crossover from diffusive to ballistic transport for a thin contact occurs when $a = 0.85\lambda$. For pure copper, the crossover contact resistance is about $0.1\ \Omega$, but it scales with the square of the resistivity, so that it can rise to a kilohm in a poor metal with $\rho \approx 1\ M\Omega$ m. The quantity $2h/e^2$ features frequently in transport theory. It has units of ohms, and the magnitude is $12{,}906\ \Omega$. The inverse $G_0 = e^2/2h$ is known as the conductance quantum. This is for both spins; the conductance for an polarized spin channel is half as large, e^2/h.

Since there is no scattering in a ballistic channel, the resistance must somehow be associated with the contacts. Ballistic nanowires with radius a comparable to the Fermi wavelength, behave as an electron waveguides with only a few transverse modes separated in k_y or k_z by π/a, but the wave-vector along the length of the wire is unconstrained. The conductance of each mode is G_0, multiplied by a transmission factor $T_i \leq 1$, giving the *Landauer formula*.

$$G = (e^2/h)\ \Sigma_i\ T_i \tag{5}$$

The third means of electron transport is *tunneling*. When two conductors are separated by a thin layer of insulator, or a vacuum gap where the electronic wave functions decay exponentially, there is a probability for the electrons to tunnel through the barrier, provided its thickness w is less than or of order the decay length of the electronic wave function. The probability of tunneling through a barrier of height ϕ and width w is given by the transmission coefficient

$$T = \exp\{-2w\sqrt{\{2m_e e\phi/\hbar\}} \tag{6}$$

The characteristic feature of tunneling transport is a conductance $G \ll G_0$ which is almost independent of temperature, and shows a nonlinear $I{:}V$ characteristic

$$I = GV + \beta V^3 \tag{7}$$

due to the deformation of the barrier in the applied electric field. Fitting this curve to an expression given by Simmonds, for example, for tunnelling through a barrier yields the effective barrier height ϕ and width w.

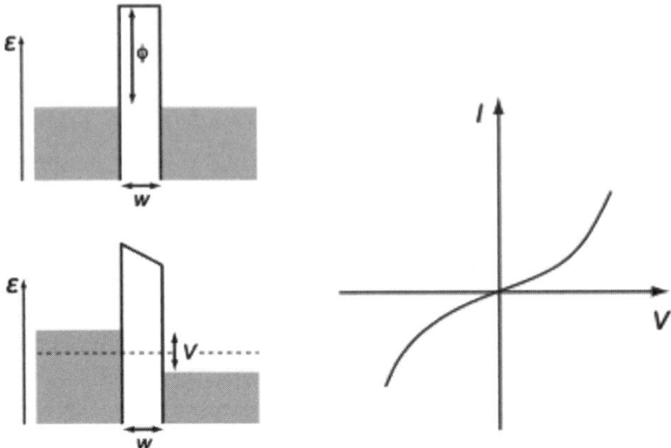

Fig. 5. A tunnel barrier for electrons is deformed when a voltage is applied, leading to a nonlinear *I:V* curve.

The fourth mode of electron transport is *hopping*. This arises when the electrons are localized, so they are described by a confined wavepacket $\psi \sim \exp\text{-}\alpha r$ rather than an extended wavefunction. Electrons move from one site to the next by thermally-assisted jumps. When the hopping to nearest-neighbour sites requires an activation energy E_a the conductivty is proportional to $\exp(\text{-}E_a/k_BT)$, much like a semiconductor. At low temperatures, the electrons may jump further afield to find a site with almost the same energy. The conductivity is then described by Mott's variable-range hopping expression

$$\sigma = \sigma_\infty \exp -(T_0/T)^{1/4} \tag{8}$$

where $T_0 = 1.5/[k_B \alpha^3 N(\varepsilon_F)]$. Electron hopping is the principal transport mechanism in defective oxides and organic conductors.

2.2. *Spin polarization*

Spin electronics depends on the creation and detection of spin polarization of mobile electrons. A simple device comprises a source of spin-polarized electrons, which are injected into a conductor of some description, where they transmit the information coded in their spin polarization to a detector. Electrical spin injection and detection is relatively straightforward for all-metal structures such as GMR spin valves, and it works well for the TMR spin valves with an insulating tunnel barrier, These spin-valve magnetoresistors are two-terminal electronic devices. Spin injection is much more problematic for semiconductors where the spin polarization of injected electrons is lost because of the impedance mismatch at the interface.

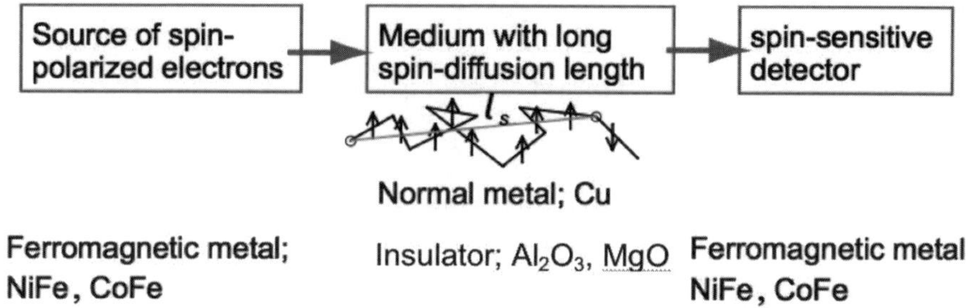

Fig. 6. A generic spin electronic device.

Spin polarization is the central concept here, so we need to understand how it is defined and measured. The definition $p = (n_\uparrow - n_\downarrow)/(n_\uparrow + n_\downarrow)$ in terms of the electron densities $n_\uparrow$ and $n_\downarrow$ is equivalent to the relative spin magnetization of the electrons, or of some specific band. When considering transport, it is appropriate to consider densities of states at the Fermi level $N_\uparrow$ and N , or within an energy eV of the Fermi level, where V is the bias voltage. However, the spin polarization deduced from different experiments depends on what precisely is being measured. A generalized definition at low bias is

$$P_n = \{ v_{F\uparrow}{}^n N_\uparrow - v_{F\downarrow}{}^n N_\downarrow \} / \{ v_{F\uparrow}{}^n N_\uparrow + v_{F\downarrow}{}^n N_\downarrow \} \tag{9}$$

where v_F is the Fermi velocity. The density of states at the Fermi level $N_\uparrow$, is weighted by the Fermi velocity raised to a power n, which is zero for electrons ejected in a spin-polarized photoemission experiment, 1 for ballistic transport and 2 for diffusive transport or tunnelling at low bias. The sense of the velocity averaging is uniaxial for ballistic transport. Furthermore, transport measurements reflect the electron mobility, so measurements in $3d$ metals are more sensitive to the polarization of $4s$ rather than $3d$ electrons. Table 2 lists some calculated spin polarizations for $3d$ ferromagnets. The interesting point is that different definitions can produce completely different results, and they do not even necessarily give the same sign. Only for a half metal, which is a metal with a gap for one spin state, so that either $N_\uparrow$ or $N_\downarrow = 0$, do all definitions agree that $P_n = 1$.

Table 2. Calculated values of spin polarization.

	Fe	Co	Ni	CrO_2	$La_{0.67}Ca_{0.33}MnO_3$
$p^\dagger$	0.27	0.18	0.06	1.00	0.91
P_0	0.52	-0.70	-0.77	1.00	0.36
P_1	0.38	-0.39	-0.43	1.00	0.76
P_2	0.36	0.11	0.04	1.00	0.92

†Spin polarization of $3d$ and $4s$ electrons.

Some methods of measuring spin polarization are indicated in Fig 7. Photoemission probes the electron density at the Fermi level directly, weighted by the photoemission cross section. Spin- dependent tunneling in a junction with two ferromagnetic electrodes and an amorphous barrier gives P_2 but spin filter effects depending on the symmetry of the transmitted wavefunctions appear with crystalline barriers. If one electrode is superconducting, spin polarization can be deduced by applying a magnetic field that is insufficient to turn the superconductor normal, as in the Tedrov-Meservey experiment. Ballistic point contacts between two ferromagnets can also be used to estimate P_1 but care is needed to avoid effects due to magnetostriction. Finally, ballistic transport across a point contact between a ferromagnet and a superconductor offers a new possibility of determining the spin polarization at low temperature, from Andreev reflection.

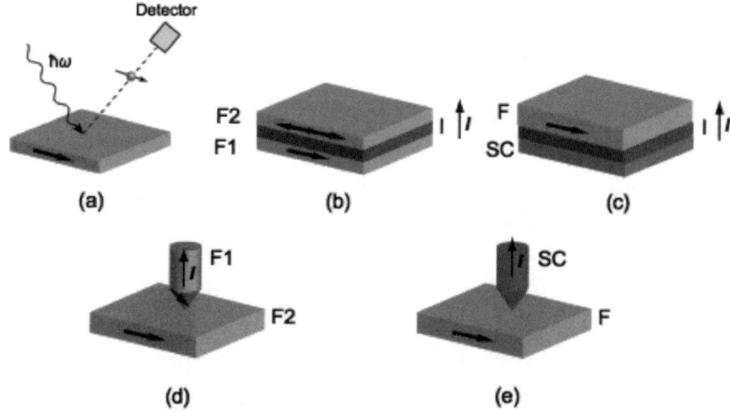

Fig. 7. Some experimental methods for deducing the spin polarization of a ferromagnet: a) photoemission with polarization analysis, b) tunnel magnetoresistance, c) Tedrov - Meservey experiment, d) ballistic point contact, e) Andreev reflection.

Table 3. Spin polarization deduced from Andreev reflection.

Fe	0.40	$Co_{50}Fe_{50}$	0.50
Co	0.40	NiMnSb	0.45
Ni	0.35	Co_2MnSi	0.55
$Ni_{80}Fe_{20}$	0.45	CrO_2	0.95

In Andreev reflection, an electron injected into the superconductor through the point contact with a normal metal must form a Cooper pair if the bias voltage is less than the superconducting energy gap Δ_{sc}. As a result, a hole with opposite spin is injected back into the metal. When V $< \Delta_{sc}$, the conductance of the contact is doubled compared with the value when V $> \Delta_{sc}$, on account of the current carried by the hole. For a half-metal, there are no vacant states with opposite spin near the Fermi level, so the conductance should be strictly zero when V $< \Delta_{sc}$. The polarization in the general case is

$$P_1 \approx (1/2) \{1 - [G(0) - G(V>\Delta_{sc})]/[G(V>\Delta_{sc})]\} \qquad (10)$$

The spin polarizations defined above are intrinsic quasi-equilibrium properties of the ferromagnet. In spin electronics, we are more often

                           J. M. D. Coey

interested in the spin polarization of electric current, which is predominantly carried by *s* electrons.

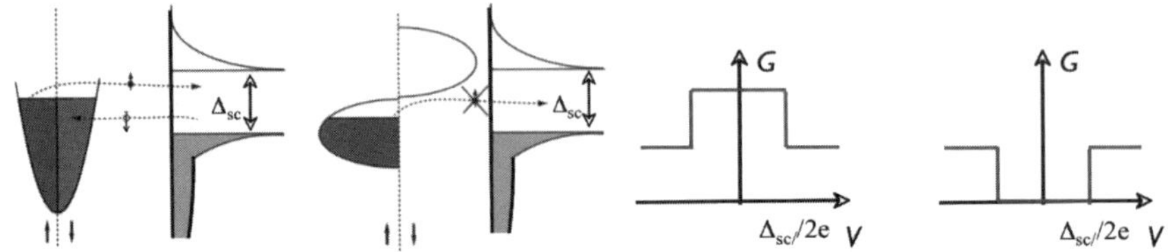

Fig. 8. Point-contact Andreev reflection, showing typical bias-dependence of the conductance for a normal metal (left) and a half-metal (right).

2.3. *Spin injection and spin accumulation*

An important concept in spin-polarized electron transport is *spin accumulation*. Spinpopulations are modified in the vicinity of a contact between ferromagnet and a normal metal, where spin-polarized electrons An important concept in spin-polarized electron transport is *spin accumulation*. Spinpopulations are modified in the vicinity of a contact between ferromagnet and a normal metal, where spin-polarized electrons diffuse into the normal metal, and unpolarized electrons diffuse into the ferromagnet. The extent of the magnetic blurring on each side of the interface is the spin diffusion length, but the equilibrium changes in spin population are very small. Significant nonequilibrium changes in spin polarization are set up when a current flows across the interface. Spin polarized electrons are injected from the ferromagnet into the normal metal, and diffuse away from it. The interface, with the associated changes of chemical potential, is illustrated in Fig. 9. Spin accumulation occurs, for example, in the copper layer of a Co/Cu/Co spin valve where current flows perpendicular to the plane (CPP). It is best sensed in a non-local transport measurement. The spin-polarized electrons are injected into the normal metal from the ferromagnet, and diffuse away from it. The potential of a second ferromagnetic electrode is measured, which is not in the current path (Fig. 10).

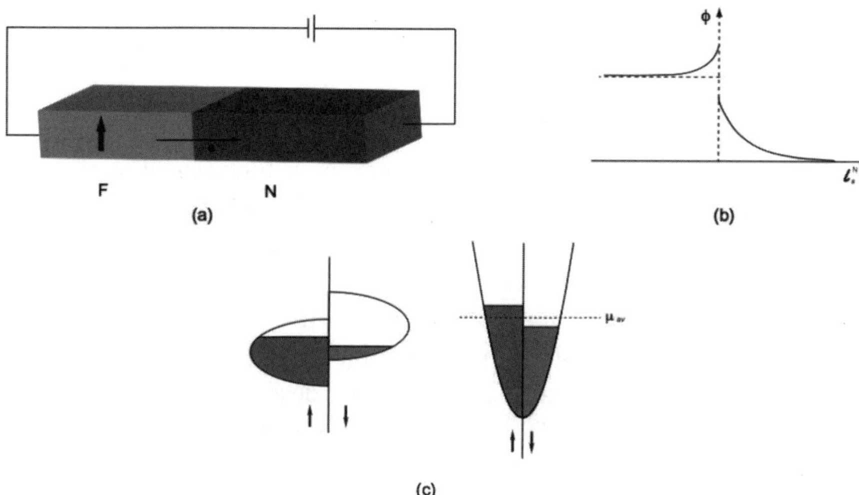

Fig. 9. Spin injection at an interface, where electrons flow from a ferromagnet into a normal metal a). Spin-polarized electrons accumulate in the normal metal b). The nonequilibrium chemical potentials near the interface are illustrated in c).

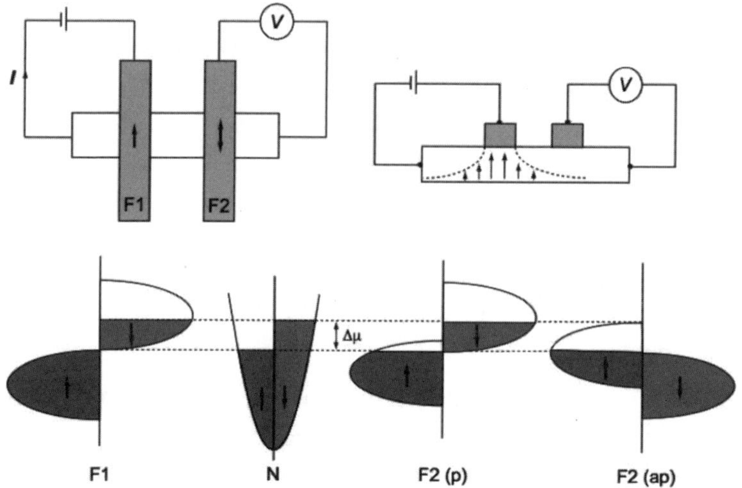

Fig. 10. A nonlocal electrical measurement to probe the spin-dependent potential, without passing a charge current through the measuring electrode. After Ref. 1.

The spin-polarized electrons drift away from the injector, and a spin-polarized charge cloud is set up on a scale of l_s. The potential of the electrode will be slightly different in the ↑ and ↓ states, because the ↑ and ↓ electrons have different chemical potentials, as indicated on the figure. By measuring the spin-dependent potential as a function of x, the spin diffusion length can be determined directly, although it was challenging to do this on account of the short length scale, which is of order 100 nm for a nonmagnetic metal at room temperature.

The decay of the *spin voltage*, $\mu^\uparrow - \mu^\downarrow$, on either side of a ferromagnet/nonmagnet (f/n) metal interface is described by the spin diffusion equation, originally developed to describe the spreading of nuclear polarization in NMR. The diffusion equations for the two spin populations are

$$n^{\uparrow,\downarrow}(x,t)/\tau_s = D_e\{\partial^2 n^{\uparrow,\downarrow}(x,t)/\partial x^2\} \qquad (11)$$

where D_e is the electronic diffusion constant, and τ_s is the spin relaxation time. The surplus spin density produced by the current flowing across the interface $m(x.t) = n^\uparrow(x,t) - n^\downarrow(x,t)$ is obtained by integrating the spin-split electronic density of states $N(\varepsilon)$ up to the chemical potential $\mu^\uparrow$ or μ , and taking the difference. Hence

$$m(x.t) = N(\varepsilon_F)[\mu^\uparrow(x,t) - \mu^\downarrow(x,t)] \qquad (12)$$

Using the differential equation for $n^\uparrow - n^\downarrow$ to obtain $m(x.t)$, the steady-state solution for the spin voltage is

$$(\mu^\uparrow - \mu^\downarrow)_x = (\mu^\uparrow - \mu^\downarrow)_0 \exp(-x/l_s) \qquad (13)$$

where l_s is given by Eq (2).

On the ferromagnetic side of the f/n junction, the spin polarization builds up over a length scale determined by the *s-d* scattering in the ferromagnet. In pure ferromagnetic metals, l_s is about 40 nm and τ_s is of order 1 ps, but the values are smaller in alloys such as Ni-Fe or Co-Fe, where the scattering is greater (Table 1).

The chemical potentials for a GMR spin valve are shown in Fig. 11 for the case where the nonmagnetic layer is thin enough for spin relaxation there to be neglected. On flipping the free layer from parallel to antiparallel, there is a change in spin-dependent potential and hence a change in the spin accumulation voltage V_{sa} at constant current — in other words, magnetoresistance.

The interface resistance per unit area R_{int} is V_{sa}/j, which is expressed in terms of the conductivity ratio $\alpha = \sigma^{\uparrow}/\sigma^{\downarrow}$

$$R_{int} = [(\alpha - 1)/2ej(1+\alpha)](\mu^{\uparrow} - \mu^{\downarrow})_0 \tag{14}$$

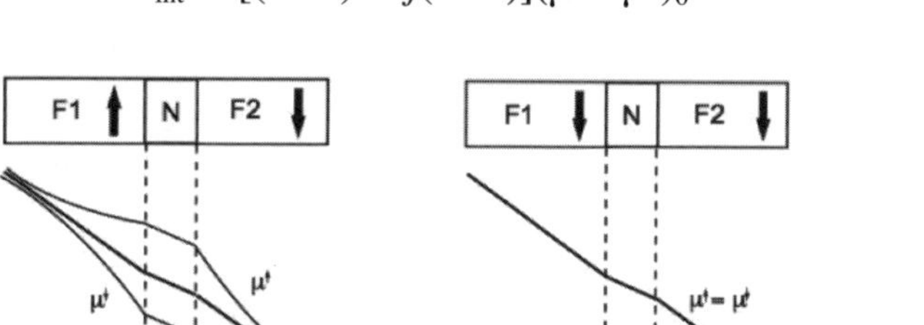

Fig. 11. Chemical potentials for $\uparrow$ and $\downarrow$ electrons in a GMR spin valve, with parallel or antiparallel alignment of the electrodes. The spacer layer is supposed to be thin compared to the spin diffusion length.

The GMR $\Delta R/R = 2R_{int}/\rho_f t_f$ where ρ_f is the resistivity of the ferromagnet and t_f is the total thickness of the two ferromagnetic layers. In the limit of a thin nonmagnetic spacer and thick ferromagnetic layers, the result is

$$\Delta R/R = = [(\alpha^2 - 1)/4\alpha](l_{sf}/t_f) \tag{15}$$

The effect is maximized when the value of α is large, and the spin diffusion length in the ferromagnet is long. If $\alpha = 5$ and $l_{sf}/t_f \approx 1$, the MR would be 20 %. The order of magnitude of the spin accumulation voltage, even at high current densities $j \sim 10^{12}$ A m^2, does not exceed a millivolt.

                    J. M. D. Coey

In this analysis, we have assumed the interfaces are transparent. In fact, spin-dependent scattering and reflection at the interface plays a crucial role in the operation of GMR spin valves. The step in chemical potential at the interface is $\Delta\mu_\uparrow^{\uparrow,\downarrow} = eI^{\uparrow,\downarrow}R_s^{\uparrow,\downarrow}$ where R_s is the interface resistance. The enormous gradient of exchange field at the interface deflects the ↑ and ↓ electrons in the adiabatic limit, as in a Stern-Gerlach experiment. In the nonadiabatic limit of perfectly abrupt interfaces, quantum reflection becomes is important.

The various mechanisms contributing to the spin polarization of electrons as they traverse a ferromagnetic interface layer into a nonmagnetic metal are depicted in Fig 12. Reflection, differential precession and the different mean free paths ensure that only a thin ferromagnetic layer is needed. Cobalt is particularly effective; one nanometer of cobalt is an effective spin polarizing layer.

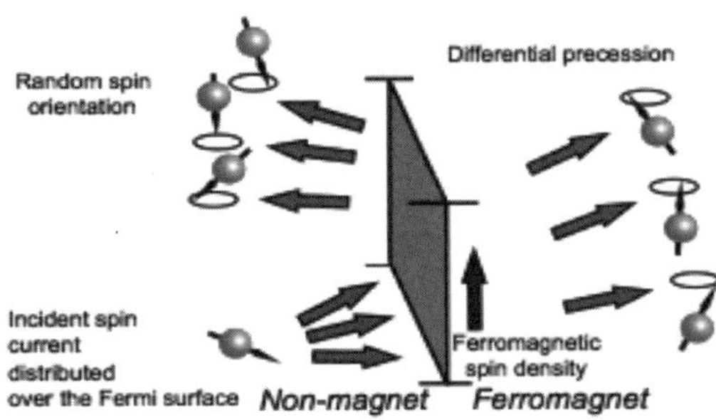

Fig. 12. Mechanisms contributing to the absorption of the transverse component of a spin polarized current which is reflected or transmitted at the interface with a ferromagnetic layer. After Ref. 2.

Spin lifetimes and diffusion lengths are much longer in semiconductors than in metals, but electrical spin injection and detection of spin-polarized electrons are problematic. The problem is the resistance mismatch between the metal and the semiconductor. Considering the two spin channel resistances adding in parallel, the metal plus contact resistance is spin dependent, but the much larger semiconductor resistance is not. This means that the electrically-injected are not significantly polarized. It is appropriate to treat scattering at interfaces in terms of resistance rather than resistivity; the scattering is integrated over the junction region where the potential changes.

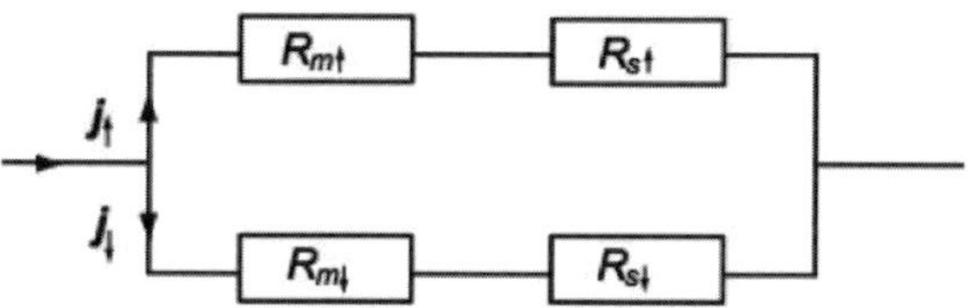

Fig. 13. The resistance mismatch problem. Here, $R_{m\uparrow} \neq R_{m\downarrow}$ but $R_{s\uparrow} = R_{s\downarrow}$ and $R_s \gg R_m$ hence $j_\uparrow \approx j_\downarrow$.

One way to overcome this problem is to introduce a Schottky barrier or a tunnel barrier at the semiconductor-metal interface, and to inject spin-polarized hot electrons over the barrier. Otherwise, optical spin injection using circularly-polarized light is good for semiconductors with a direct bandgap, such as GaAs. The spin polarization of the conduction electrons P_c defined as $(n_{c\uparrow} - n_{c\downarrow})/(n_{c\uparrow} + n_{c\downarrow})$ can also be detected optically, using the Faraday or Kerr effect.

2.4. *Spin transfer torque*

Besides the flow of charge $j = nev_d$ C m^{-2} s^{-1}, a spin-polarized electric current also carries with it a flow of angular momentum

$$j_s = n(\hbar/2)p_c v_d = = j\hbar/p_c/2e \tag{16}$$

Units are J m^{-2}. Unlike charge, which is conserved in the total electric current, the angular momentum can be absorbed by spin-dependent scattering and other processes in a lattice, leading to a torque Γ which is equal to the rate of change of angular momentum in the conducting lattice. The total angular momentum of the system electrons + lattice has to be conserved, so any loss of angular momentum of the electric current has to be balanced by a gain of angular momentum of the lattice. Spin transfer torque is the rate of change of angular momentum Even an initially-unpolarized current can exert torque, if it becomes spin polarized via a spin-dependent scattering process in a ferromagnetic lattice. Spin transfer torque is able to excite magnons and microwaves, move domain walls and reverse the magnetization of nanoscale magnetic free layers. The theory has been developed by Luc Berger and John Slonczewski. In the nanoworld, it is more effective to exert torque by spin transfer rather than by magnetic fields due to currents in nearby conductors, known in this context as Oersted fields. Manipulation of magnetization by spin-polarized currents is possibly the most exciting development in contemporary magnetism.

The field at the surface of a current-carrying wire of radius r has magnitude $H = j/2r$ so the Oersted fields become increasingly ineffective, falling off as $1/r$ as the dimensions are reduced. The angular momentum associated with the magnetization of the free layer in a nanopillar, where it has thickness t and radius r, is $\gamma M \pi r^2 t$. The rate of flow of angular momentum associated with the current density is $j\hbar p_c \pi r^2 / 2e$ Hence the switching effect, which depends on the ratio of these quantities, varies as jp_c/Mt independent of r. Spin transfer torque is *scalable*, although it only becomes effective at nanoscale dimensions. Scaleability is an essential requirement of any new technology that hopes to make its way into mainstream electronics.

To understand the physics underlying spin transfer torque, consider in more detail the spin-polarized electrons entering the ferromagnetic thin film of Fig 12, which is magnetized in the z direction. If the electrons are moving in the x-direction and are initially polarized in a direction making an angle θ with $\mathbf{e}_x$ they may be either reflected or transmitted at the interface. For example, at a Cu/Co interface, the majority-spin $\uparrow$ electrons are more easily transmitted than the minority-spin $\downarrow$ electrons. Reflection at the interface is a source of spin torque, especially in the nonadiabatic case. The electrons that make it into the cobalt are then subject to a huge exchange field B_{ex} in the z-direction, which is of order 10^4 T. (The exchange splitting of the $3d$ band is of order 1 eV). They precess at the Larmor frequency, completing many turns as they diffuse from site to site. By the time they leave the cobalt film, the *xy*-component of the moments will be completely dephased, as the electrons will have followed paths of different length, but their *z*-component of magnetization will be unchanged provided the thickness is less than the spin diffusion length. Angular momentum is therefore transferred from the electron current to the cobalt at a rate $(j\hbar p_c \sin\theta)/2e$ J m^{-2}. All this has little effect on a thick cobalt layer, but it can greatly modify the magnetization of a thin one, which will tend to rotate towards the direction of polarisation of the incoming current. The transverse angular momentum can be absorbed in a distance which is less than the mean free path.

Another mechanism is also effective at transferring angular momentum from the electron current. The mean-free paths are different for $\uparrow$ and $\downarrow$ electrons in a ferromagnet (Table 1); often with $\lambda_\uparrow > \lambda_\downarrow$. The minority spin electrons share their angular momentum with the lattice within the first nanometer or so of the interface, while their majority-spin colleagues travel much further before they are scattered. Spin-selective scattering at the interface can also contribute to the torque. Scattering must be spin-selective, but spin-flip scattering is not required. The spin diffusion length l_s is not a relevant lengthscale.

By whatever mechanism, an unpolarized electron current becomes spin polarized after traversing about a nanometer of cobalt. The ability of ultra-thin cobalt layers to effectively impart spin polarization is widely exploited in thin films stacks used for spin electronics. A ballistic electron traveling at 10^6 m s^{-1} in a random direction completes a precession in a layer which is of order a nanometer thick, so the transverse component of angular momentum is very soon absorbed by the precession mechanism.

Next consider electron flow across the structure including a thick and a thin ferromagnetic layer, shown in Fig. 14. Think first of electron flow in the direction from the thick layer to the thin one. The charge current is in the opposite direction, of course, because the electron has a negative charge.

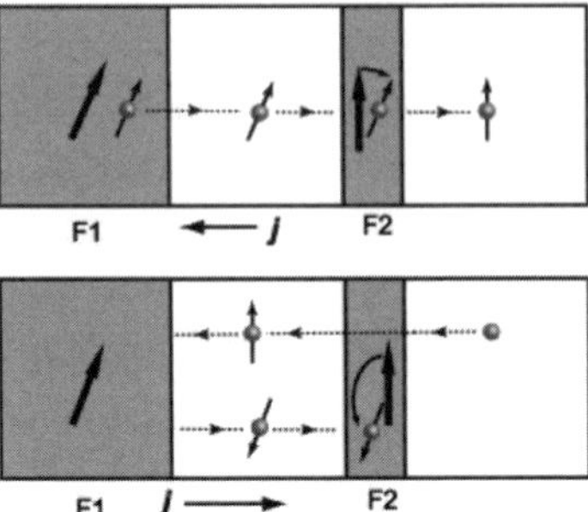

Fig. 14. Spin transfer torque associated with flow of angular momentum in a nanowire or nanopillar with thick and thin ferromagnetic layers. The sense of the torque on the thin free layer depends on the current direction.

The initially unpolarized electron current is polarized in the direction of magnetization of the thick layer when it emerges. The spin polarisation is not perfectly efficient; for a Co/Cu stack p_c is about 35 %. The transverse component of angular momentum is later absorbed in the thin layer as we have just explained. A torque acts on F_2, tending to turn themagnetization towards the orientation of the incoming spins. A parallel configuration of F_1 and F_2 is stabilised.

Next consider what happens when the electrons flow in the opposite direction, from the thin layer to the thick one. They cross F_2 acquiring spin polarisation, and most of them go on to enter F_1 where they ineffectually exert torque on the thick pinned layer. But some, predominantly those with spin opposite to F_1, are reflected, and travel back to F_1 where their transverse component of an angular momentum is absorbed, this time tending to stabilise an antiparallel configuration of F_1 and F_2. In summary, depending on the direction of the current, the spin transfer torque tends to stabilise or destabilise the parallel configuration of the two ferromagnetic layers.

The shortest possible time t taken to switch a free layer of thickness t can be estimated from the rate of flow of angular momentum. Equating the angular momentum delivered in time t to the angular momentum $M\mathcal{A}t\hbar/2\mu_B$ associated with the magnetic moment of the free layer of cross section area $\mathcal{A}$ gives

$$t = Mte/j\mu_B \tag{17}$$

This is independent of A but switching speed is effectively limited by the maximum current density j that can be pumped through the device, which is ultimately limited to about 10^{12} A m^{-2} by electromigration. In practice, the free layers must be only a few nanometers thick, which is well within the scope of modern thin film technology.

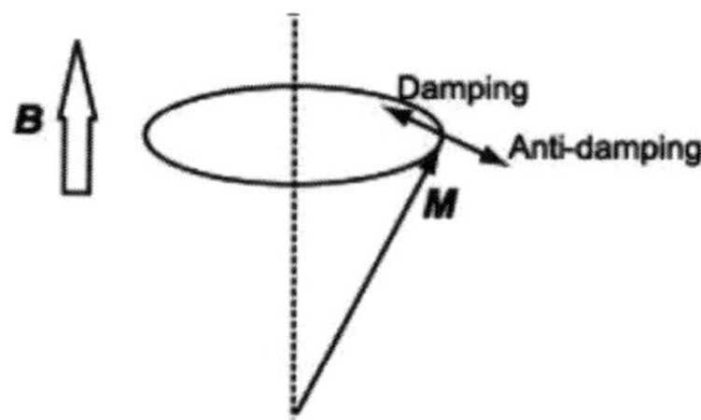

Fig. 15. Precession of magnetization with anti-damping due to spin-transfer torque.

To treat the dynamics of the free layer in more detail, allowing for the presence of an applied magnetic field, we make the *macrospin approximation*, and assume that the free layer rotates coherently like a giant spin at $T = 0$ The precession is described by the Landau-Lifschitz-Gilbert equation with an extra damping-like term α' added to account for the spin tranfer torque.

$$d\mathbf{m}/dt = \gamma\mu_0\mathbf{m} \times \mathbf{H}_{\text{eff}} - (\alpha + \alpha')\mathbf{e}_m \times d\mathbf{m}/dt \tag{18}$$

The effective field $\mathbf{H}_{\text{eff}}$ is the sum of the applied and anisotropy fields. According to the direction of the current, the spin transfer torque adds to the damping torque, or subtracts from it. When the current flows in the $-x$ direction, the spin transfer term simply reinforces the damping. However, when the current flows in the $+x$ direction, a dynamic instability arises when the spin transfer term overcomes the damping. Instead of relaxing towards the z-axis, the magnetization begins to spiral away from it.

The extra "damping" term α' due to spin transfer torque may be of either sign, according to the direction of spin-polarized current in the structure of Fig. 14. The macrospin dynamics when $-\alpha' > \alpha$ can take several forms. In one case, the free layer may spiral all the way to $\theta = \pi$, and get damped in the reverse direction. Switching by spin transfer torque is particularly valuable for magnetic memory because of the scalability of the effect.

Another possibility is that the free layer may not spiral all the way to $\theta = \pi$, but precess continually at some intermediate angle. The current generates steady-state magnetic oscillations in the GHz frequency range with a frequency proportional to j (Fig. 16). Easily-tunable on-chip microwave sources offer new perspectives for magnetism and interchip communications.

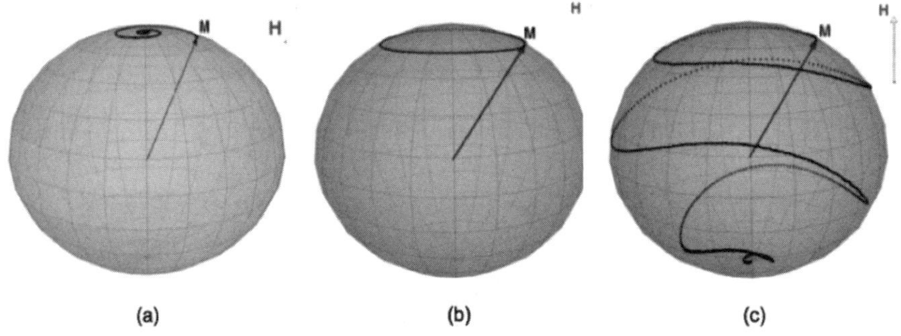

Fig. 16. Precession of a macrospin with damping a), no net damping b) and antidamping c).

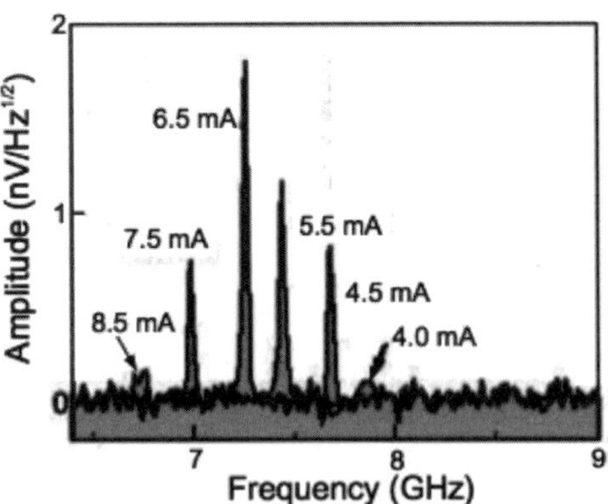

Fig. 17. Excitation of microwaves from a GMR spin valves by spin-transfer torque.

2.5. *Spin currents*

When an electric current passes through a conductor that contains spin-orbit scatterers, there is a tendency for spin to accumulate at the surface of the conductor, as shown in the figure. Intrinsic spin-orbit scattering may have a similar effect. This is the *spin Hall effect*. The principle is the same as for the Mott detector, which senses the polarization of an electron beam by passing it through a gold foil where it undergoes spin-orbit scattering, which separates the ↑ and ↓ beams. The effect is difficult to detect in a wire, because the Oersted field created by the current has the same symmetry as that due to the accumulated spin. However, the spin accumulation has been detected optically in a flat slab of slab of GaAs, using the Faraday effect.

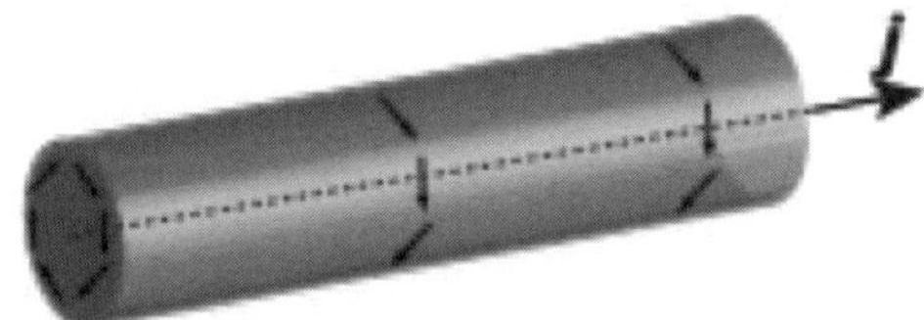

Fig. 18. Spin Hall effect in a wire. The current produces a circular magnetization pattern by spin-orbit scattering.

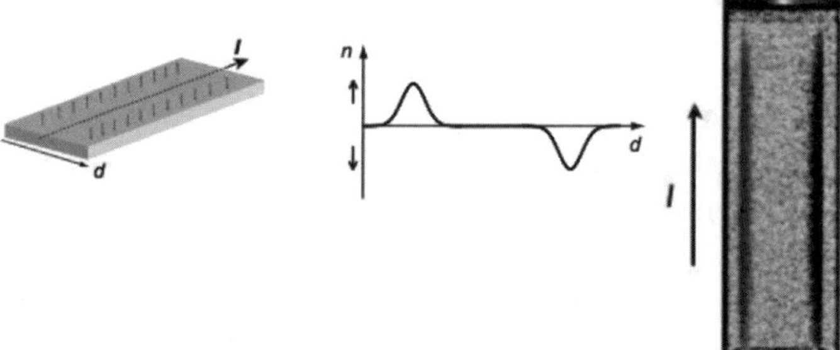

Fig. 19. An experiment to detect the spin Hall effect in a slab of GaAs. The spin-up and spin-down electron densities are indicated by different colours.

The spin current in these examples, and the one that flows when the device in Fig. 10 is switched on, is driven by the charge current, although, unlike a normal spin-polarized electric current, it flows in the transverse direction. In the example of the flat slab, the transient spin current flows across the slab, and and equilibrium concentration of ↑ and ↓ electrons accumulate at opposite edges.

It is possible for spin currents to be completely divorced from charge currents when domain wall motion occurs; AC spin currents are associated with the propagation of spin waves. However, these pure spin currents can only be transient or alternating. The maximum possible transfer of angular momentum is constrained by flipping all the spins in the system. It would be a challenge to devise a DC spin battery, which would drive a steady flow of angular momentum unassociated with a steady flow of charge. New concepts of spin-based computation seek to eliminate the charge current entirely, and thereby bypass the mounting energy dissipation associated with CMOS processors.

3. Materials for Spin Electronics

The generic spin electronic device of Fig. 6 consists of a source of spin-polarized electrons, a transport medium and an analyser. The ferromagnetic polarizer and analyser are often Co, Fe, Co-Fe or Ni-Fe. Half-metallic Heusler alloys with high T_C such as Co_2MnSi ($T_C = 985$ K) are also used. Other half metals of interest include oxides such as LSMO and CrO_2 but room-temperature device operation with these oxide materials is unlikely.

In principle, the spin transport medium can be almost anything that transmits electrons. without completely destroying their polarization. In all-metal thin film stacks, Cu or Al is often used. Heavy metals introduce spin-flip scattering, due to their strong spin-orbit interaction. Semiconductors are very effective spin transport media because there are few impurities or other electrons to scatter the spin-polarized carriers. Electrical spin injection and detection are problematic in semiconductors, because of the problem of resistivity mismatch between the metal and the semiconductor. Optical injection and magneto-optic detection are feasable for semiconductors with a direct bandgap, such as III-V materials.

Insulators are useful as spin transport media only when they are thin enough to permit tunnelling. Tunnel barriers are commonly made of amorphous Al_2O_3 or 001–oriented MgO. Ballistic conductors such as graphene, and organics where spin transport is by hopping are also potentially interesting for spin electronics. The organics have very poor mobility, but there is a very long spin lifetime which may find an application. Relevant electronic properties of some representative materials are summarized in Table 4.

Table 4. Some materials for spin electronics.

	conduction	l_s (nm)	μ (m^2 V^{-1}s^{-1})	v_F (m s^{-1})	τ_s(s)
Cu	diffusive	200	$4\ 10^{-3}$	$1.6\ 10^6$	$3\ 10^{-12}$
Co	diffusive	40	$2\ 10^{-3}$	$1.3\ 10^6$	$4\ 10^{-13}$
Si	diffusive	10^5-10^6	0.1	†	10^{-9} - 10^{-7}
GaAs	diffusive	10^5	0.3	†	10^{-9} - 10^{-7}
CNT	ballistic	$5\ 10^4$	10	$1\ 10^6$	$3\ 10^{-8}$
graphene	ballistic	1500	0.2	$1\ 10^6$	10^{-10}
rubrene ($C_{42}H_{28}$)	diffusive	13	$2\ 10^{-3}$	†	-
Hexathiophene	hopping	200	10^{-9}	-	10^{-6}
Alq$_3^*$	hopping	45	$10^{-14}- 10^{-12}$	-	10^{-2} - 1

*tris-(8-hydroxyquinoline) aluminium.   † Depends on the carrier concentration.

The thin film stacks on which spin electronic devices are based frequently incorporate layers of antiferromagnetic metal in contact with one of the ferromagnetic layers. $IrMn_3$ and $PtMn_3$ are often used. The antiferromagnet imparts a unidirectional anisotropy to the adjacent ferromagnet known as *exchange bias*, which serves to pin its direction of magnetization. The other ferromagnetic layer in the stack is then free to rotate, changing the resistance of the structure. A common building block in these stacks is the synthetic antiferromagnet (SAF), which consists of two ferromagnetic layers separated by a very thin spacer layer, typically 0.8 nm of Ru, which serves to couple the magnetization of the two ferromagnetic layers antiparallel. By using a SAF as the pinned layer, it is possible to eliminate the stray field acting on the free layer.

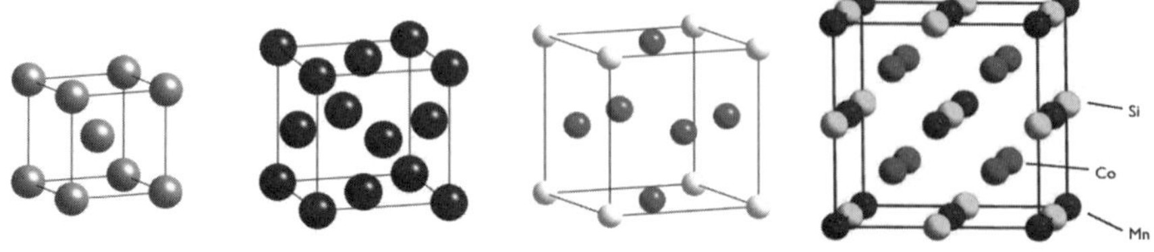

Fig. 20. Crystal structures of some metallic alloys used in spin electronics. a) Co-Fe, b) Ni-Fe, c) IrMn$_3$ and d) Co$_2$MnSi.

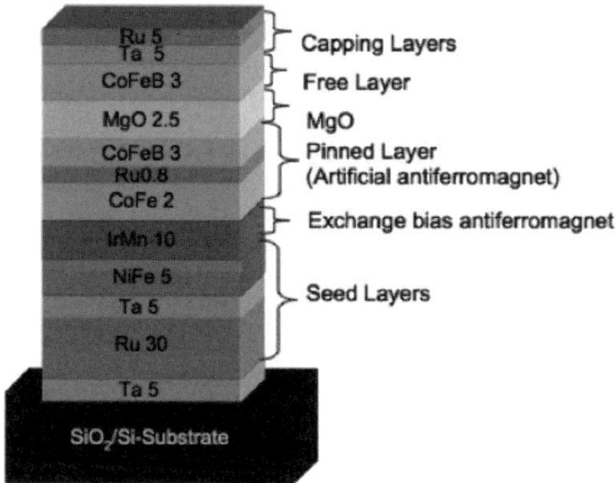

Fig. 21. An example of a thin film stack incorporating TMR spin valve with an MgO tunnel barrier and a synthetic antiferromagnetic layer pinned by coupling to an antiferromagnetic exchange-bias layer.

4. Magnetic Sensors

A variety of different physical effects are exploited in magnetic sensors, which are passive devices that detect the presence of a magnetic field. They deliver an electrical voltage that is monotonic, and preferably proportional to one component of the field B acting at the sensor. The sensors are usually magnetoresistive structures or Hall bars, but occasionally it is advantageous to use magnetoimpedance sensors or magneto-optic sensors, addressed with light. The great advantage of magnetic sensing is that it avoids the need for direct contact between the sensor and the object sensed. Well over a billion magnetic field sensors are manufactured each year, about half of them destined for hard disc drives. We are most interested here in structured thin film sensors based on giant magnetoresistance (GMR) or tunneling magnetoresistance (TMR), which are the simplest, two terminal spin electronic devices. We focus on aspects of thin film sensor design, and consider the critical issue of noise, which ultimately limits their sensitivity.

Characteristics of the main sensor types are summarized in Table 5.

Table 5. Characteristics of magnetic field sensors.

Sensor	Principle	Detects	Frequency Range	Field (T)	Noise	Comments
Coil	Faraday's law	$d\Phi/dt$	$10^{-3}-10^9$	$10^{-10}-10^2$	100 nT	bulky, absolute
Fluxgate	Saturable ferromagnet	$d\Phi/dt$	$0-10^3$	$10^{-10}-10^{-3}$	10 pT	discrete
Hall generator	Lorentz force	B	$0-10^5$	$10^{-5}-10$	100 nT	thin film
Classical magnetoresistance	Lorentz force	B^2	$0-10^5$	$10^{-2}-10$	10 nT	thin film
Anisotropic magnetoresistance (AMR)	Spin-orbit scattering	H	$0-10^7$	$10^{-9}-10^{-3}$	10 nT	thin film
Giant magnetoresistance (GMR)	Spin accumulation	H	$0-10^9$	$10^{-9}-10^{-3}$	10 nT	thin film
Tunnelling magnetoresistance (TMR)	Tunelling	H	$0-10^8$	$10^{-9}-10^{-3}$	1 nT	thin film
Giant magnetoimpedance (GMI)	permeability	H	$0-10^4$	$10^{-9}-10^{-2}$		wire
Magneto-optic	Faraday/Kerr effect	M	$0-10^8$	$10^{-5}-10^2$	1 pT	bulky
SQUID (4 K)	Flux quantization	Φ	$0-10^9$	$10^{-15}-10^{-2}$	1 fT	cryogenic
SQUID (77K)	Flux quantization	Φ	$0-10^4$	$10^{-15}-10^{-2}$	30 fT	cryogenic
Nuclear precession	Larmor precession	$|B|$	$0-10^2$	$10^{-10}-10$	1 nT	bulky, very precise

Fig. 22. A barber pole.

If a thin film sensor, which depends for its operation on the direction of magnetization of a responsive ferromagnetic layer, is to deliver a magnetoresistive signal that is monotonic in field, the magnetization has to be able respond continuously, rather than flop between bistable configurations ("high" and "low" resistance states).

In the case of a single film anisotropic magnetoresistance (AMR) sensor, which is usually made of permalloy, the magnetoresistive effect $\Delta\rho(\Pi) = [\rho(\Pi) - \rho\,(\pi/2)]$ is equal to $\Delta\rho\cos^2\Pi$ where Π is the angle between M and j. Sensitivity is greatest when $\Pi = \pi/4$. Ways of setting this angle are either to induce in-plane anisotropy by depositing or annealing the permalloy film in a magnetic field, or else to make a *barber pole* pattern overlaying the permalloy with stripes of a more conducting metal such as gold, which defines equipotentials across which the current flows. In either case, the sensor responds to the transverse component of the field in the plane of the film. Using the equilibrium condition that there is no net torque on the magnetization due to the internal field in the film, the demagnetizing field $NM\sin\Pi$ is equal to the transverse applied field H_0 For small rotations, $\delta\Pi = H_0/NM$ and the variation of resistance with field is

$$\Delta\rho(H) \approx \Delta\rho H_0/NM \tag{19}$$

The response is linear for small fields.

A planar Hall sensor also uses a single film of an AMR material such as permalloy. Here a transverse voltage is measured when a magnetic field is applied in-plane, perpendicular to the current. The effect is closely related to AMR, but the sensor design is a little simpler.

The anomalous Hall effect, measured in the standard Hall geometry, can be exploited for detection of higher fields, applied perpendicular to the plane of the sensor, in which case $N \approx 1$. Since the anomalous Hall voltage is proportional to the out-of-plane magnetization $M_\perp$, and the demagnetizing field is $-NM_\perp$ it follows from the condition that there is zero torque due to the internal field that $H_0 - NM_\perp = 0$. The signal is linear, and proportional to H_0 up to saturation.

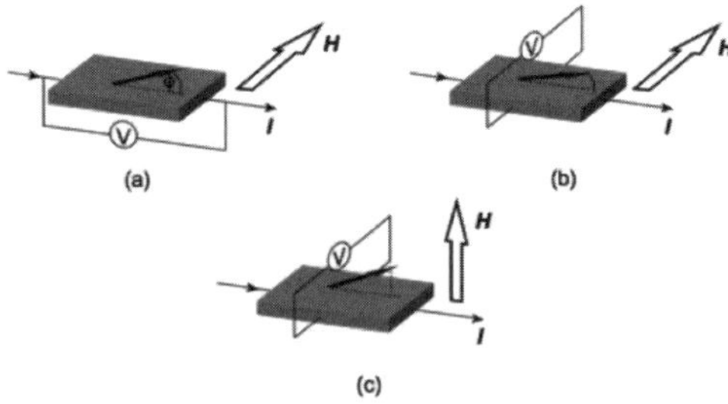

Fig. 23. a) AMR, b) planar Hall and c) anomalous Hall effect sensor configurations. The easy direction is shown by the dashed line.

Fig. 24. Spin valve sensor with an antiferromagnetic layer for exchange bias.

In GMR structures, the spacer is a metal, usually copper, and the current may flow either in-plane or perpendicular to the plane. In a TMR structure, the spacer is an insulator, and the current flows perpendicular to the plane. The magnetizations of the free and pinned layers are perpendicular, in zero field. The linearization of a GMR or TMR sensor is slightly different from the linearization of an AMR sensor. The free layer of a spin valve is normally free to rotate in the plane. The rotation in a well-designed sensor is coherent, and the free layer remains single-domain. The MR in these cases varies as $\sin^2\phi/2$, where ϕ is the angle between the magnetization of the free and pinned layers. Hence, the sensitivity is optimized for the crossed anisotropy configuration where $\phi = \pi/2$. This can be achieved by setting the exchange bias direction of the pinned layer perpendicular to the easy axis of the free layer. The GMR or TMR sensors are only linear for small changes of ϕ, $\Delta\phi \approx 10^0$ so only a fraction of the total magnetoresistance is actually exploited in sensors such as read heads.

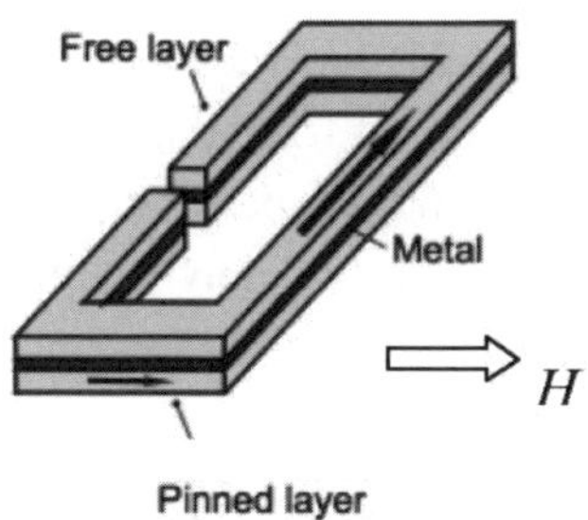

Fig. 25. A yoke-type spin valve sensor. In this design, the direction of the free layer is set by shape anisotropy, and the pinned layer is set in a perpendicular direction by exchange bias.

Giant magnetoimpedance (GMI) offers excellent sensitivity in low magnetic fields. In wire sensors, the circumferential anisotropy can be set by controlling the magnetostriction, or annealing a current-carrrying wire, which creates a circumferential magnetic field. Excellent results, up to 1000 %, are achieved using wires with a nonmagnetic copper core, and an electroplated permalloy sheath, about a micron thick. The noise sensitivity is comparable to that of good tunnel junctions. The sensors are widely used for direction sensors in interactive computer games, for example.

4.1. *Noise*

Noise in a magnetic sensor can be electrical or magnetic in origin. It is manifest as uncontrolled random fluctuations of the voltage $V(t)$ across the device. The frequency spectrum of the fluctuations $V(\omega)$ is the Fourier transform of $V(t)$. The average power (per unit resistance) dissipated in fluctuations during a measuring time t_m which tends to infinity is

$$P = (1/t_{\mathrm{m}}) \int_{-\infty}^{\infty} |V(\mathrm{t})|^2 = (1/2\pi t_{\mathrm{m}}) \int_0^{\infty} |V(\omega)|^2 \, d\omega \tag{20}$$

The *power spectral density* of the fluctuating process is therefore

$$S_{\mathrm{V}}(\omega) = \lim_{tm \to \infty} 2|V(\omega)|^2 / t_{\mathrm{m}} \tag{21}$$

The voltage spectrum can be measured directly in a spectrum analyzer. It is conveniently quoted for a bandwidth $\Delta f = \Delta \omega / 2\pi$ of 1 Hz.

Electrical noise is of four main types. The first is thermal or Johnson[3] noise, which is a characteristic of any resistor R. The spectrum is frequency-independent

$$S_{\mathrm{V}}(f) = 4k_{\mathrm{B}}TR \tag{22}$$

and the mean square voltage fluctuation measured across a resistor, with no imposed current, is $V_{\mathrm{av}}^2 = 4k_{\mathrm{B}}TR \, \Delta f$. For example, the root mean square (RMS) voltage fluctuation across a 1 MΩ resistor in a 1kHz bandwidth at room temperature is 4 µV. Hence the interest in making electrical measurements in a narrow bandwidth at the frequency of the signal of interest, in order to improve the signal/noise ratio. This is the principle of the synchronous, phase-sensitive detector (lock-in amplifier), which is almost universally used for precise electrical measurements.

Shot noise is a non-equilibrium phenomenon associated with electric currents, which is related to the discrete nature of electrical charge. It was first observed in vacuum tubes, where the power spectral distribution is flat in frequency $S_I(f) = 2eI$. The current noise in a bandwidth Δf is

$$I_{sn} = \sqrt{(2eI\Delta f)} \tag{23}$$

Shot noise can be observed in tunnel junctions when sufficient current passes through the junction to exceed the equilibrium Johnson noise. It may then determine the ultimate sensitivity of high-frequency sensors such as read heads. Operating at too high a bias reduces the TMR, but it also increases the shot noise. Carrier lifetime may influence shot noise at high frequency.

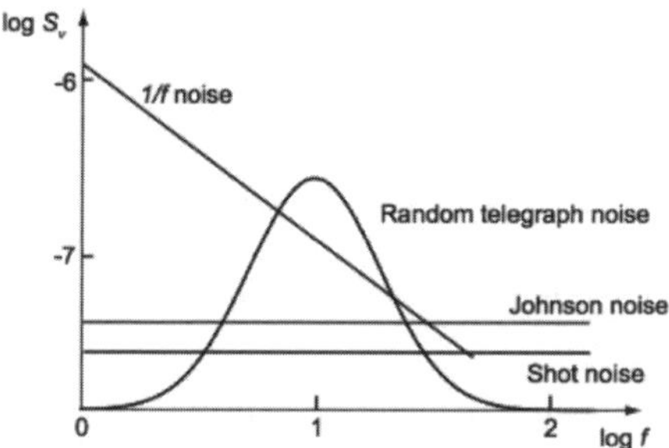

Fig. 26. Frequency dependence of various contributions to the noise in a sensor.

Signal/noise ratio is critical for sensor applications. At high frequency, the noise contributions that limit performance are Johnson noise and shot noise, varying as $\sqrt{R}$ and $\sqrt{I}$ respectively. Both are *white noise* with a flat frequency spectrum. Another contribution called *pink noise*, or $1/f$ noise dominates at low frequencies, where the power spectral density S_V varies as $1/f^\alpha$ where $\alpha \approx 2$. The fascination of $1/f$ noise is its ubiquity. There is an astonishing range of different natural and man-made examples, ranging from the human heartbeat ($1/f$ below 0.3 Hz) to the water level of rivers to the musical output of radio stations. The $1/f$ noise in electronic systems may be several orders of magnitude greater than thermal noise at 1 Hz. In magnetic sensors, it is related to resistance fluctuations $R(t)$ which translate into voltage fluctuations when a constant current is passed. The current does not create the fluctuations, it just reveals them above the white noise. The square of the voltage fluctuations $V(t)$ varies as I^2, which permits the $1/f$ noise coming from the sample to be distinguished from that arising from other sources, such as the preamplifier. In magnetoresistive sensors, there is a contribution to $1/f$ noise that is associated with domain walls.

The $1/f$ resistance fluctuations are conventionally parameterized in terms of a phenomenological expression due to Hooge.

$$S_V(f) = \gamma_H V_a^2 / N_c f \tag{24}$$

where V_a is the applied voltage, N_c is the number of charge carries in the noisy volume ($N_c = n\Omega$ where n is the carrier density and Ω is the volume) and γ_H is a dimensionless number known as the Hooge coefficient, which allows a normalized comparison of different systems. Values of $\gamma_H \sim 10^{-3}$ are found for well-crystallized metal films, and for semiconductors. The values of γ_H for conducting magnetic systems including GMR and TMR sensors, may be orders of magnitude greater.

The origin of the $1/f$ noise remains an open question, but the noise can be modeled in terms of an ensemble of elementary thermally-activated fluctuators with a broad range of energy barriers. The task for an engineer concerned with magnetic sensors it to avoid it or reduce it as far as possible. One approach is to ensure that thin films of excellent crystalline quality are used to reduce the electrical contribution. A common solution is to modulate the signal in the kHz range, with a microcantilever for example, in order to shift the detected signal beyond the range where $1/f$ noise dominates.

Finally a type of noise, which is sometimes encountered in conducting magnetic thin films, is *random telegraph noise*, where a particular two-level system is activated in a certain temperature window. This leads to a broad feature in the noise spectrum. An example of the low-frequency noise spectrum in a magnetic tunnel junction is shown in Fig. 27.

The *magnetic noise*, or electrical noise of magnetic origin in AMR or spin-valve sensors is related to fluctuations of the magnetization, which are reflected in the magnetoresistance signal. These fluctuations can be severe when domain walls are present, and they are mitigated by ensuring that the ferromagnetic layers are single-domain, and that the rotation of the free layer is coherent.

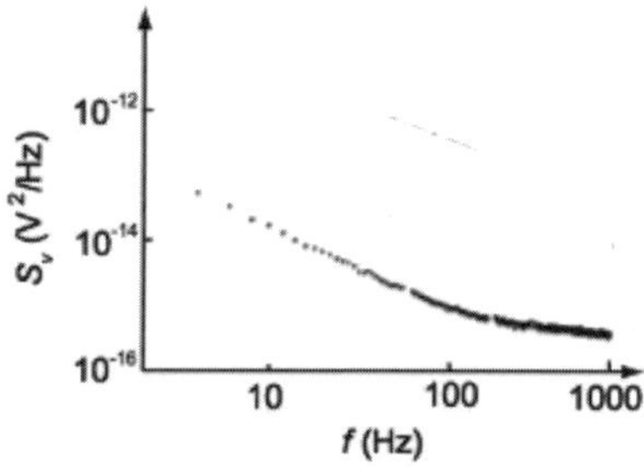

Fig. 27. $1/f$ and white noise in a CoFeB/MgO/CoFeB magnetic tunnel junction.

One way to achieve this is to build a permanent magnet into the sensor structure, which creates a small bias field. Otherwise, domain formation can be discouraged by reducing the demagnetizing field, in a yoke-type configuration like that shown in Fig. 25. When the sensor dimensions are very small, collective fluctuations of the magnetization may be a significant source of noise.

5. Magnetic Memory and Logic

Magnetic memory is as old as digital electronics. It has the incontrovertible advantages of being nonvolatile, and indefinitely rewritable. Mass memory for computers has been provided by hard discs for over fifty years. A series of schemes have been advanced for other, faster schemes which can be addressed electrically at random, rather than mechanically in sequence. These included ferrite core memory which was dominant in the 1950s and 60s until it was superseded by semiconductor memory, permalloy plated wire memory in the 1960s, magnetic bubble memory in the 1970s and early 1980s, and magnetic random-access memory (MRAM) which has been under development since the mid 1990s. In principle, MRAM can combine the speed of SRAM with the density of DRAM and the nonvolatility of flash, with radiation hardness and reduced power consumption. The former has always been an attraction of magnetic memory for military and space applications - the computers in the Challenger space shuttle and the Hubble space telescope included magnetic memory. With increasing miniaturization, it could become an issue for civilian applications because of the problems associated with unshieldable background radiation.

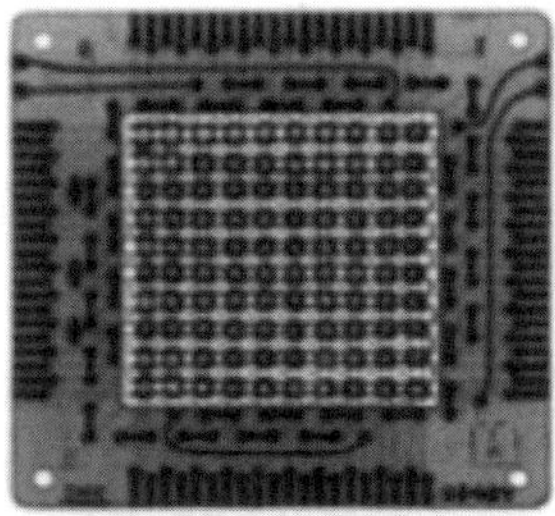

Fig. 28. A 256-bit ferrite core memory from 1960.

5.1. *Magnetic random-access memory*

The tiny ferrite toroids a few hundred microns in diameter in the original core memory were threaded on write and sense lines; they operated on the *half-select principle*. The same principle has been used in MRAM. Schemes that have been investigated include those with current-in-plane based on GMR spin valves and current-perpendicular-to-plane based on magnetic tunnel junctions, which produce much larger signals. The requirement is for a bistable device where the free layer can flip very fast between the parallel and antiparallel states, which represent the binary data, "0" and "1". A thin-film MRAM based on TMR and Oersted-field switching is illustrated in Fig. 29.

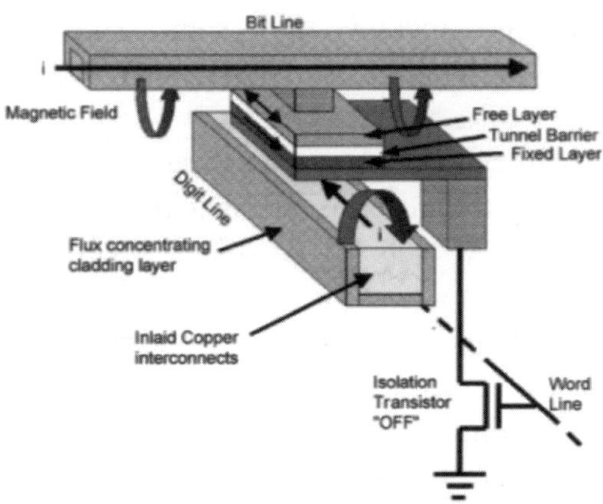

Fig. 29. A TMR spin valve memory cell operating on the half-select principle, with orthogonal bit and word lines. Illustration from Ref. 3.

Half-select switching uses current pulses in orthogonal bit and word lines to generate magnetic field pulses H_p. A single pulse does not disturb the state of the free layers of a line of elements, but when two pulses arrive simultaneously along the two perpendicular lines that cross at the selected element, the resultant field $\sqrt{2}H_p$ is sufficient to induce switching. The margin of error for this switching is rather small, and immense reliability is demanded for a working memory device. An improved pulse sequence known as toggle-mode switching greatly enhances switching reliablility; it was devised by Motorola for the first commercial MRAM chips, which appeared in 2006.

Unfortunately, half-select switching with Oersted fields does not scale favourably as the dimensions are reduced. Possible ways around this difficulty are i) thermally-assisted switching where the current passing through the memory cell heats it sufficiently to reduce the anisotropy, making it easier to switch and ii) spin-transfer-torque MRAM, where the switching is accomplished by a spin-polarized current pulse surging through the spin valve stack. TMR devices with MgO barriers offer useful voltage changes (> 0.1 V) and a compact footprint of a single transistor switch, with the magnetic tunnel junction added in a back-end metallization step. The magnetoresistance characteristic of the exchange-biased TMR stack of Fig. 21 is shown in Fig. 30.

Nanopillar structures made from these stacks with MgO barriers thin enough to allow for spin-transfer torque switching may be able to replace the six-transistor SRAM cell. It is still a challenge to reduce the switching current density to a level that permits sustained, reliable operation, but the remarkable feat of covering a 200 mm or 300 mm wafer with a thin film stack that includes a uniform MgO barrier layer barely one nanometer thick has been achieved.

A different scheme for magnetic memory, proposed by Stuart Parkin, is based on the movement of a pattern of magnetized domains around a permalloy track, which constitutes a magnetic shift register. The head-to head and tail-to-tail domain walls would move in opposite directions in an applied field, thereby annihilating the data, but the pattern of magnetization can be pushed around a track by spin transfer torque. A current passing through a magnetized track adopts the polarization of the domain, so the walls are all pushed in the same direction. A string of bits can be moved at velocities of order 100 m s^{-1} by a train of current pulses in the *magnetic racetrack memory.* (MRM) illustrated in Fig. 31. The scheme promises non-volatile solid state memory, without the mechanical inconvenience of a mechanical disc drive. Very high storage densities should be achievable if the registers can be built in the vertical direction. The scheme has yet to be demonstrated in practice.

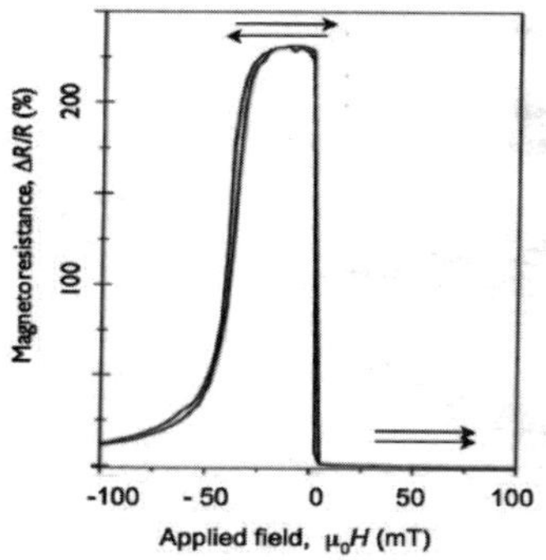

Fig. 30. Magnetoresistance of the exchange-biased TMR spin valve with and MgO barrier and a synthetic antiferromagnetic pinned layer shown in Fig 21. When used as a memory element, the axes of the free and pinned layers are parallel, giving an abrupt switch near zero field, as shown. When used as a sensor, the easy axes are perpendicular, leading to a linear $R(H)$ transfer curve, as the free layer rotates coherently in the applied field. Nanopillars patterned from stacks with the MgO barrier layer thinner than 1 nm are being developed for spin-transfer switched MRAM.

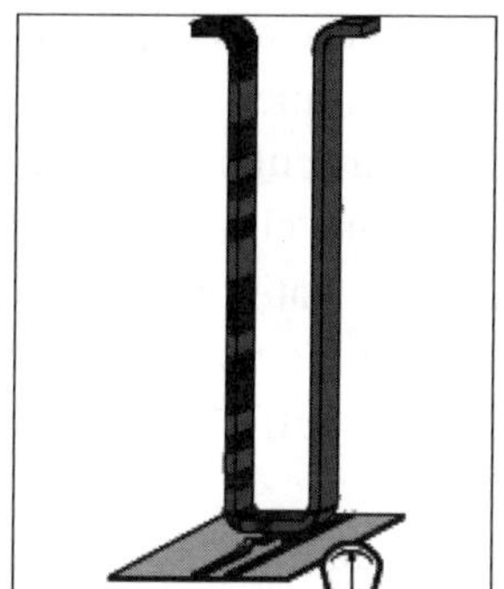

Fig. 31. Magnetic racetrack memory (after S. S. P. Parkin) The domains are driven around the track by current pulses, and the stray field is detected by a sensor at the bottom.

5.2. *Logic*

Ferromagnetic thin-film elements can be pressed into service for logic as well as memory. Useful output voltages can be achieved from logic gates based on MgO barrier magnetic tunnel junction. An alternative implementation uses a switchable ferromagnetic film grown on a semiconductor Hall sensor. The stray field from Co-Fe acting on a two-dimensional electron gas Hall cross creates a useful Hall voltage which follows the Co-Fe magnetization, reversing sign as the magnetization switches.

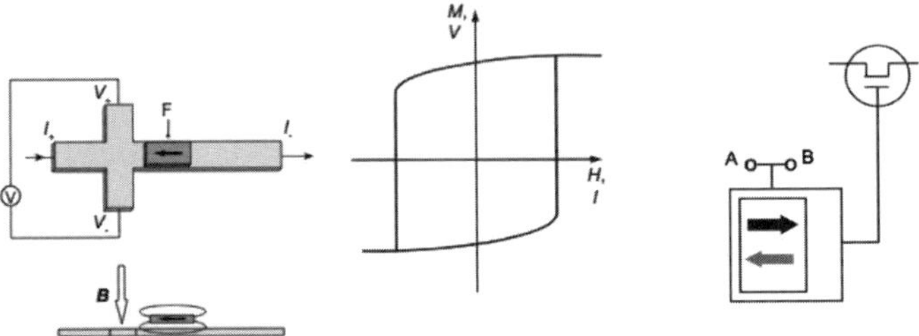

Fig. 32. A nonvolatile magnetic switch based on a switchable thin film and a Hall cross. With half-select pulses, the switch becomes an AND gate (right).[3]

A generic nonvolatile magnetic switch may be applied in field-reprogrammable gate arrays. These are applications-specific CMOS logic circuits, which can be reprogrammed at will to carry out some specific set of logic operations for a particular application. This avoids the need to design and produce a custom chip for each purpose. The magnetic switch delivers its output to the gate of an FET. Half-select input pulses at the terminals A and B determine the state of the device. The output of the switch, of order 0.1 volt, is sufficient to pinch off the current in the channel of the FET.

The magnetic switch can also be applied for Boolean logic. As it stands, the half-select switch acts as an AND gate, switching only when it receives simultaneous pulses at A and B. Reset, and the other Boolean operations OR, NAND and NOR can be conducted by adding a third terminal C, which is used for clock and reset pulses. The reset pulses can be positive or negative, setting the switch in either the high (1) or the low (0) state. All the operations can be achieved in just two clock cycles.

Other magnetic logic schemes can be envisaged. One, due to Russel Coburn, uses magnetic domain walls which are driven around specially shaped structures on a permalloy track. Another exploits the phase of spin waves, which have the advantage that they carry pure alternating spin currents. The phase of the spin wave is shifted when it traverses a domain wall. By passing the spin waves around parallel arms of a narrow loop and detecting the signal at a point where they recombine, it is possible to make a logic gate based on the presence of $360°$ domain walls in either or both of the arms. An attractive feature is that no transfer of charge is involved, and so there is no energy loss due to Joule heating although there may be energy dissipation associated with Gilbert damping. Furthermore, the spin waves may be generated on-board the chip using spin transfer torque.

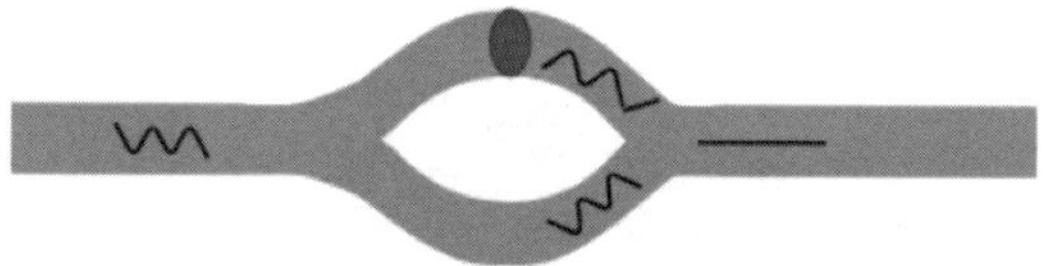

Fig. 33. Spin wave logic. The spin wave running through the two branches acquires a phase shift if it encounters $360°$ a domain wall.

Another proposed logic scheme is based on three single-electron nanodots, coupled by antiferromagnetic exchange (Fig. 39). There is a weak bias field to define an easy direction. The two outer dots provide the input, and the central dot is the output. The four possible inputs illustrated in Fig. 28a) show that the device operates as a NAND gate, from which the other logic operations can be constructed.

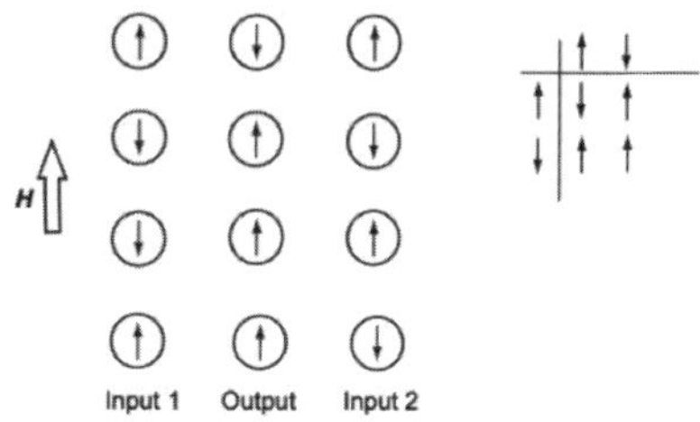

Fig. 34. Logic scheme based on exchange-coupled nanodots.[4]

Finally, among the many candidate systems that have been considered *for quantum computing*, entangled q-bits based on weakly-interacting spins match many of the requirements. A single electron spin on a quantum dot can be regarded as a qbit in a state $\alpha \mid\uparrow> + \beta\mid\downarrow>$. The qbits are initialized in the $\mid\uparrow>$ state in a large magnetic field, and the individual bits are then manipulated by varying the Zeeman splitting. The two-q-bit operations, which are required for universal quantum computing are performed by tuning the weak exchange coupling between two adjacent q-bits. The coherence of the state tends to be destroyed by spin orbit coupling and hyperfine interactions which introduce longitudinal, spin flip T_1 relaxation processes and transverse, T_2 processes which describe the decay of the superposition state.

Fig. 35. A two q-bit structure composed of two weakly interacting single-electron spins on adjacent quantum dots.

5.3. *Spin transistors*

A key requirement for any implementation of logic is *fanout*. The digital output from one logic gate must be sufficient to apply to the inputs of one or more contiguous gates in the next stage. Normally, fanout requires power gain, which is achieved with conventional CMOS. Spin electronics will enter a new phase when it becomes possible to make a three-terminal devices with spin gain, which allows the intrinsic fanout required in logic circuitry. There have been several suggestions and demonstrations of physically interesting devices. The first three-terminal spin electronic device was the Johnson transistor Fig. 36a. Essentially a GMR spin valve, with the nonmagnetic metal as base electrode, the base-emitter current injects polarized spins into the base region. The floating collector samples a small voltage change depending on the relative orientation of the two ferromagnetic emitter and collector electrodes. The device is equivalent that used in the nonlocal measurement of spin accumulation.

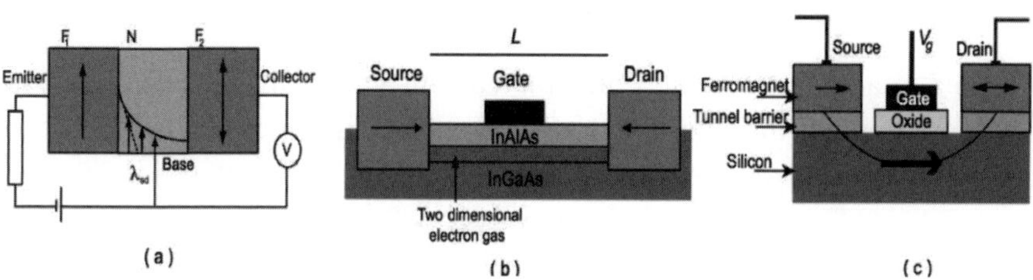

Fig. 36. Some possible three-terminal spin devices. a) the Johnson transistor, b) the Datta-Das transistor, c) the spin MOSFET.

An influential device proposal has been the Datta-Das transistor. This is an FET-like structure, with ferromagnetic source and drain. Spins are injected into the 2-dimensional electron-gas channel, where transport is supposed to be ballistic. They are subject to an electric field from the gate, which looks like a magnetic field $B^* = (v \times E)/c^2$ to the electron moving with velocity v. This is the Rashba effect. The electrons undergo Larmor precession, and the source-drain current depends on the number of turns they have made by the time they arrive at the drain, which can be controlled by the gate potential. Neither the Datta-Das device, nor the spin MOSFET where the semiconductor channel is diffusive, made of Si or GaAs, and which promises a combination of power amplification and non-volatile memory via the switchable ferromagnetic drain electrode, have been realized in practice, although there has been recent progress towards realizing electrical spin injection and detection in these semiconductors.

A way to circumvent the resistance mismatch problem is to inject hot electrons via a Schottky or tunnel barrier, which subsequently lose energy by spin-dependent inelastic scattering in the base region of a device, Fig. 37. In the Monsma transistor, the base is a GMR multilayer with two Schottky barriers, at the source and drain. Another device is the magnetic tunnel transistor, where the base is a ferromagnetic electrode and there is injection from a ferromagnetic source via a tunnel barrier, and a semiconductor collector with a Schottky barrier serves as the drain. The device produces a significant magnetocurrent, defined as

$$\mathrm{MC} = (I_{\mathrm{Cp}} - I_{\mathrm{Cap}})/\, I_{\mathrm{Cap}} \qquad (25)$$

where I_{Cp} and I_{Cap} are the collector current with parallel or antiparallel configurations of the two electrodes. However, in no case has gain of the spin-polarized current yet been achieved.

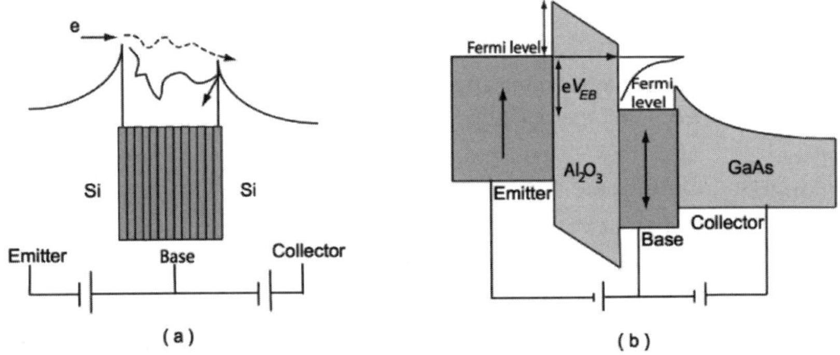

Fig. 37. Hot-electron spin transistors a) the Monsma spin valve transistor and b) the magnetic tunnel transistor.

6. Conclusion

First-generation spin electronics has been remarkably successful in producing two-terminal devices, essentially magnetoresistors, for use as sensors and bistable elements. These sensors have been critical for the continuing progress of magnetic recording, on which digital computing and the internet depend. There are numerous opportunities to use these GMR and TMR magnetic sensors for other purposes. Spin-torque switching is scaleable, and competitive at small lengthscales. Spin-torque MRAM based on magnetic tunnel junctions remains a serious candidate for nonvolatile memory, but it is still unclear to what extent it will be implemented on a large scale by the electronics industry.

Two major challenges facing spin electronics are now:

i) to realize a spin amplifier — a device which can sense a spin-polarized current, and then enhance the product of its intensity and spin polarization, of whichever sign.

ii) to separate somehow the spin and charge currents so that logic operations can be performed using pure spin currents, without the ohmic dissipation of charge currents.

Success in meeting either of these challenges would lead to a second generation of spin electronics

It seems very likely that spin-transfer torque will play an important part in future developments of spin electronics, not least because it because it enables scaleable switching. There are opportunities to realize new magnetic alloys, and alloy combinations, which will deliver the optimal performance in the new thin-film devices.

References

1. F. J. Jedema et al, Nature 416, 713 (2002).
2. D. Stiles and J. Miltat, *Spin Transfer Torque and Dynamics in Spin Dynamics in Confined Spaces* Ed. A Thiaville (Springer, Berlin, 2006) pp.225-308
3. M. Johnson (Ed.), *Magnetoelectronics*, (Elsevier, Amsterdam, 2004)
4. S. Bandyopadhyay and M. Cahay, *Introduction to Spintronics* (CRC Press, 2007).

Further reading

J. M. D. Coey *Magnetism and Magnetic Materials*, Cambridge 2009, 600 pp.

M. Ziese and M. Thornton (Eds.), *Spin Electronics*, Springer, Berlin 2001, 493 pp.

U. Hartmann (Ed.) *Magnetic Multilayers and Giant Magnetoresistance*, Springer, Berlin 1999, 321 pp.

S. Maekawa (Ed.), *Concepts in Spin Electronics*, Oxford 2006, 398pp.

J. G. Gregg and C. Dennis, *The Art of Spintronics*, Oxford 2009

CHAPTER 12

SPINTRONICS

Myung-Hwa Jung

Department of Physics, Sogang University
1 Shinsu-dong Mapo-gu, Seoul 12-74, South Korea
E-mail: mhjung@sogang.ac.kr

Spintronics is the spin electronics in which both charge and spin degrees of freedom play the determining role in solid state environment. The goal of the spintronic application is to realize functions controlling electronic properties by spin or magnetic field as well as controlling magnetic properties by current or electric field. In practical sense, metal-based spintronics (i.e. giant magnetoresistance; GMR) is used as hard disk read heads and semiconductor-based spintronics (i.e. tunnel magnetoresistance; TMR) is used as magnetic random access memory (MRAM) devices. Spin transfer torque (STT) technology is applied to the existing MRAM devices of magnetic tunnel junctions with MgO barriers. STT-MRAM has the advantages of better selectivity and better scalability over conventional MRAM.

1. Introduction

This chapter is about spintronics, mostly for senior undergraduate and beginning graduate students of science and engineering. I will try to explain spintronics based on magnetic materials. The first part describes spin-dependent transport, which is the origin of spintronics. The typical physical property of spin-dependent transport is magnetoresistance (MR). So, I will briefly explain four typical magnetoresistance phenomena; ordinary MR (OMR), anisotropic MR (AMR), giant MR (GMR), and tunneling MR (TMR). Spintronics can be classified into three categories

405

based on spintronic materials. One is metal-based spintronics, the other is semiconductor-based spintronics, and another is insulator-based spintronics. In the second part, I will treat metal-based spintronics. The examples of metal-based spintronics are GMR head in hard disk drive, which is awarded a Novel prize in 2007, and spin transistor. The third part describes semiconductor-based spintronics; spin field effect transistor and Hall effect device. And in the fourth part, I will report insulator-based spintronics, which has TMR magnetic tunnel junction with powerful applications for magnetic random access memory (MRAM) devices. Especially, the progress for MRAM applications is discussed in terms of conventional MRAM, longitudinal spin transfer torque(STT)-MRAM, and perpendicular STT-MRAM. Basically, these magnetic memory devices have magnetic tunnel junction with tunnel barriers such as Al_2O_3 and MgO. Finally, I will summarize in the last part.

2. Spin-Dependent Transport

Before entering the main topic of spintronics, two fundamental physical states should be mentioned. One is electronic states and the other is magnetic states. There are five electronic states; superconductor, metal, semimetal, semiconductor, and insulator. These electronic states are classified by the electrical resistance level from zero to infinity. In more strictly speaking, these electronic states are governed by the density of states at the Fermi level in the band structure. Metal has a large number of charge carriers, approximately Avogadro number. Insulator has a large gap between the valence and conduction bands. Between them, there is a semiconductor, which has a narrow band gap. But the small number of charge carriers can be generated by doping. Interestingly, there are also five different magnetic states; ferromagnet, ferrimagnet, antiferromagnet, paramagnet, and diamagnet. Ferromagnet has all parallel moments, antiferromagnet has all antiparallel moments, and ferrimagnet has two different moments in size. In zero magnetic field, both paramagnetic and diamagnetic materials have randomly aligned moments. But when you apply magnetic field, the magnetic moments of paramagnetic materials are aligned in the same direction of the magnetic field, while the magnetic moments of diamagnetic materials are aligned in the opposite direction of the magnetic field. These magnetic states are basically governed by double or super exchange interactions in d-electron system and RKKY or Kondo interactions in f-electron system. From the viewpoint of band structure, nonmagnetic material has equal number of spin-up and spin-down electrons (Fig. 1(a)). So, the net magnetization should be zero. On the other hand, ferromagnetic material has different number of spin-up and spin-down electrons. The imbalance between the spin-up and spin-down bands occurs by exchange splitting, as shown in Fig. 1(b). Then, the material can be ferromagnetic with non-zero net magnetization. Figure 1(c) shows the typical magnetization curve, the so-called hysteresis loop of ferromagnetic material. In high magnetic field, the magnetization is saturated, which is the saturation magnetization M_s. As reducing the field to zero, the magnetization remains, which is the remanent magnetization M_r. The coercive field H_c is defined as the field that requires the net magnetization to zero. Usually, ferromagnetic materials are d- or f-electron systems including Fe, Co, Ni with 3d electrons and lanthanide with 4f electrons. The origin of ferromagnetism lies in the behavior of the 3d and 4f electrons.

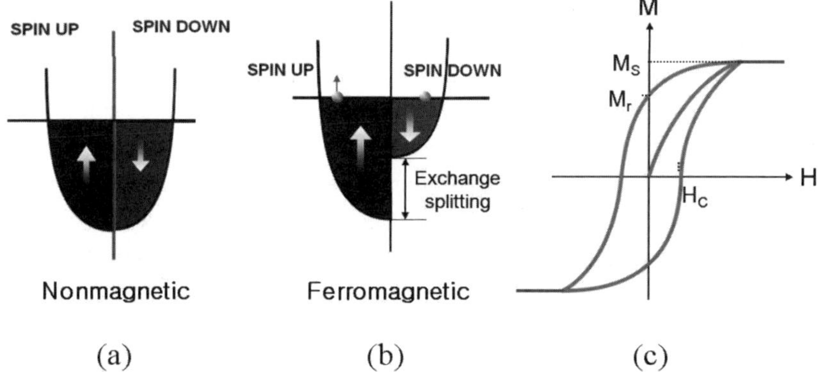

Fig. 1. a) and b) Schematic band structures of nonmagnetic and ferromagnetic material, respectively. c) Typical magnetization curve of ferromagnet. M_S is the saturation magentization, H_C is the coercive field, and M_r is the remanent magnetization.

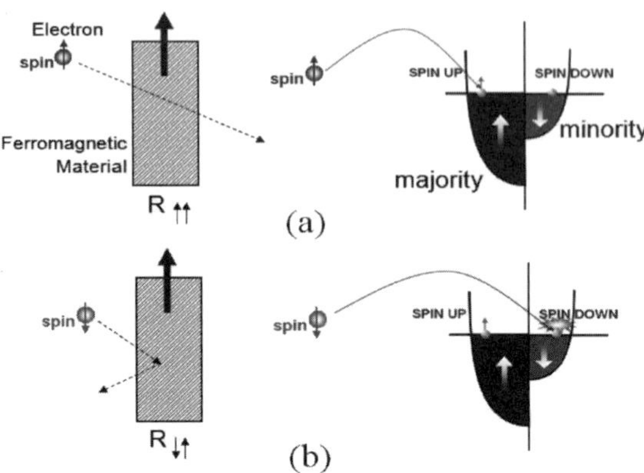

Fig. 2. Schematic diagram a) when the spin-up electron easily passes through a ferromagnet with majority spin-up band and b) when the spin-down electron is hard to pass through the ferromagnet.

These two fundamental physical states, electronic and magnetic states, are in a close relation each other. This means that the spin states affect the electrical transport properties. This is the so-called spin-dependent transport, which is the origin of spintronics. As shown in Fig. 2, let us consider one ferromagnetic material that is perfectly polarized with spin-up electrons and apply current through the material. The spin-up electrons easily pass through the ferromagnet (Fig. 2(a)), while the spin-down electrons are hard to pass (Fig. 2(b)). In a viewpoint of band structure, when the spin-up electrons transfer to the ferromagnet, there is a relatively high density of states for transport. This makes low resistance state. On the other hand, the spin-down electrons have a relatively low density of states for transport. This makes high resistance state. These are the reason why the electrical transport is affected by the spin states. This spin-dependent transport is the origin of spintronics. The electron has two degrees of freedom; one is charge and the other is spin. And electronics is based on electron charges and their energy, while magnetics is based on electron spins and their coupling. This combination of electronics and magnetics is magnetoelectronics, simply spintronics, that uses two fundamental properties of electron, charge and spin, at the same time. This is the reason why the spintronics has attracted a lot of attention.

Existing electronics expresses data as binary digits; 0 and 1. Here, "0" represents the absence of electron charge and "1" represents the presence of electron charge (Fig. 3). And it operates by applying voltages. It has been generally accepted that electronics follows Moore's law. This indicates that microprocessors will double in power every 18 months. According to this Moore's law, the size of individual bits approaches the dimension of atoms. This means the end of silicon road map. So, we need a new class of device based on another property of electron, that is "spin", in addition to charge.

Fig. 3. Spintronics uses two fundamental properties of electron, namely its charge and its spin simultaneously.

As like electronics, spintronics should also express data as binary digits; 0 and 1. Electrons can rotate in the clockwise or counterclockwise direction. Then, counterclockwise ones give up spins and clockwise ones give down spins. These spin-up and spin-down electrons play a role of "0" and "1" states. And the spintronics operates by applying magnetic fields. So, there are several advantages of spintronics, compared with electronics. For example, much smaller device size can be expected and the smaller device needs less electricity with higher density. And the spin movement is very fast, so higher speed can be expected. Furthermore, because spintronics utilizes two degrees of freedom, charge and spin, at the same time, more powerful functionality can be expected.

Spintronics applications can be found in magnetoresistance effects, which are shown in Fig. 4. There four examples of typical magnetoresistance (MR) phenomena; ordinary MR (OMR), anisotropic MR (AMR), giant MR (GMR), and tunneling MR (TMR). OMR is caused by the curving of the carrier trajectories in a magnetic field as a result of Lorenz force. When the current and magnetic field are applied in perpendicular, the Lorentz force gives cyclotron motion of electrons. In most case, OMR has quadratic dependence. Usually, OMR ratio is observed to be very small about 0.1%. In AMR case, it depends on the relative direction of the current and the magnetization. If there is an angle θ between them, the resistance depends on the square of $\cos\theta$. So, if there is an angle zero, then the resistance is a maximum. If there is an angle 90 degree, then the resistance is a minimum. Usually, AMR ratio is observed to be about 2%. GMR and TMR effects are observed in sandwich structures composed of two ferromagnets separated by nonmagnetic space layer. If the space layer is conductor such as Cr, Ru, Cu, Au, etc. then it is GMR. If the space layer is insulator such as Al_2O_3 and MgO, then it is TMR. The ferromagnets can be 3d metals and their alloys such as Fe, Co, Ni, CoFe, NiFe, CoFeB, etc. One ferromagnet with high coercive field acts as a pinned layer or fixed layer because it is hard to flip the spin direction. And the other ferromagnet with low coercive field acts as a free layer because it is relatively easy to flip the spin direction.

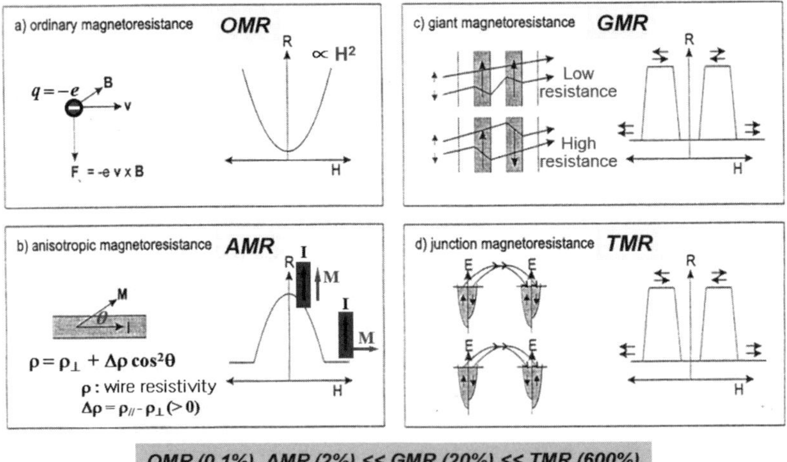

Fig. 4. Four typical magnetoresistance phenomena; a) OMR, b) AMR, c) GMR, and d) TMR.

3. Conductor-Based Spintronics

3.1. *GMR effect*

By applying magnetic field, two different spin alignments of parallel and anti-parallel are generated (Fig. 5). According to the spin-dependent transport mechanism, spin-up and spin-down electrons move independently. In the parallel case, the spin-up electrons easily pass through the two ferromagnets, but the spin-down electrons cannot. The spin-up electrons have a current path with two low resistors and the spin-down electrons with two high resistors. On the other hand, in the anti-parallel case, the spin-up electrons pass through the first ferromagnet but not through the second ferromagnet. And the spin-down electrons are reverse. So, both spin-up and spin-down electrons have a current path with one high resistor and another low resistor. Thus, the parallel case gives low resistance and the antiparallel case gives high resistance.

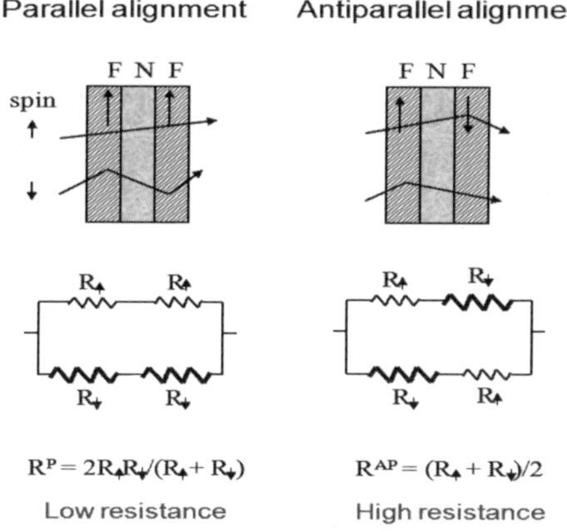

Fig. 5. Basic concept for GMR and TMR. F represents ferromagnet and N represents normal metal for GMR case and insulator for TMR case.

Usually, GMR ratio is observed to be about 20%. In a viewpoint of band structure, same story can be applicable. The injected spin-up electrons in the first ferromagnet transfer its spin information into the second spin-up band. Since the parallel spin alignment corresponds to the spin transfer from majority to majority spin bands, the parallel spin alignment is in a low resistance state. On the other hand, the anti-parallel spin alignment corresponds to the spin transfer from majority to minority spin bands. This generates a high resistance state.

GMR effect was discovered by Peter Grünberg and Albert Fert in 1988. As shown in Fig. 6, Peter Grünberg reported 1.5% GMR ratio at room temperature in Fe/Cr/Fe trilayer and Albert Fert reported 50% GMR ratio at 4.2 K in FeCu multilayers. Their discovery of GMR is considered as the birth of spintronics and is commercialized as GMR head in hard disk drive. For this reason, their contributions to the field of spintronics and computer hard drives were awarded a Nobel Prize in physics in 2007.

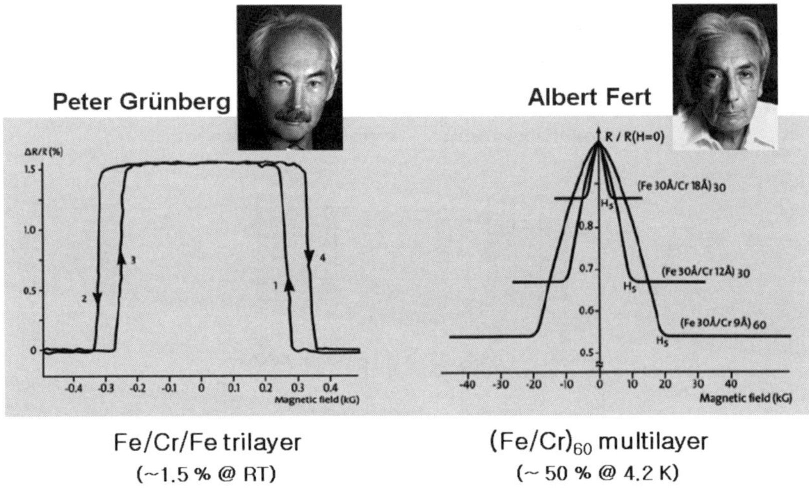

Fig. 6. The discovery of GMR of Fe/Cr/Fe trilayer reported by Peter Grünberg and of $(Fe/Cr)_{60}$ multilayer reported by Albert Fert in 1988.

There are some remarks on the Nobel prize. The discovery of GMR opened the door to a new field of science, magnetoelectronics (or spintronics), where two fundamental properties of the electron, namely its charge and its spin, are manipulated simultaneously. The GMR devices have spin valve structures. The standard structure of GMR devices can be trilayer such as Fe/Cr/Fe as Grünberg introduced or multilayer such as $(Fe/Cr)_{60}$ as Fert intoduced. But in order to enhance the GMR ratio, very complicated multilayer structures have been developed.

3.2. *Spin transistor*

All-metal spin transistor in Fig. 7 was suggested by Mark Johnson at the Naval Research Laboratory in 1994. He designed three electrodes with two ferromagnets, collector and emitter, separated by nonmagnetic conducting layer, base. By applying voltage with battery between emitter and base, the current between base and collector can be measured. When the spin alignment is parallel, a positive current is measured. However, when the spin alignment is antiparallel, a negative current is measured. However, there are several problems in all-metal spin transistor. Metal-based transistor cannot amplify signals, needs extremely small device using nanolithographic technique, and is difficult to be integrated into existing semiconductor-based circuits.

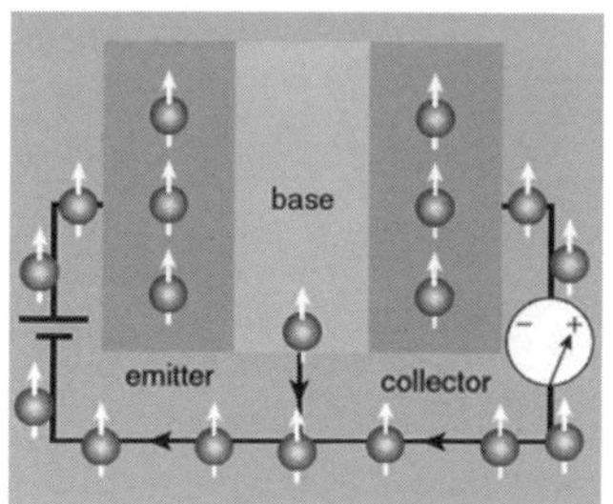

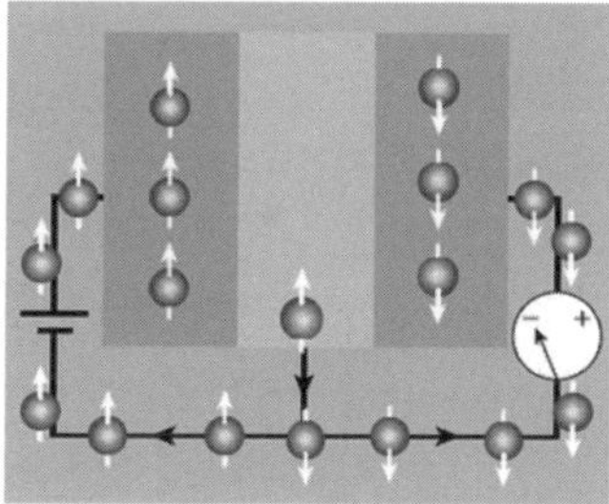

Fig. 7. Schematic diagram of spin transistor suggested by Mark Johnson in 1994.

4. Semiconductor-Based Spintronics

4.1. *Spin FET*

Spin FET(field effect transistor) in Fig. 8 was suggested by Datta and Dass at Purdue University in 1990. The device has nearly same structure with the Johnson's spin transistor. But they used 2DEG(2-dimensional electron gas) for space layer, so that the spins can move in the 2DEG channel. In the left-hand side, emitter electrode emits electrons with polarized spins. In the right-hand side, collector electrode accepts electrons with the same spin only. In the middle of them, gate produces an electric field that forces the spins to precess in the 2DEG channel. So, if the spin alignment is parallel, a current is generated. However, if the spin alignment is antiparallel, no current is generated. However, there are several problems also in spin FET, interface scattering and conductivity mismatch between the ferromagnetic metal and semiconductor. And also this spin field effect can include complicated phenomena by AMR, Hall effect, etc. So, usually very low efficiency for spin injection is observed less than 1%. In order to solve these problems, some solutions were suggested such as inserting tunnel contacts at the interface, adapting ferromagnet/superconductor junction, and using dilute magnetic semiconductors as the injector. Nevertheless, there are still problems to be solved for spin transistor.

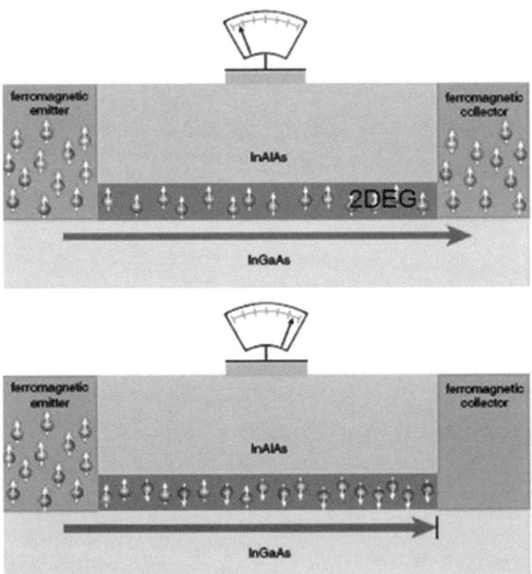

Fig. 8. Schematic diagram of spin FET suggested by Datta and Dass in 1990.

4.2. *Hall effect device*

Figure 9 shows the comparison between the Hall and spin Hall effects. When the current is applied in the x direction and the magnetic field in the z direction, the Hall voltage is generated in the y direction, because of the Lorentz force. This is the normal Hall effect. On the other hand, without magnetic field, by inducing the electric field spin-up and spin-down electrons can be separated, if the material has strong spin-orbit coupling. This is called as spin Hall effect. Recently, inverse spin Hall effect has been suggested. In the spin Hall case, the applied charge current induces the spin current by separating the spin-up and spin-down electrons. On the other hand, in the case of inverse spin Hall effect, the unbalance between the spin-up and spin-down electrons can generate charge current from the spin current.

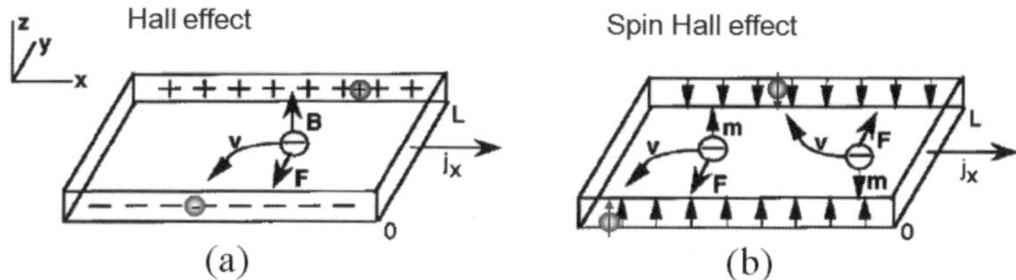

Fig. 9. Schematic diagrams for a) Hall effect and b) spin Hall effect devices.

In Fig. 10, the inverse spin Hall device has a bilayer structure of ferromagnet NiFe and nonmagnetic normal metal Pt. By inducing the microwave from ferromagnetic resonance, spin pumping occurs. Then,

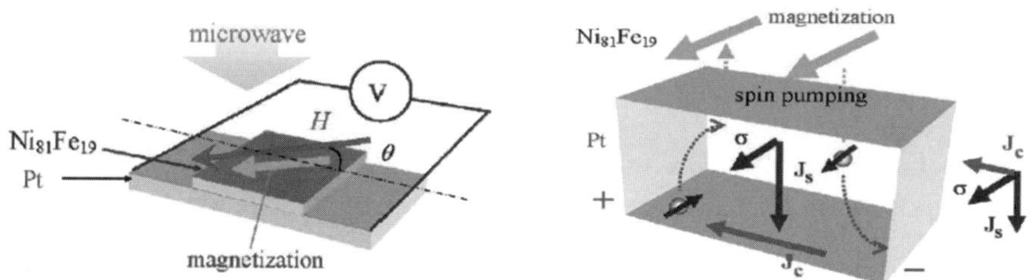

Fig. 10. Schematic diagrams for inverse spin Hall effect device.

the electrons polarized in the same direction of the ferromagnet moments are accumulated in the ferromagnet side and generate a pure spin current out of plane in the Pt layer. Then, finally the charge current is measured in the plane of Pt layer. This is because the charge current is the vector product of the spin current and the spin direction.

More recently, spin Seebeck effect has been suggested. There are two different metals A and B connected to each other. If they have different Seebeck coefficients, then the voltage between the output terminals is generated to be proportional to the temperature difference between the ends of the couple. This is the normal Seebeck effect, which is applied for thermocouple. On the other hand, in the case of spin Seebeck effect, spin-up and spin-down electrons have different spin Seebeck coefficients. So, by generating a temperature gradient in a single ferromagnetic metal, a spin voltage at the ends of ferromagnet is detected to be proportional to the temperature difference. This is called as spin Seebeck effect, which is the generation of spin voltage as a result of a temperature gradient.

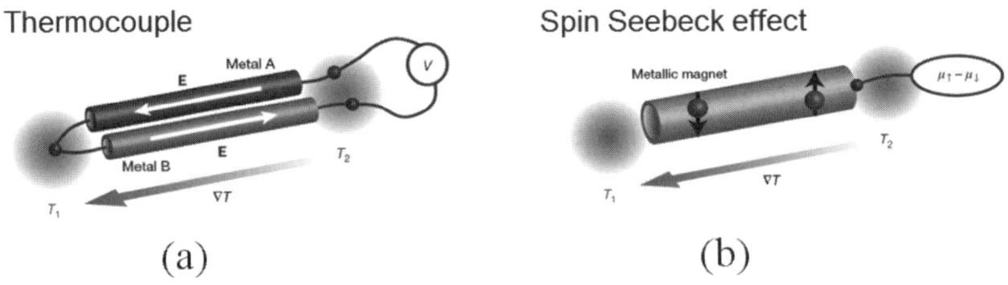

Fig. 11. Schematic diagram for thermocouple and spin Zeebeck effect device.

5. Insulator-Based Spintronics

5.1. *Comparison between GMR and TMR*

Figure 12 shows the basic concept of GMR and TMR effects. A trilayer is composed of two ferromagnets separated by nonmagnetic material. The parallel spin alignment corresponds to low resistance state and the antiparallel spin alignment corresponds to high resistance state. By changing the direction of magnetic field, two different resistances are achieved. The parallel state gives low resistance and the antiparallel state gives high resistance. This is the basic mechanism of magnetic memory.

The magnetic memory devices can have GMR or TMR structures. Basically, GMR device has spin valve structure. Two ferromagnets are separated by conductor such as Cu, Cr, etc. And usually, the conductor has short spin relaxation length. So, the GMR ratio is observed to be low about 20 %.

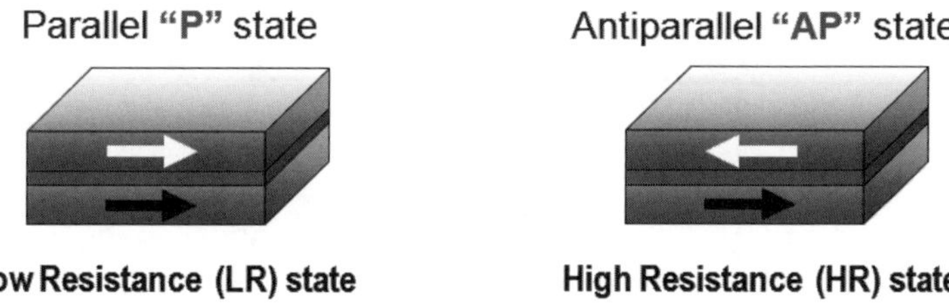

Fig. 12. Basic concept of GMR and TMR effects. Parallel (antiparallel) state corresponds to low (high) resistance state.

On the other hand, the TMR device has magnetic tunnel junction (MTJ) structure, where the space layer is insulator such as Al2O3 and MgO. And usually, insulator has long spin relaxation length. So, the TMR ratio is observed to be high up to 600%. However, in both cases, there is an antiferromagnetic layer, which is called pinning layer. Usually, the pinning layer makes exchange bias.

5.2. *Conventional MRAM*

The first generation of MTJs is based on Al_2O_3 tunnel barrier. The maximum value of TMR ratio for Al_2O_3 is about 80%. For MgO as tunnel barrier, the TMR ratio is strongly enhanced up to 600% at room temperature and 1144% at low temperature. Figure 13 shows the advance of the TMR ratio. The increment of TMR ratio for Al_2O_3-based MTJs is slow. By the use of MgO tunnel barrier, the TMR ratio is suddenly developed. There are two main reasons why the TMR ratio is large in MgO tunnel barrier. One is that MgO can be epitaxially grown by annealing with the ferromagnets of CoFeB. Then, more spins are transferred without loosing its spin information. The other is MgO has anisotropic band structure. Only Δ_1 electrons can pass through crystallized MgO tunnel barrier, implying that the polarization of Δ_1 band is almost 100%. The first magnetic memory starts from AMR effect, and the second generation is GMR effect. The TMR effect was introduced from the Al_2O_3-based MTJ in the third generation. Now, we are in the fourth generation of MgO-based magnetic memory.

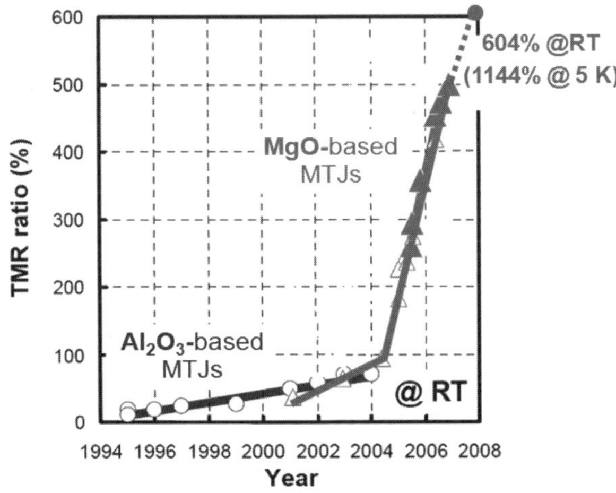

Fig. 13. TMR ratio of Al_2O_3- and MgO-based MTJs

By applying magnetic field, parallel or antiparallel spin alignment is achieved. According to the spin-dependent transport mechanism, the parallel/antiparallel spin alignment gives rise to low/high resistance state. These two states behave as binary digits 0 and 1 for low and high resistances, respectively.

The magnetic memory used for magnetic random access memory MRAM operates like this. You have trilayer with insulating tunnel barrier Al_2O_3 or MgO. When you apply magnetic field pointing out the right-hand side, which is positive field by definition in this case, then the spins of the free layer switch the direction of the magnetic field. Then, you have antiparallel spin alignment, which gives high resistance state. On the other hand, when you apply magnetic field pointing out the left-hand side, negative field, the spins of the free layer switch the direction of the magnetic field. Then, you have parallel spin alignment, which gives low resistance state. So, if you measure the magnetoresistance, you have high resistance in the positive field. And low resistance in the negative field. As you know, computer expresses data as binary digits as 0 and 1. So, in this case, these two resistance states behave as binary digits 0 and 1 for low and high resistances, respectively. Here, you have to focus the two resistance states at zero field. This means that this magnetic device can memorize even in no power. This is the reason why we are paying lots of attentions to the MRAM because it is nonvolatile. MRAM is nonvolatile as like flash memory and it has high speed as like SRAM because MRAM utilizes spins instead of charges. It is also possible to have high density as like DRAM because it could be nanosized. When the HDD functionality is involved in addition to these properties, MRAM can be a universal memory for next generation. However, the commercialized MRAM has only 4 Mbits memory. One reason is selectivity issue, because the magnetic field cannot be localized. The MRAM cell could be affected by the magnetic field in neighbor. The other reason is scalability issue, because more power is needed with reducing the cell size.

5.3. *Longitudinal STT-MRAM*

Magnetization acts on the current, which is the magnetoresistance, used for GMR head on hard disk drive and conventional MRAM. Then, what is the reaction? Does the current act on magnetization? The answer is YES. Spin transfer torque (STT)-MRAM is that case. Usually, STT can be expressed by the LLG equation. The time derivative of magnetization is given by three terms. The first term is precession term by effective field. The second term is the field-induced damping torque term with Gilbert damping parameter. And the third term is the current-induced spin transfer torque term. By applying magnetic field, the spins have a tendency to be aligned in the same direction of magnetic field. This is the field-induced damping torque. And also, by applying current, the spins have another tendency to be switched to the opposite direction of magnetic field. This is the current-induced spin transfer torque. Then, if the spin transfer torque is larger than the damping torque, then the magnetization switches. However, if the spin transfer torque is comparable with the damping torque, then stationary precession is generated from their balance.

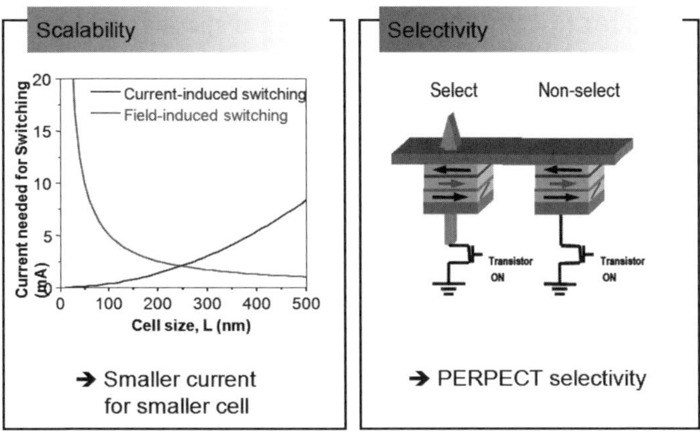

Fig. 14. Comparison between current- and field-induced switching devices. The two main issues of scalability and selectivity are solved for current-induced case.

STT-MRAM has been proposed in order to solve the selectivity and scalability issues in the conventional MRAM. Compared with the conventional MRAM, STT-MRAM is operated by current. It should be noted that the ferromagnet must have single domain structure, meaning the requirement of nanosized device.

When the current starts from pinned layer, i.e. the electrons flow from free layer to pinned layer, the longitudinal spin components are injected into the pinned layer, while the transverse spin components are not injected but reflected at the interface of the pinned layer. Then, the spin direction of the free layer is switched into the opposite direction to the pinned layer. This is referred to as spin transfer torque (STT) effect. It gives rise to high resistance state. On the other hand, when the current starts from free layer, the polarized electrons in the pinned layer are injected into the free layer. Then, the spin direction of the free layer is switched into the same direction to the pinned layer, resulting in low resistance state.

As mentioned before, conventional MRAM is operated by magnetic field. This field-induced switching needs more power with reducing the cell size. The current needed for field-induced switching is exponentially increased with reducing the cell size. However, STT-MRAM is operated by current. This current-induced switching needs less power with reducing the cell size. Thus, by using STT effect, the scalability issue is solved. And the selectivity issue is also solved by applying current, instead of magnetic field, because the current can be localized to select a cell with no effect on the next cell.

Nevertheless, there are some key issues of STT-MRAM; how to reduce the critical current density and how to enhance the thermal stability. In order to reduce the critical current, we need low saturation magnetization, high spin polarization, and low damping parameter. And in order to enhance thermal stability, we need optimal materials and structures to achieve $kV/k_BT > 60$ at 30 nm node without reducing spin polarization. Recently, a new type of STT-MRAM using perpendicular magnetic anisotropic (PMA) materials has been suggested.

6. Perpendicular STT-MRAM

The most benefit of perpendicular magnetic anisotropy materials is that it has less sensitivity to shape variations. In general, the longitudinal magnetic anisotropy materials have the magnetization curling at edge, leading to very complicated magnetic domain structure. On the other hand, the perpendicular magnetic anisotropy materials have no magnetization curling at edge. This results in homogeneous magnetic domain structure with low saturation magnetization. Furthermore, the thermal stability for perpendicular STT-MRAM was suggested to be improved.

The known PMA materials are Co hcp alloy, multilayer of ferromagnet and nonmagnetic metal, L10 alloys, and rare-earth based materials

(Table 1). There are some important physical parameters reported in PMA materials. However, we have still limited information of spin polarization and damping parameter. Pt doping in Fe, Co, and Ni leads to an increase of damping parameter. For example, the damping parameter of FePt is about 0.05, which is much larger than that (=0.0005) of CoFeB. New PMA materials and structures with damping parameter less than 0.01 should be developed for practical MRAM applications. In amorphous TbFeCo, the thermal stability is estimated to be 32. This implies that the magnetic properties of TbFeCo can be easily degraded during processing at temperatures above 300°C. We need optimal materials and structures to achieve the thermal stability higher than 60 at 30 nm node.

Table 1. Suggested perpendicular magnetic anisotropic materials.

Examples of some Uniaxial Ferromagnetic Materials					
Material	$Ku \times 10^7$ (ergs/cc)	Ms (emu/cc)	Tc (K)	Hk (kOe)	Vp (nm3)
Co hcp alloy					
CoCrPt	0	330		18	966
Co3Pt	2	1100		36	145
CoPt3	1	300	600	33	580
Co	0	1422	1393	6	644
Multilayer					
Co2/Pt9	1	360	500	56	290
Co2/Pd9	1	360	500	33	483
L10					
FePd	2	1100	760	33	161
FePt	7	1140	750	116	44
CoPt	5	800	840	123	59
MnAl	2	560	650	61	170
Rare Earth					
Fe14Nd2B	5	1270	585	72	63
SmCo5	20	910	1000	440	14
YCo5	6	844	903	130	53
CeCo5	7	692	737	208	40
PrCo5	8	891	912	173	38

The known PMA materials are Co hcp alloy, multilayer of ferromagnet and nonmagnetic metal, L10 alloys, and rare-earth based materials. There are some important physical parameters reported, but we have still limited information of spin polarization and damping parameter. So, we are still looking for new PMA materials and structures for the STT-MRAM applications.

7. Summary

We classified the spintronic devices into three categories in terms of the space layer; conductor-, semiconductor-, and insulator-based spintronics. Even though the conductor-based spintronics has several advantages, we have serious problems such as no amplification of the signals and low GMR ratio. It is also hard to be integrated into the semiconductor-based circuits. So, in order to solve these problems, we needed the semiconductor-based spintronics, such as spin-FET and Hall effect device. However, this semiconductor-based spintronics has also some problems, such as interface scattering and conductivity mismatch between the ferromagnetic metal and semiconductor and low efficiency of spin injection. So, we needed another spintronics, which is based on the insulator. Even though we have high TMR ratio, conventional MRAM operated by magnetic field has two main problems of scalability and selectivity issues. To solve these problems, a new physics was involved, which is spin transfer torque. Early works on MRAM are concentrated on magnetic tunnel junction (MTJ) with longitudinal magnetic anisotropy. The longitudinal STT-MRAM meets problems of high critical switching current and low thermal stability. Recently, a new type of STT-MRAM using perpendicular magnetic anisotropic (PMA) materials has been suggested. For better MRAM applications, we are now in progress for optimal materials and structures.

CHAPTER 13

THERMOELECTRIC MATERIALS

Silke Paschen

Institute of Solid State Physics, Vienna University of Technology
Wiedner Haupstrasse 8-10, 1040 Vienna, Austria
E-mail: paschen@ifp.tuwien.ac.at

Thermoelectric materials can be used for refrigeration and power generation. To date efficient materials for applications in the vicinity of room temperature exist and thermoelectric modules fabricated from these are commercially available. However, both at low and high temperatures, efficiencies are still too low and much effort is put worldwide in optimizing known and finding new materials. After a brief introduction to thermoelectric phenomena and a presentation of the state-of-the-art in traditional Bi-Te based thermoelectrics, focus will be on material classes considered more recently for thermoelectric applications. Special emphasis will be on "cage compounds" as one important representative of the class of "complex metallic alloys".

1. Introduction

Thermoelectric (TE) materials convert heat into electric current and vice versa and can, as such, be exploited for power generation or refrigeration. Potential areas of application for the former are all kind of industrial processes, automotive exhaust, domestic heating, and many others. The use of TE power generators would not only increase the overall efficiency but also actively counteract global warming. Thus, a breakthrough in the efficiency of TE materials is a key contribution to an environment friendly and sustainable energy system for the coming generations.

An equally important issue is "green" cooling. When used in solid state Peltier coolers thermoelectric materials cool without any harmful

425

liquid refrigerants, without movable parts, and completely silently. Promising application areas are the replacement of compression based refrigerators, cooling of CPUs and other normal or even superconducting electronic components, of infrared detectors, and so on. With the miniaturization of electronic devices the dissipated power density increases to an extent which will shortly require cooling efficiencies that exceed those of presently available technologies.

Currently broad applications of thermoelectric materials are inhibited by their limited efficiency. The optimization of materials for higher efficiency is challenging since the three physical quantities determining the efficiency (thermopower, thermal conductivity, electrical conductivity) are generally interdependent in a way that optimizing one will adversely affect another. Thus, the strategy governing the search for better materials has long been to find the best compromise. In the mid-1990's the field was given new impetus by the advent of nanotechnology which could, in some cases, decouple the physical quantities from each other. Today thermoelectric materials build a highly active research field both in academia and industry.

A number of thermoelectric materials are well established and are already today used in commercial applications. At temperatures around room temperature Bi_2Te_3-based materials are universally used in thermoelectric refrigeration. Between about 100 and 600°C lead tellurides are the materials of choice and, at even higher temperatures, SiGe is employed. The thermoelectric figure-of-merit of these and other less established materials is shown in Fig. 1.

Section 2 deals with a relatively new class of materials with promising thermoelectric properties at high temperatures, cage compounds. Also, attempts to improve their properties at low temperatures by the introduction of effects of strong electronic correlations will be discussed shortly (section 3). Cage compounds are considered to be members of a larger class of materials called "complex metallic alloys (CMAs)", which are the topic of the European Network of Excellence CMA. Thermoelectric properties of non-caged CMAs are covered in Chapter 3, Volume 2 of the CMA Series. Other promising materials classes such as oxides or low-dimensional systems are beyond the scope of this contribution.

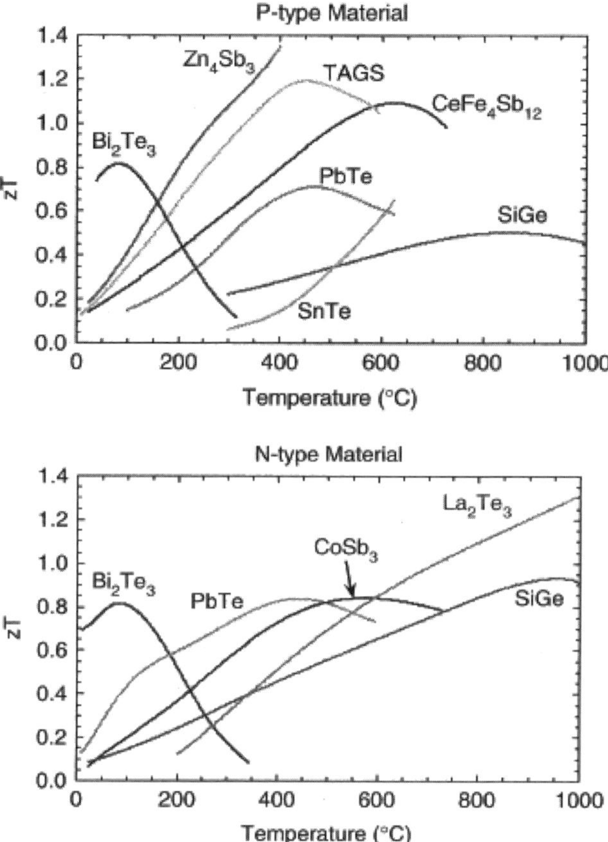

Fig. 1. Dimensionless thermoelectric figure-of-merit of established and more recently investigated *p*- and *n*-type materials versus temperature[1] (upper panel and lower panel, respectively).

2. Cage Compounds

2.1. *Definitions*

What are cage compounds? This notion is not yet well established in the literature and various definitions are conceivable which shall be explained in more detail below.

2.1.1. *Guest/host atoms*

Guests are situated in cages made up by the host species. An example is shown in Fig. 2.

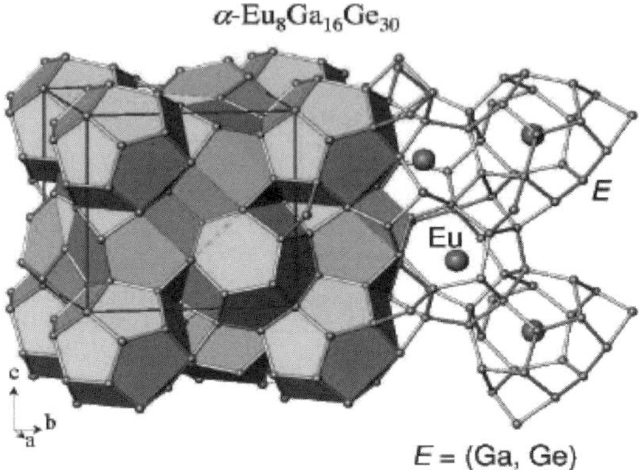

Fig. 2. Cage compound defined as a guest – host system[2] (Eu are guest atoms).

2.1.2. *Coordination number (c.n.)*

In a cage compound the guest atom has a large c.n. with the host atoms. In the clathrate β-Eu$_8$Ga$_{16}$Ge$_{30}$, for instance, there are two guest sites. Guest 1 in the smaller cage has c.n. = 20 while guest 2 in the larger cage has c.n. = 24 (cf. Fig. 3).

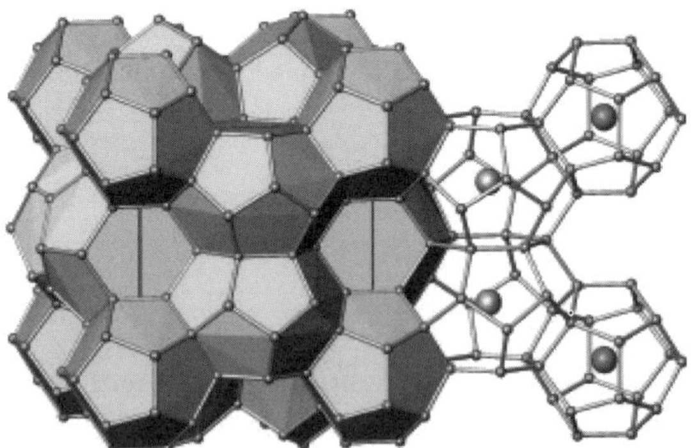

Fig. 3. Large coordination numbers in the cage compound β-Eu$_8$Ga$_{16}$Ge$_{30}$.

2.1.3. *Bond length/strength*

Bonds between guest and host species are longer (bond strength weaker) than in related non-caged compounds. While in β-$Eu_8Ga_{16}Ge_{30}$ the shortest Eu–Ga/Ge distance is 3.41 Å or 3.59 Å for the smaller and larger cage, respectively (cf. Fig. 3), the shortest Eu–Ge distance in $EuGe_2$ is 3.11 Å (cf. Fig. 4).

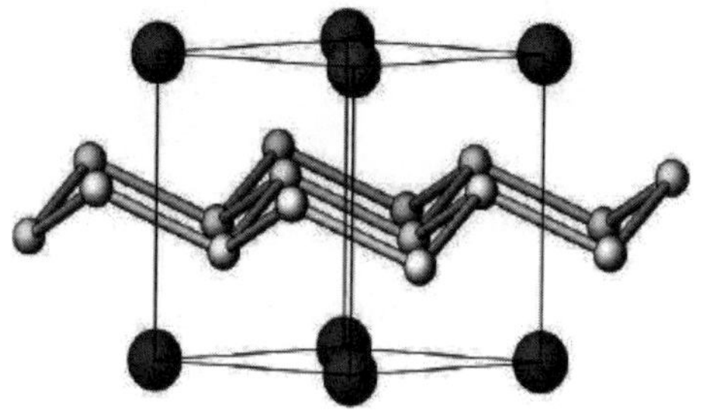

Fig. 4. Crystal structure of $EuGe_2$.[3]

2.1.4. *Empty host*

The host can, at least in some cases, exist without the guest. This is typical for the skutterudites which are well known both for their empty (Fig. 5) and filled (Fig. 6) form.

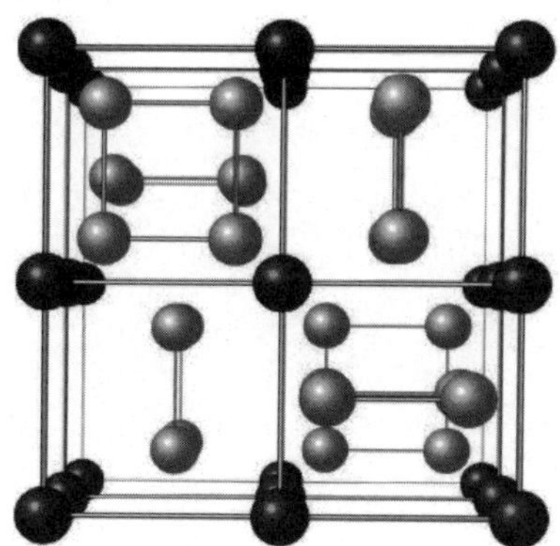

Fig. 5. Empty skutterudite, e.g. $CoAs_3$ (Co dark spheres, As light spheres).

                                   S. Paschen

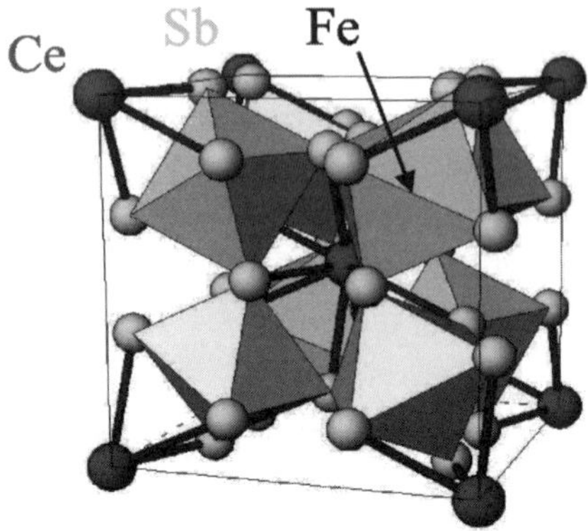

Fig. 6. Filled skutterudite, e.g., $CeFe_4Sb_{12}$, where Ce is the guest.

In clathrates the existence of empty variants appears to be much more subtle. While in so-called type-I clathrates no partial filling of the cages has ever been observed, the type-II structure is more disposed to exist in empty form (Fig. 7).

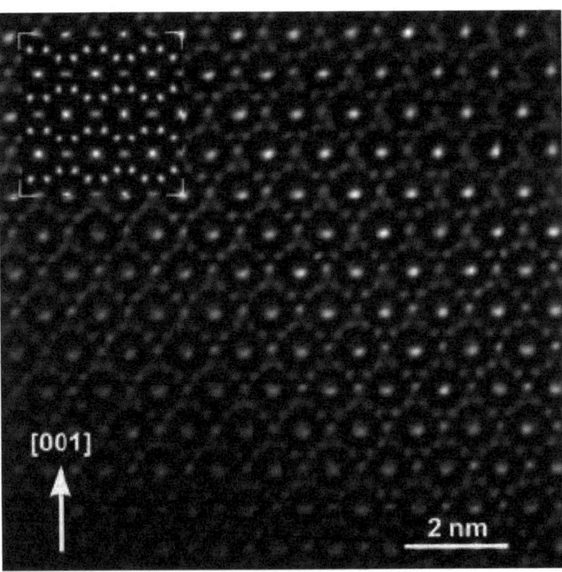

Fig. 7. TEM image of the empty type-II clathrate $\Upsilon_{24}Ge_{136}$. (Υ denotes a void).[4]

2.2. Examples

Cage structures occur in various classes of materials. The below list is certainly not exhaustive.

2.2.1. Filled skutterudites

The structure of filled skutterudites results from the one of binary (or empty) skutterudites (cf. Fig. 5). These latter consist of eight formula units per unit cell and have a cubic crystal structure of space group $Im\overline{3}$. The large voids in $2a$ position can accommodate interstitially various kinds of atoms, which may be represented by the notation $\Upsilon_{24}T_8X_{24}$, where Υ = void. The transition metal atoms T occupy the $8c$ site, the atoms X the $24g$ site. The interstitial ions are enclosed in an irregular dodecahedral cage of X atoms.

Skutterudites of many different compositions exist. The prototypical binary skutterudite is $CoAs_3$ (Fig. 5). Other binary skutterudites can be created by replacing Co and As by the isoelectronic elements Rh and Ir (Fig. 8) or P and Sb (Fig. 9), respectively, resulting in the possible compositions given in the first column of the upper table of Table 1. Many other compositions, in particular ternary and quatenary compounds, can be designed by respecting the overall charge balance, e.g., if one atom per formula unit is replaced by another one with one extra valence electron then another atom has to be replaced by an atom with one less valence electron (Table 1).

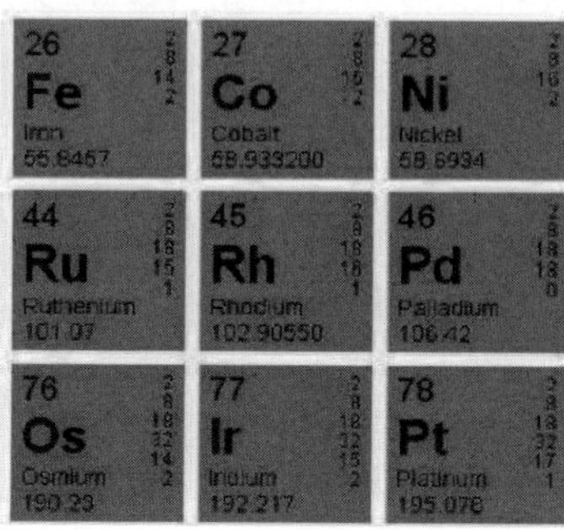

Fig. 8. Neighbours of Co in the periodic table.

Table 1. Possible compositions of skutterudites (*by courtesy of A . Grytsiv, unpublished*).

unfilled skutterudites						
binary			ternary			
T_4X_{12}	T'_4X_{12}	T''_4X_{12}	$T_2X'_6X''_6$	$T'_2T''_2X_{12}$	$T'_4X_8X''_4$	$T''_4X_8X'_4$
Co_4P_{12}	"Fe_4Sb_{12}"	Ni_4P_{12}	$Co_4Ge_6Te_6$	$Fe_2Ni_2Sb_{12}$	$Fe_4Sb_8Se_4$	$Ni_4P_8Ge_4$
Co_4As_{12}		Pd_4P_{12}	$Co_4Sn_6Se_6$	$Fe_2Ni_2As_{12}$	$Fe_4Sb_8Te_4$	$Ni_4Bi_8Ge_4$
Co_4Sb_{12}			$Co_4Sn_6Te_6$	$Fe_2Pd_2Sb_{12}$	$Ru_4Sb_8Se_4$	$Pt_4Sb_{7.2}Sn_{4.8}$
Rh_4P_{12}			$Co_4Ge_6S_6$	$Fe_2Pt_2Sb_{12}$	$Ru_4Sb_8Te_4$	$Ni_4Sb_8Sn_4$
Rh_4As_{12}			$Co_4Ge_6Se_6$	$Ru_2Ni_2Sb_{12}$	$Os_4Sb_8Te_4$	$Ni_4As_8Ge_4$
Rh_4Sb_{12}			$Rh_4Ge_6S_6$	$Ru_2Pd_2Sb_{12}$		
Ir_4P_{12}			$Ir_4Ge_6S_6$	$Ru_2Pt_2Sb_{12}$		
Ir_4As_{12}			$Ir_4Ge_6Se_6$			
Ir_4Sb_{12}			$Ir_4Sn_6S_6$			
			$Ir_4Sn_6Se_6$			
			$Ir_4Sn_6Te_6$			

filled skutterudites		
ternary	Quaternary	
EpT_4X_{12}	$EpT'_4(X_{1-x}X'_x)_{12}$	$Ep(T_{1-x}T'_x)_4X_{12}$
T = Fe, Ru, Os **X = P, As, Sb** Ep=Ca, Sr, Ba, La, Ce, Pr, Nd , Na, Sm, Eu, Gd, Tb, Yb, Th, U	**T' = Co, Ir** **X = Sb; X' =Ge, Sn** Ep = La, Nd, Sm, Tl $LaIr_4Sb_9Ge_3$,$NdIr_4Sb_9Ge_3$,	**T = Fe; T' = Co** **X = Sb** Ep = Tl $TlFeCo_3Sb_{12}$
$LaFe_4P_{12}$, UFe_4P_{12}, $ThOs_4As_{12}$, $YbFe_4Sb_{12}$, ...	$SmIr_4Sb_9Ge_3$,$TlCo_4Sb_{11}Sn$, $La_{0.9}Co_4Sb_{10.3}Sn_{2.44}$, ...	

metastable partially filled skutterudites:

$Ep_{1-y}Fe_4Sb_{12}$; Ep = Na,Y, Hf, Sn, Lu $Ep_{1-y}Co_4Sb_{12}$; Ep = Sn, Pb

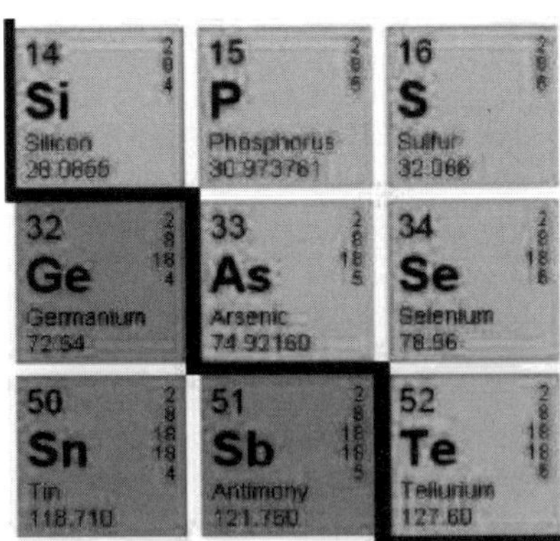

Fig. 9. Neighbours of as in the periodic table.

2.2.2. *Intermetallic clathrates*

Intermetallic clathrates are periodic solids in which tetrahedrally bonded atoms form a space-filling framework of cages that enclose guests. Clathrates have been classified with respect to the type of cage-forming polyhedra (Table 2).

The bonding situation may, in a first approximation, be discussed with the help of the Zintl concept: The more electropositive guest atoms (cations) donate electrons to the more electronegative cage atoms (anions) such that the latter complete their valence requirements (octet rule) when forming a covalently (sp^3-like) bonded space-filling framework (Fig. 10). E.g., for the clathrate of type-I $Ba_8Ga_{16}Ge_{30}$ this may be written as

$$(Ba^{2+})_8(Ga^{1-})_{16}(Ge^0)_{30} \tag{1}$$

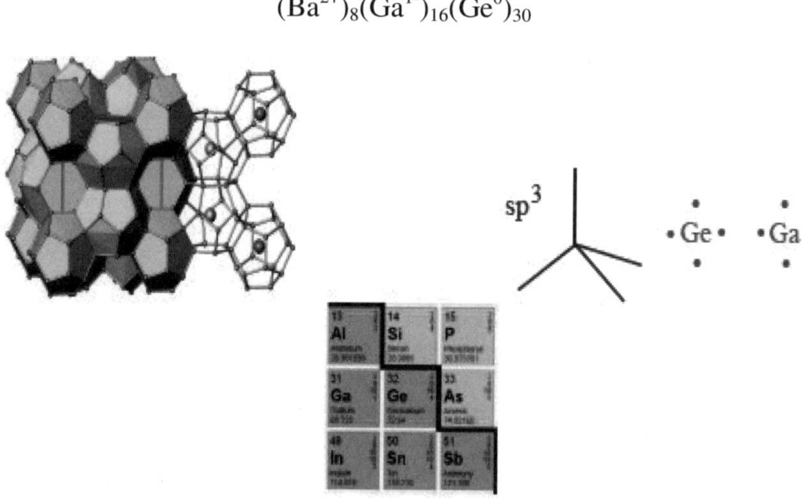

Fig. 10. Bonding situation in intermetallic type-I clathrates.

2.2.3. *Clathrate-like compounds*

Clathrate-like compounds do not fulfill the strict structural requirements of clathrates but have nevertheless a cage structure with atoms situated inside the cages (guests). Examples are, e.g., $Ce_2Ni_{36}P_{15}$ (Fig.11) or $Ce_3Pd_{20}Si_6$ (Fig. 12).

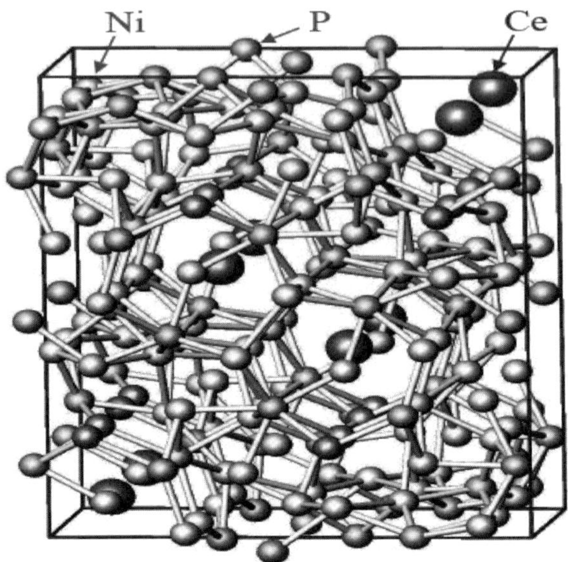

Fig. 11. Orthorhombic *Pbam* crystal structure of $Ce_2Ni_{36}P_{15}$. The cages consist of 16 Ni and 6 P atoms.[6]

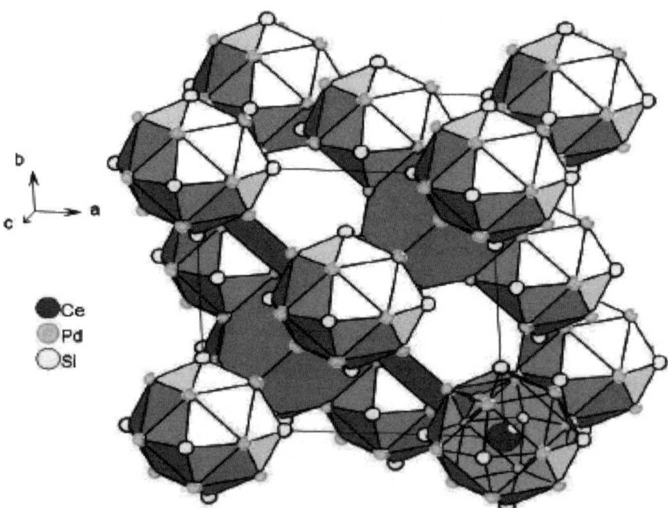

Fig. 12. $Ce_3Pd_{20}Si_6$. Cubic $Fm\bar{3}m$. Cage 1: 16 Pd, cage 2: 12 Pd, 6 Si.[7]

Table 2. Structure types of intermetallic and hydrate clatrathes: X, Y= guest atoms, T = host atoms.[5]

Type	Ideal unit cell formula	Polyhedra	Space group	Intermetallic Clathrate
I	$6X^*2Y^*46T^a$	$[5^{12}6^2]_6[5^{12}]_2$	Pm-3n	K_8Ge_{46-x}, $Ba_8Al_{16}Ge_{30}$, $Eu_2Ba_6Cu_4Si_{42}$
II	$8X^*16Y^*136T$	$[5^{12}6^4]_8[5^{12}]_{16}$	Fd-3m	Na_xSi_{136}, $Cs_8Na_{16}Ge_{136}$
III	$20X^*10Y^*172T$	$[5^{12}6^2]_{16}[5^{12}6^3]_4[5^{12}]_{10}$	P4$_2$/mnm	$Cs_{30}Na_{(1.33x-10)}Sn_{(172-x)}$. x = 9.6
IV	$8X^*6Y^*80T$	$[5^{12}6^2]_4[5^{12}6^3]_4[5^{12}]_6$	P6/mmm	$\sim K_7Ge_{40-x}$, x $\sim 2^c$
V	$4X^*8Y^*68T$	$[5^{12}6^4]_4[5^{12}]_8$	P6$_3$/mmc	-
VI	$16X^*156T$	$[4^35^96^27^3]_{16}[4^45^4]_{12}$	I-43d	-
VII	$2X^*12T$	$[4^66^8]_2$	Im-3m	-
VIII	$8Y^*46T$	$[3^34^35^9]_8^d$	Im-3m	$Ba_8Ga_{16}Sn_{30}$, $Eu_8Ga_{16}Ge_{30}$
IX	$16Y^*8X^*100T$	$[5^{12}]_8 + [4^{70}]_4 + ...^e$	P4$_1$32	$Ba_6Ge_{21}In_4$, Ba_6Ge_{25}

2.2.4. *Oxides*

Also in oxides cage structures can occur. An example is KOs_2O_6 which crystallizes in the β-pyrochlore structure and where K is the guest atom in cages built of Os and O (Fig.13).

Fig. 13. β-pyrochlore structure; K (largest spheres), Os (medium sized spheres), O (smallest spheres). K in cages.[8]

In $Na_2V_3O_7$ instead of approximately spherical cages there exist nanotubes formed by VO_5 pyramides. Only part of the Na atoms are captured within these tubes (Fig.14).

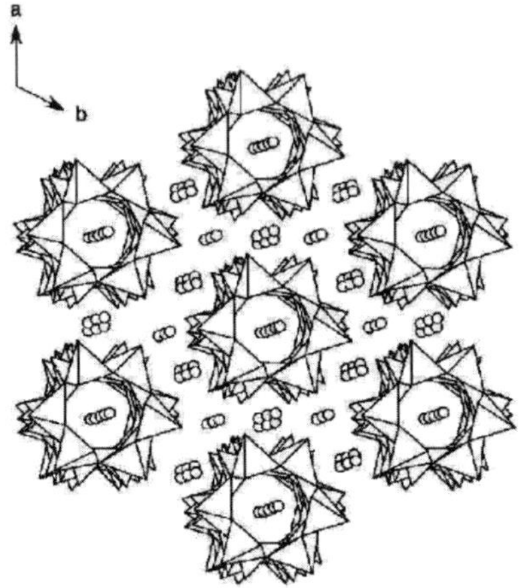

Fig. 14. $Na_2V_3O_7$ nanotubes. Square pyramides: VO_5, circles: Na. Cage (or ring) diameter ~ 5 Å.[9]

2.3. *Characteristic properties of cage compounds*

Cage compounds may as well be identified by the occurrence of characteristic properties.

2.3.1. *Rattling/tunneling*

Rattling refers to one mass (the guest) moving incoherently with respect to a framework (host) just as peas in a baby rattle. Tunneling is a purely quantum mechanical phenomenon where a particle can traverse an energy barrier which, classically, can only be overcome by thermal activation across the barrier. In what follows the evidence for rattling and/or tunneling from different experiments is presented.

First, neutron scattering experiments on a clathrate are presented. The neutron is a subatomic particle with no net electric charge but a magnetic moment. Neutron radiation is commonly employed in neutron scattering facilities, where the radiation is used in a similar way one uses X-rays for the analysis of condensed matter. Since they are neutral they can deeply penetrate into matter and since they have a spin they are sensitive to magnetism.

In Fig. 15 atomic displacement parameters (U_{eq}) and nuclear densities, both determined by neutron scattering experiments, are shown for the guest atoms X = (Ba), Sr and Eu in $X_8Ga_{16}Ge_{30}$. Eu has the largest U_{eq}. The (linear) extrapolation to zero temperature suggests that at the absolute zero in temperature Eu still has a large U_{eq}. Since there are no temperature induced oscillations at 0 K this indicates that so-called split sites exist in $Eu_8Ga_{16}Ge_{30}$, as indeed concluded from the nuclear density maps.

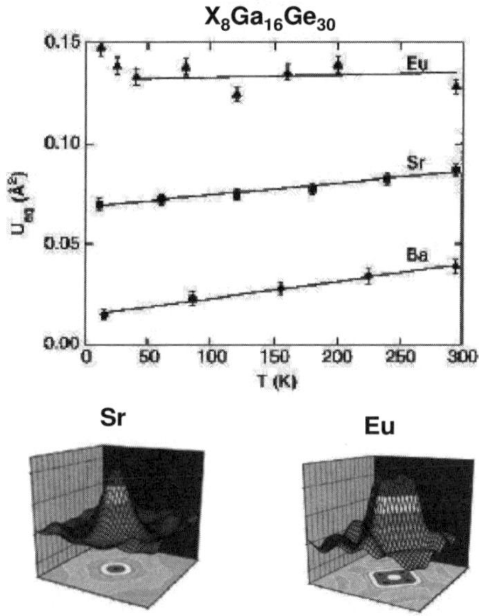

Fig. 15. Atomic displacement parameters and nuclear densities.[10]

Very direct evidence for rattling comes from Raman scattering experiments. In Raman spectroscopy monochromatic light, usually from a laser, is inelastically scattered from a solid, thereby exciting certain phonon modes or other degrees of freedom in the system. Figure 16 displays Raman spectra of various filled skutterudite compounds. Arrows show the rattling modes. With increasing cage size (from top to bottom), the rattling frequency of second-order phonons decreases. The peak intensity decreasing with lowering the temperature (not shown) indicates that (thermally activated) rattling is dominant and not (temperature independent) tunneling.

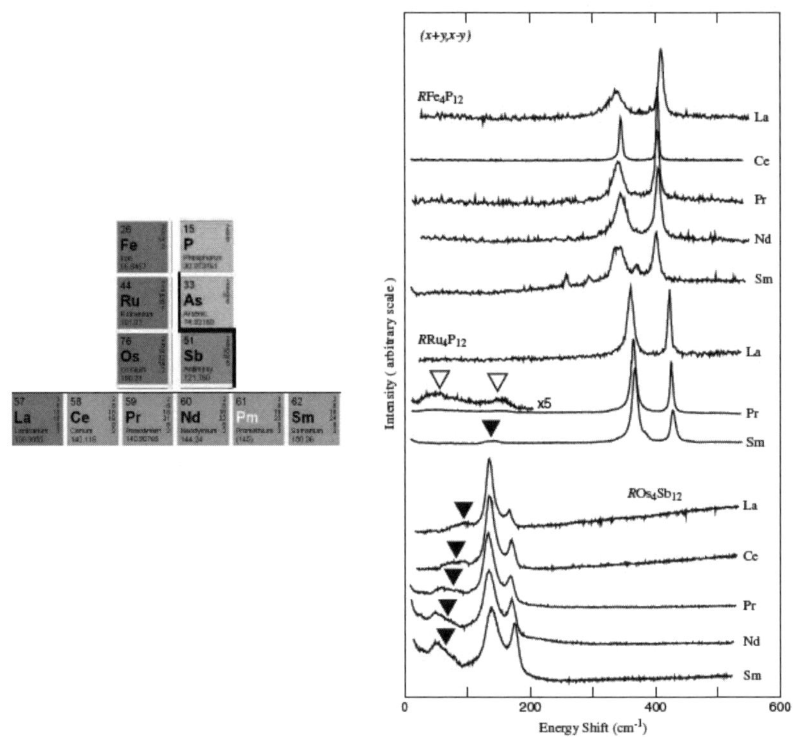

Fig. 16. Raman scattering spectra of filled skutterudites[11] and sections of the periodic table.

The results of resonant ultrasound spectroscopy (Fig. 17), on the other hand, need to be compared to theoretical curves in order to be conclusive. Theoretical curves (full lines) agree well with the experiment (symbols) if a four-well potential (top left) is assumed for the guest atoms in the clathrate cages.

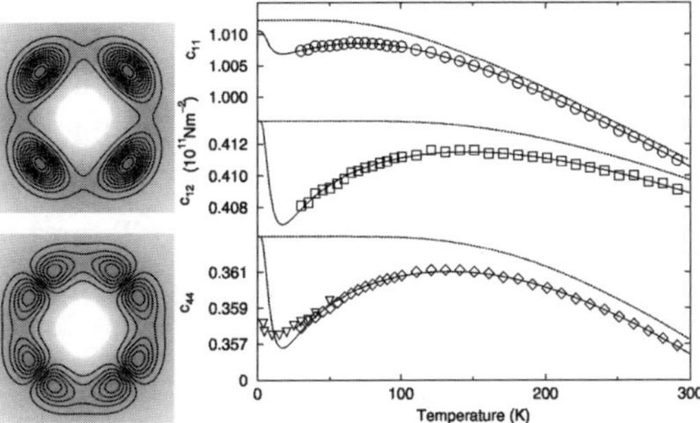

Fig. 17. Four-well potential and elastic constants of $Eu_8Ga_{16}Ge_{30}$.[12]

Direct evidence for tunneling of guest atoms between split sites is provided by Mössbauer spectroscopy on $Eu_8Ga_{16}Ge_{30}$ (Fig. 18).

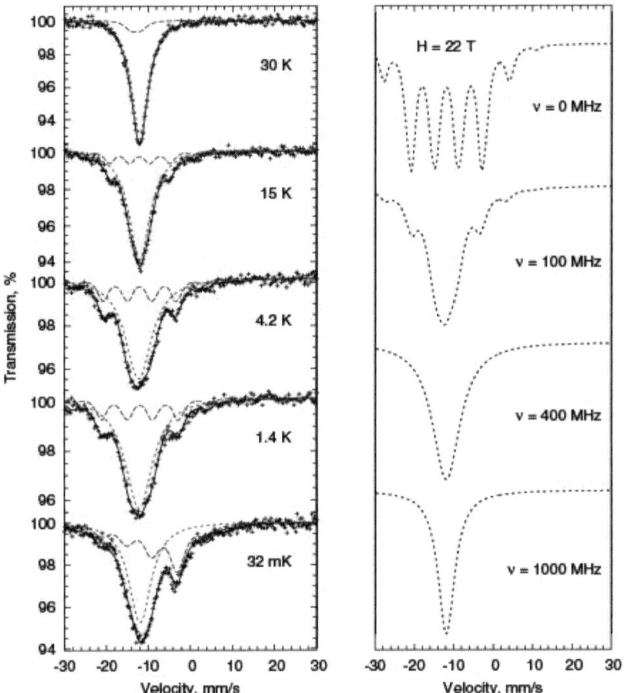

Fig. 18. Mössbauer spectroscopy on $Eu_8Ga_{16}Ge_{30}$.[13]

The panels on the right hand side are calculated spectra for divalent Eu in a hyperfine field of 22 T, fluctuating at different frequencies ν. The panels on the left hand side show the experimental results at different temperatures together with the fits representing spectra of two Eu atoms in small cages in a static hyperfine field of 20 T, and six Eu atoms in large cages tunnelling at 450 MHz between 4 split sites.

2.3.2. *Phonon glass-electron crystal*

A phonon glass–electron crystal (PGEC) (Fig. 19) is a material which conducts heat like a glass, i.e., very poorly, but electricity like a (single) crystal. It is clear from the expression of the dimensionless thermoelectric figure-of-merit: $\propto$

$$ZT = (S^2 \cdot \sigma/\kappa) \cdot T \tag{2}$$

that such a material would be an efficient thermoelectric since the ratio σ/κ is maximized.

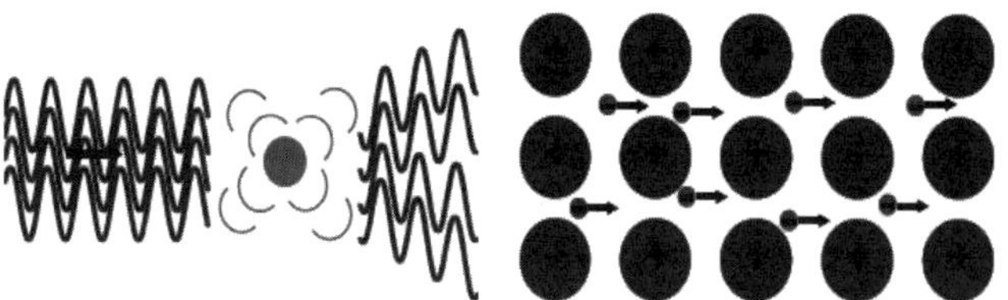

Fig. 19. Schematic representation of a "phonon glass" (κ small, left) and an "electron crystal" (σ large, right).

Since the electronic thermal conductivity κ_e is related to the electrical resistivity via the Wiedemann-Franz law ($\kappa_e \propto T/\rho$) large $\sigma = 1/\rho$ and small κ seemingly contradict each other. This is, however, not the case in materials where the thermal conductivity $\kappa = \kappa_L + \kappa_e$ is dominated by its lattice contribution κ_L, as in the clathrate $Ba_8Ga_{16}Ge_{30}$ (Fig. 20). The temperature dependence of κ_L of $Ba_8Ga_{16}Ge_{30}$ ressembles the one of vitreous SiO_2, – simple window glass, which is known for its very low thermal conductivity. The fit (full line) accounts for phonon-electron scattering at low temperatures and resonant scattering off the rattling guest atoms at higher temperatures.

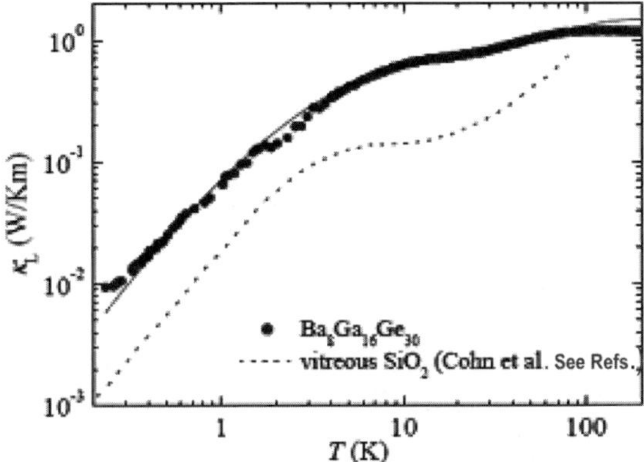

Fig. 20. Lattice thermal conductivity of the clatrathe $Ba_8Ga_{16}Ge_{30}$.[14]

The drastic lowering of the thermal conductivity by rattling guest atoms is demonstrated by Fig. 21. Upon increasing the Nd filling x in the skutterudite $Nd_xFe_4Sb_{12}$ the peak in the thermal conductivity at somewhat below 50 K is successively weakened and suppressed altogether at $x \approx 1$.

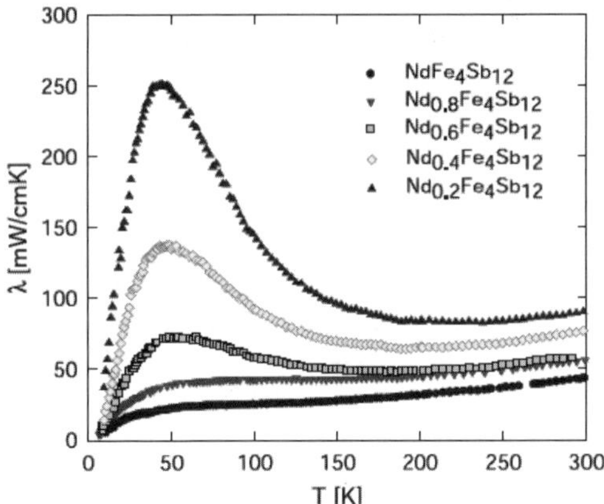

Fig. 21. Thermal conductivity of the skutterudite $Nd_yFe_4Sb_{12}$ (by courtesy of E. Bauer, unpublished).

Very instructive is also the comparison of thermal (Fig. 22) and electrical (Fig. 23) conductivities of the clathrates α- and β-Eu$_8$Ga$_{16}$Ge$_{30}$. The α-phase is a type VIII clathrate, the β-phase is of type I. Both are n-type conductors. The effective masses as deduced from electrical transport are $m^*(\alpha) \approx 1.5 m_0$ and $m^*(\beta) \approx 3.2 m_0$. $\kappa_L(T)$ appears to be dominated by phonon-electron scattering below 10 K, and by resonant scattering off the rattling guest atoms above 10 K. $\kappa_L(T)$ being, at low temperatures, much smaller in β- then in α-Eu$_8$Ga$_{16}$Ge$_{30}$ was argued to be due to the larger effective mass of the electrons, leading to enhanced phonon-electron scattering rates.

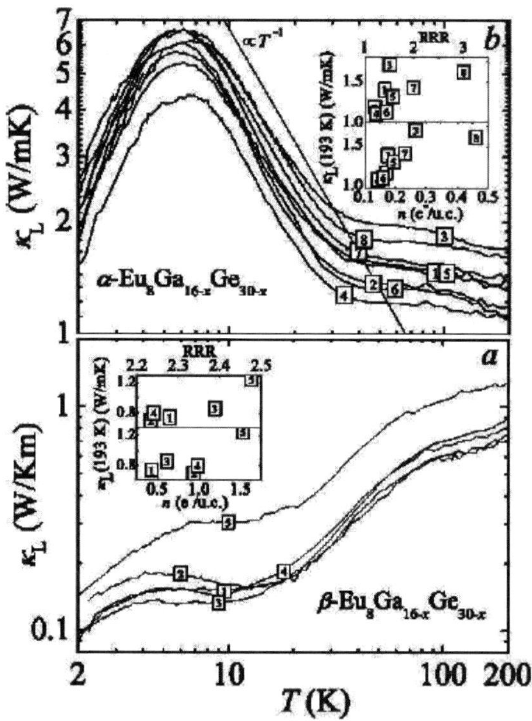

Fig. 22. Thermal conductivity of the clatrathes α- and β-Eu$_8$Ga$_{16}$Ge$_{30}$.[15]

Combining electrical resistivity and Hall effect data the electronic mean free path has been deduced for all α- and β-Eu$_8$Ga$_{16}$Ge$_{30}$ samples of

Fig. 23 at 2 K: α4 : 22 Å, α8 : 83 Å, β1 : 29 Å, β5 : 37 Å. These are very short compared to typical values for clean crystalline metals/semiconductors (which can be of the order of μm). There are several possible reasons for these short mean free paths and corresponding low charge carrier mobilities: Ga/Ge disorder, deviations from the ideal Ga:Ge stoichiometry, rattling of the guest atoms. Further investigations are needed to determine the dominating effect. I any case, the low mobilities of the clathrates α- and β-$Eu_8Ga_{16}Ge_{30}$ show that the "electron crystal" has to be used with some caution for these systems.

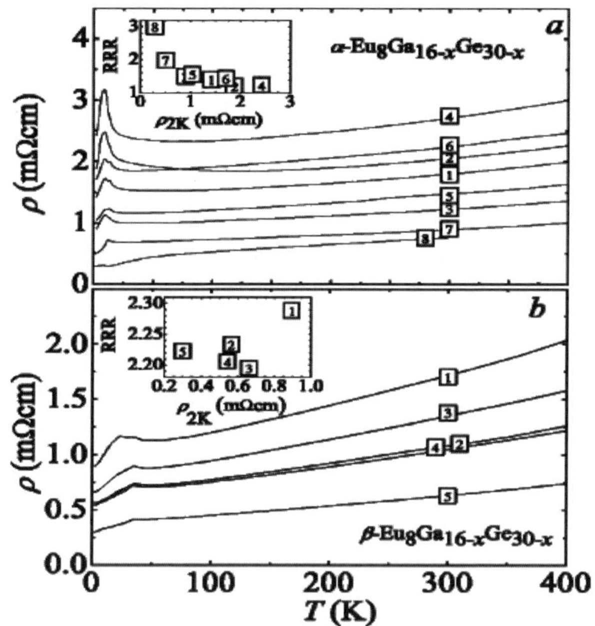

Fig. 23. Electrical resistivity of the clatrathes α- and β-$Eu_8Ga_{16}Ge_{30}$.[15]

2.4. *Tuning for optimized performance*

Many efforts worldwide aim at improving the thermoelectric properties of known materials. The attempts to do so can be subdivided as follows:

2.4.1. Stoichiometry

A precise control of the stoichiometry is important for many thermoelectric materials. An example is $Eu_8Ga_{16-x}Ge_{30+x}$ where x can be deliberately changed by annealing at different temperatures. The variation of x translates into a variation of the charge carrier concentration n and the thermoelectric properties, as shown in Fig. 24.

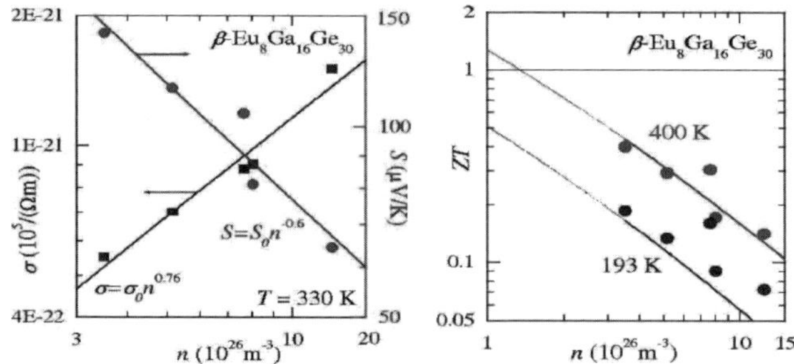

Fig. 24. Resistivity and thermopower (left) and thermoelectric figure-of-merit (right) after annealing $Eu_8Ga_{16-x}Ge_{30+x}$ at different temperatures: $n \approx x$.[15, 16]

2.4.2. Doping

In many cases doping is used to maximize ZT. Figure 25 shows such a study on the filled skutterudite $Nd_xCo_{4-y}Ni_ySb_{12}$.

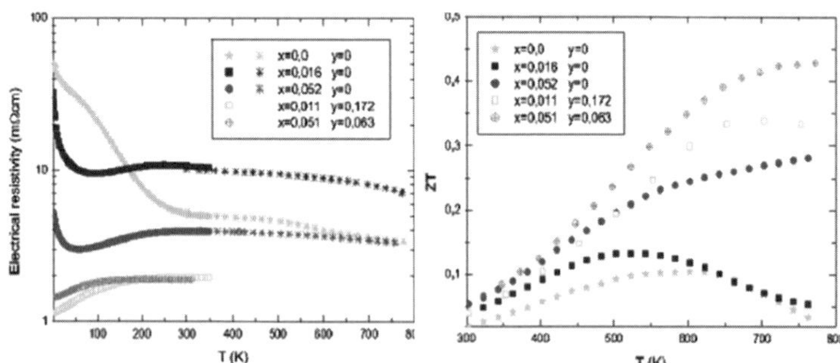

Fig. 25. Electrical resistivity (left) and thermoelectric figure-of-merit (right) of the filled skutterudite $Nd_xCo_{4-y}Ni_ySb_{12}$.[17]

2.4.3. *Substitution*

Also substitutions such as the partial replacement of Sr by Yb in the double-filled skutterudite $(Sr,Yb)_yCo_4Sb_{12}$ (Fig. 26) can lead to sizable *ZT* enhancements.

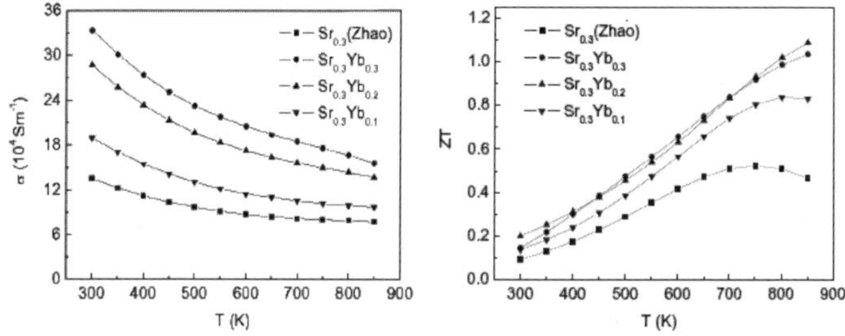

Fig. 26. Thermoelectric figure-of-merit.[17]

2.4.4. *Micro/nanostructuring*

A technique that becomes increasingly popular is the micro- and/or nanostructuring of materials. The concept behind is that electrons are not as strongly scattered by grain boundaries as phonons and that thus the (lattice) thermal conductivity is more strongly suppressed than the electrical conductivity. Figure 27 shows, with the example of the unfilled skutterudite $CoSb_3$ processed by spark plasma sintering, that most efficient in this respect are grains of different length scales.

Fig. 27. Micro- and nanograins (left) and thermoelectric figure-of-merit (right) of the unfilled skutterudite $CoSb_3$ processed by spark plasma sintering.[18]

3. Strongly Correlated Cage Compounds

3.1. *The concept of strongly correlated cage compounds*

The idea to exploit the emerging class of materials, strongly correlated cage compounds, for thermoelectric applications (Fig. 28) and the underlying physical concepts (Fig. 29) shall be discussed in this section.

The motivation for using cage compounds is the phonon glass – electron crystal concept explained above and symbolized by the "vibrating" guest atoms in the clathrate structure in Fig. 28.

In addition, a new concept for increasing the thermopower – strong electronic correlations – is introduced. The symbol in Fig. 28 represents a localized magnetic moment (flash) that interacts with an ensemble of conduction electrons.

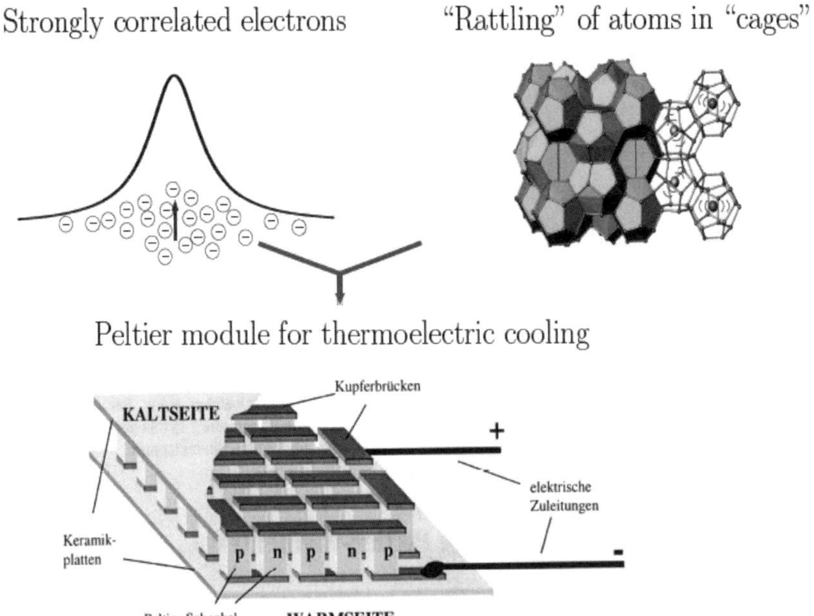

Fig. 28. Exploitation of strongly correlated cage compounds for thermoeleactric cooling.

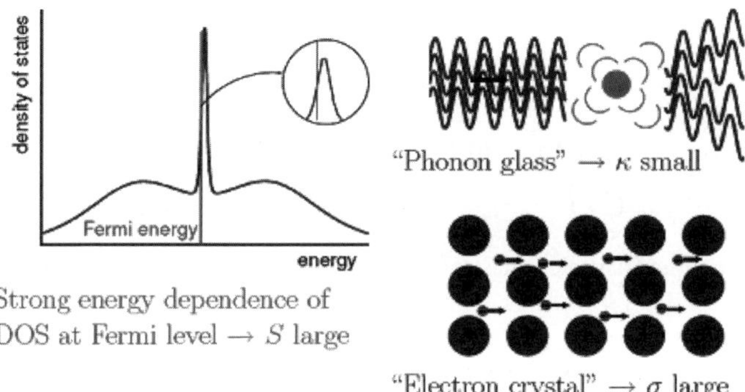

$$ZT = (S^2 \cdot \sigma/\kappa) \cdot T$$

maximized for

strongly correlated electrons in cage compounds

Fig. 29. The concept of large thermoelectric figure-of-merit in strongly correlated cage compounds.

3.2. *Brief introduction to strongly correlated electron systems*

Strongly correlated electron systems (SCES) are characterized by electrons strongly interacting with each other. As a consequence their effective mass is strongly enhanced (renormalized). In so-called heavy fermion systems, it is two different types of electrons that interact with each other, localized ones originating from uncompletely filled localized f (or d) shells, and itinerant ones originating from much broader bands of s, p (or d) character. At low temperatures the Kondo interaction leads to the formation of a "composit quasiparticle" consisting of the local magnetic moment collectively screened by a conduction electron cloud (Fig. 30). These quasiparticles may be described by Landau's Fermi liquid theory. However, for metallic heavy fermion systems, the prefactors in the temperature dependencies of the specific heat C, the magnetic susceptibility χ, and the electrical resistivity ρ, γ_0, χ_0, and A, respectively (see Fig. 30), are strongly remormalized compared to the typical values in simple metals (where $m^* \approx m_e$).

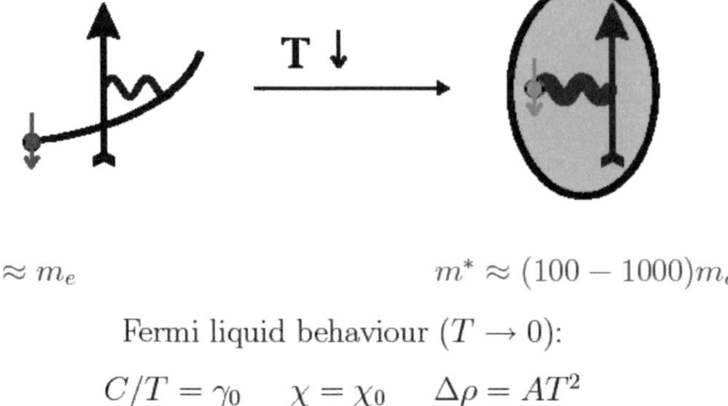

Fermi liquid behaviour $(T \to 0)$:

$$C/T = \gamma_0 \qquad \chi = \chi_0 \qquad \Delta\rho = AT^2$$

Fig. 30. Heavy quasiparticles in strongly correlated electron systems. (Graphics from Ref. 19.)

As for simple (uncorrelated) materials, strongly correlated metals generally face the problem of too high (electronic) thermal conductivities. But also in SCES there exist materials with semimetallic or semiconducting properties (so-called Kondo insulators/semiconductors/semimetals) which are better candidates for thermoelectric applications. The origin of the (pseudo) energy gap formation in these systems is the correlation effects itself. An example of a semimetallic SCES is the Kondo semimetal CeNiSn. Its pseudo energy gap ΔE is approximately $k_B \cdot$ 14 K, a very small value as compared to typical energy gaps in semiconductors (Si, Ge: $\Delta E \approx k_B \cdot$ 10 000 K).

Figure 31 shows its crystal structure, which is clearly not cage like, and the thermopower at low temperatures. The thermopower has a large peak below the gap temperature. The value at 4 K is more than an order of magnitude larger than in a typical thermoelectric clathrate. On the other hand due to the absence of a cage structure, the thermal conductivity of CeNiSn is about an order of magnitude higher.

An example of a strongly correlated semiconductor is FeSi. Its energy gap ΔE is approximately $k_B \cdot$ 500K. Angle resolved photoemission spectroscopy (ARPES) experiments (Fig. 32) have revealed extremely sharp features in the electronic density of states near the Fermi level, along certain directions in reciprocal space.

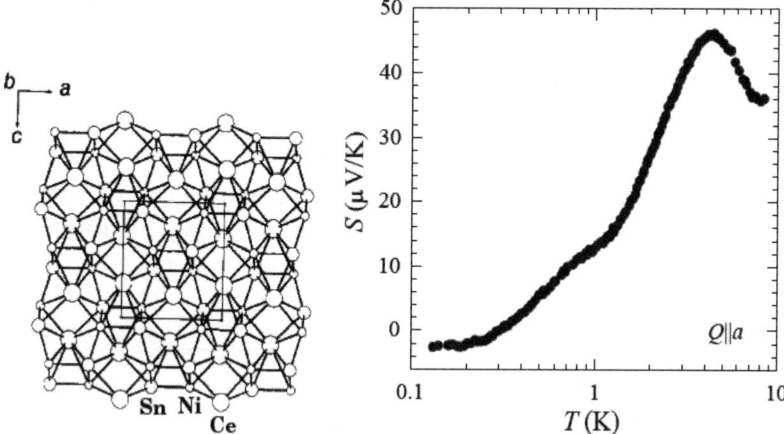

Fig. 31. Crystal structure and temperature dependence of thermopower of the Kondo semimetal CeNiSn.[20]

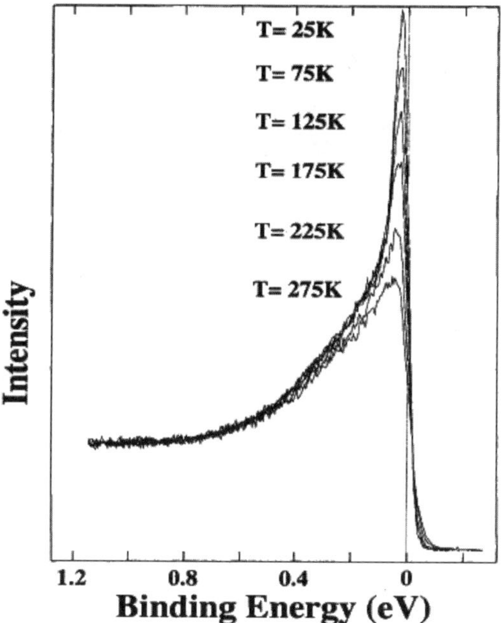

Fig. 32. Electronic density of states of FeSi determined in ARPES experiments.[21]

As expected, the thermopower of FeSi is extremely large (Fig. 33). However, since σ/κ is small, ZT reaches only a modest value of less than 0.01 at 50 K. Thus, to bring SCES to practical use, σ/κ has to be enhanced. Since in both CeNiSn and FeSi κ is dominated by the phonon contribution, it is this contribution which has to be suppressed. One possible route might be to combine, in a single substance, the advantages of cage compounds (large σ/κ) and of SCES (large S). First attempts to obtain such "strongly correlated cage compounds" are described below.

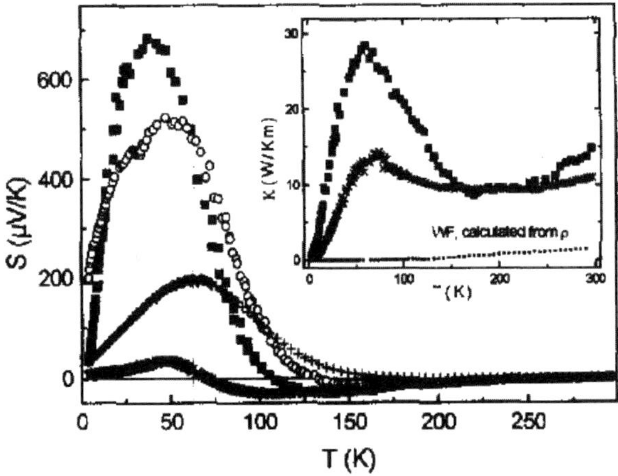

Fig. 33. Thermopower versus temperature for FeSi (full squares) and doped variants (other symbols).[22]

3.3. *Attempts to obtain strongly correlated cage compounds*

A possible route to strongly correlated cage compounds is to use rare earth elements tending to hybridization with conduction electron bands as guest atoms in cage compounds. Few such attempts are presented below.

3.3.1. *Substituted $Eu_8Ga_{16}Ge_{30}$*

Even though there are examples of Eu based compounds showing effects of strong correlations this is the exception rather than the rule.

In $Eu_8Ga_{16}Ge_{30}$ no signs of strong correlations are observed. Eu is divalent in the entire temperature range and the system orders ferromagnetically at low temperatures. Worldwide efforts to (partially) replace Eu by other rare earth elements have failed.[23,24] Also the seemingly successful incorporation of Ce into a Si-based type-I clathrate, $Ba_6Ce_2Au_4Si_{42}$,[25] was subsequently shown to have failed.[26]

3.3.2. *The clathrate-like compound* $Ce_2Ni_{36}P_{15}$

$Ce_2Ni_{36}P_{15}$ has a complex crystal structure (Fig. 34), with Ce encapsulated as guest atom in cages of Ni and P. Electrical resistivity and specific heat show clear signs of correlation effects, with non-Fermi liquid temperature dependencies ($\rho \propto T$, $\gamma \propto -\ln(T)$).[27] However, due to a very small Kondo temperature and the metallic character of the system the thermoelectric performance is rather poor.

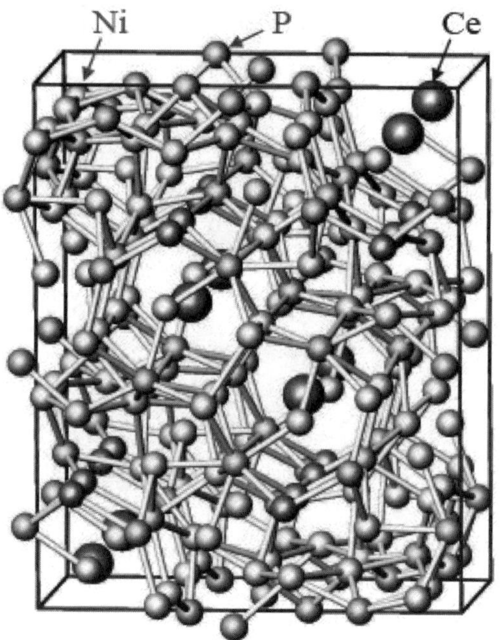

Fig. 34. Crystal structure of $Ce_2Ni_{36}P_{15}$.[27]

3.3.3. *The cage compound CeRu₄Sn₆*

In $CeRu_4Sn_6$ strong hybridization between the guest atom Ce and the host atoms leads to the opening of an energy gap of approximately 40 K. The peak in the thermopower at temperatures somewhat below, as measured on polycrystalline samples[28], is surprisingly low. One may speculate whether this is an extinction effect of thermopowers of different sign along different crystallograhpic directions. Measurements on single crystals are needed to clarify the situation. The thermal conductivity is not as strongly glass like as in some clathrates. This might be due to the smaller size of the cages.

3.3.4. *The cage compound Ce₃Pd₂₀Si₆*

Finally, the cage compound $Ce_3Pd_{20}Si_6$ shall be discussed in some more detail. In the cubic crystal structure of space group $Fm\overline{3}m$, Ce is situated in two types of cages. The polyhedron 1 consists of 6 Si and 12 Pd atoms, polyhedron 2 of 16 Pd atoms (Fig. 35).

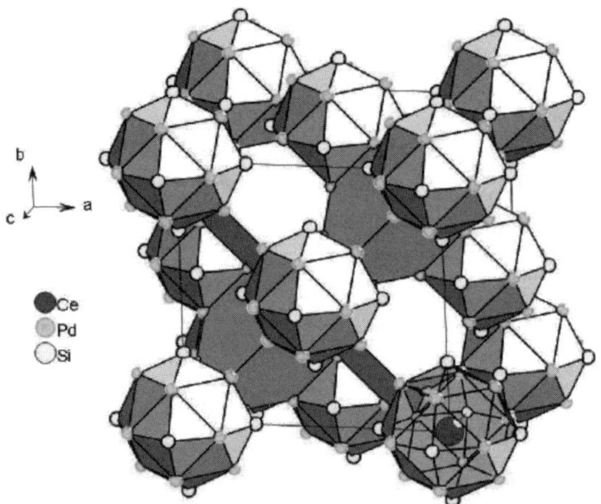

Fig. 35. Cubic crystal structure of $Ce_3Pd_{20}Si_6$.[7]

The magnetic susceptibility is Curie-Weiss-like at high temperatures but tends to saturate to a high value at low temperature where, finally, a very weak signature of (presumably) magnetic ordering is observed

(Fig. 36), behaviour typically observed also in heavy fermion compounds. Also the electrical resistivity (not shown) with a T^2 temperature dependence with a large prefactor at very low temperatures is typical for heavy fermion compounds.[29]

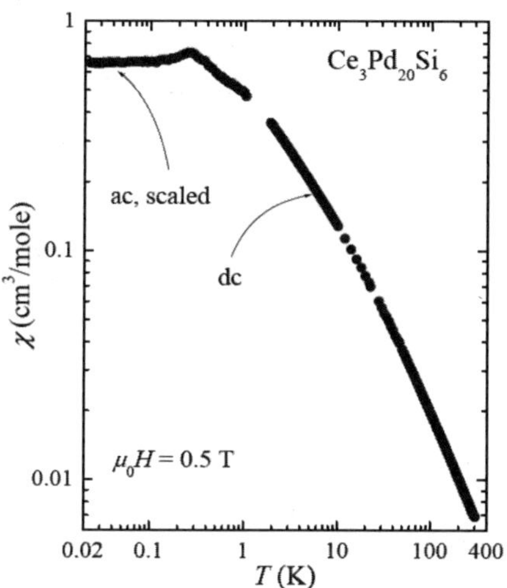

Fig. 36. Magnetic susceptibility versus temperature.[29] Similar to earlier findings.[30]

Thermopower does indeed show a relatively large value of about 45 μV/K at somewhat above 20 K (Fig. 37).

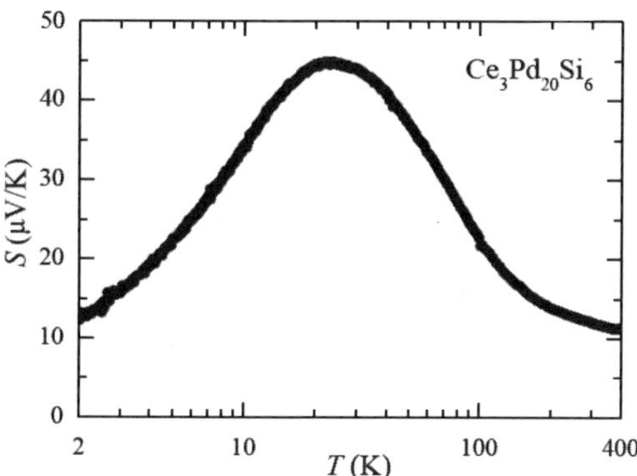

Fig. 37. Thermopower versus temperature of $Ce_3Pd_{20}Si_6$.[31]

However, similar to the case of $CeRu_4Sn_6$, also here the thermal conductivity (Fig. 38) does not show strongly pronounced glass-like features. At 20 K it has already reached a value of 3 W/Km (one third of which is due to electrons) which is distinctly higher than typically found in clathrates. Thus, also here the "rattling" of the guest atoms apprears less pronounced than in clathrates. This results in a relatively modest but nevertheless promising thermoelectric figure-of-merit, with a peak value of approximately 0.03 at 25 K (Fig. 39).

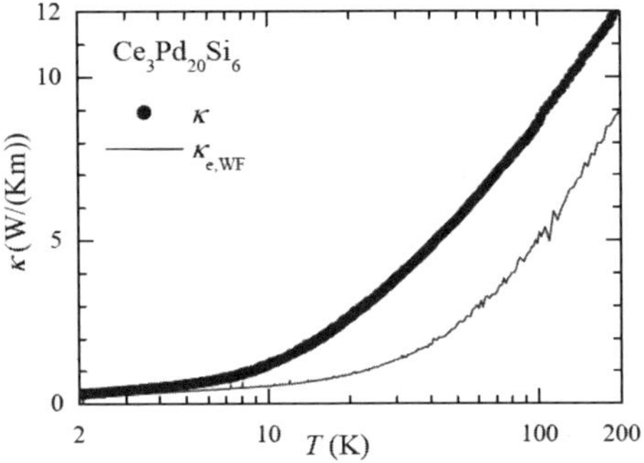

Fig. 38. Thermal conductivity versus temperature of $Ce_3Pd_{20}Si_6$. The electronic contribution as estimated from the Wiedemann-Franz law, $\kappa_{e,WF}$, is shown as solid line.[31]

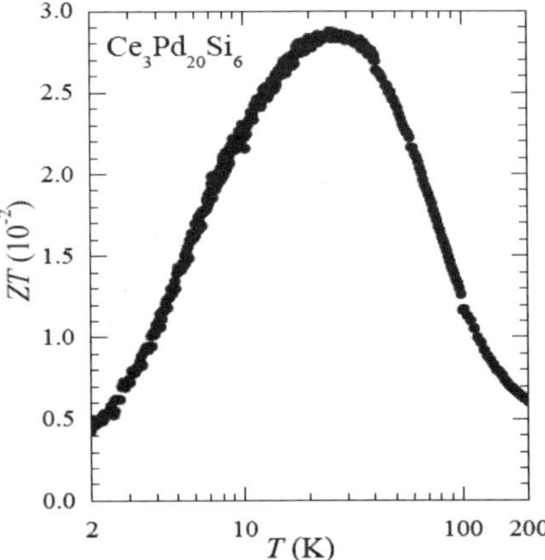

Fig. 39. Dimensionless figure-of-merit versus temperature of $Ce_3Pd_{20}Si_6$.[31]

Finally, Fig. 40 compares the figure-of-merit of $Ce_3Pd_{20}Si_6$ with other compounds (see legend). While ZT of $Ce_3Pd_{20}Si_6$ is highest below 40 K ZT clearly needs further improvement in the temperature range well below room temperature and the concept of strongly correlated cage compounds needs further exploration.

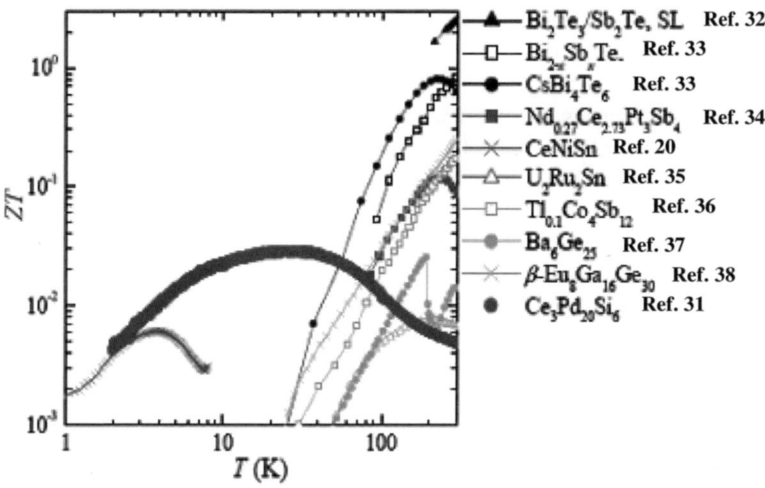

Fig. 40. Figure-of-merit: A comparison. Top three symbols (black): Bi-Te based materials; next three symbols (dark grey): Kondo insulators; next four (light grey and black): Cage compounds.

References

1. G. J. Snyder and E. S. Toberer, *Nature Materials* **7**, 105 (2008) and Refs. herein.
2. S. Paschen, W. Carrillo-Cabrera, A. Bentien, V. H. Tran, M. Baenitz, Y. Grin, and F. Steglich, *Phys. Rev.* **B 64**, 214404 (2001).
3. S. Bobev et al., *J. Solid State Chem.* **177**, 3545 (2004).
4. A. M. Guloy, R. Ramlau, Z. Tang, W. Schnelle, M. Baitinger and Y. Grin, *Nature* **443**, 320 (2006).
5. P. Rogl, in *Formation and crystal chemistry of clathrates*, in CRC Handbook of Thermoelectrics, Ed. D. M. Rowe (CRC Press, Boca Raton, 2005), Chap. 32.
6. S. Budnyk, Ph.D. thesis, National University Lwiw, Ukraine; TU Dresden, Germany, 2003.

7. A. V. Gribanov, Y. D. Seropegin, and O. J. Bodak, *J. Alloys & Comp.* **204**, L9 (1994).
8. J. Kunes, T. Jeong and W. E. Pickett, *Phys. Rev.* **B 70**, 174510 (2004).
9. J. Choi, J. L. Musfeldt, Y. J. Wang, H.-J. Koo, M.-H. Whanghoo, J. Galy and P. Millet, *Chem. Mater.* **14**, 924 (2002).
10. B. C. Sales, B. C. Chakoumakos, R. Jin, J. R. Thompson, and D. Mandrus, *Phys. Rev.* **B 63**, 245113 (2001).
11. N. Ogita, T. Kondo, T. Hasegawa, Y. Takasu, M. Udagawa, N. Takeda, K. Ishikawa, H. Sugawara, D. Kikuchi, H. Sato, C. Sekine and I. Shirotani, *Physica* **B 383**, 128 (2006).
12. I. Zerec, V. Keppens, M. A. McGuire, D. Mandrus, B. C. Sales and P. Thalmeir, *Phys. Rev. Lett.* **92,** 185502 (2004).
13. R. P. Hermann, V. Keepens, P. Bonville, G. S. Nolas, F. Grandjean, G. J. Long, H. M. Christen, B. C. Chakoumakos, B. C. Sales and D. Mandrus, *Phys. Rev. Lett.* **97**, 017401 (2006).
14. S. Paschen, V. Pacheco, A. Bentien, A. Sanchez, W. Carrillo-Cabrera, M. Baenitz, B. B. Iversen, Yu. Grin, and F. Steglich, *Physica* **B 328**, 39 (2003).
15. A. Bentien, V. Pacheco, S. Paschen, Yu. Grin, and F. Steglich, *Phy. Rev.* **B 71**, 165206 (2005).
16. V. Pacheco, A. Bentien, W. Carrillo-Cabrera, S. Paschen, F. Steglich, and Y. Grin, *Phy. Rev.* **B 71**, 165205 (2005).
17. V. Da Ros, B. Lenoir, A. Dauscher, C. Candolfi, C. Bellouard, C. Steve, E. Müller and J. Hejtmanekand, in *Proc. 22nd Int. Conf. on Thermoelectrics* (168, Vienna, Austria, ISBN 1-4244-0810-5, 2006).
18. J. X. Zhang, Q. M. Lu, X. Zhang and Q. Wei, in *Proc. 22nd Int. Conf. on Thermoelectrics* (168, Vienna, Austria, ISBN 1-4244-0810-5, 2006).
19. A. Schofield, *Contemporary Physics* **40** 95 (1999).
20. S. Paschen, B. Wand, G. Sparn, F. Steglich, Y. Echizen, and T. Takabatake, *Phys. Rev.* **B 62**, 14912 (2000).
21. C. H. Park, Z. X. Shen, A. G. Loeser, D. S. Dessau, D. G. Mandrus, A. Migliori, J. Sarrau and Z. Fisk, *Phys. Rev.* **B 52**, 16981 (1995).
22. B. Buschinger, C. Geibel, F. Steglich, D. Mandrus, D. Young, J. L. Sarrao, and Z. Fisk, *Physica* **B 230-232**, 784 (1997).
23. S. Paschen, S. Budnyk, U. Köhler, Y. Prots, K. Hiebl, F. Steglich, and Y. Grin, *Physica* **B 383**, 89 (2006).
24. S. Paschen, C. Gspan, W. Grogger, M. Dienstleder, S. Laumann, P. Pongratz, H. Sassik, J. Wernisch, and A. Prokofiev, *J. Cryst. Growth* **310,** 1853 (2008).
25. T. Kawaguchi, K. Tanigaki, and M. Yasukawa, *Phys. Rev. Lett.* **85**, 3189 (2000).

26. V. Pacheco, W. Carrillo-Cabrera, V. H. Tran, S. Paschen, and Yu. Grin, *Phys. Rev. Lett.* **87**, 9601 (2001).

27. A. Bentien, Ph.D. thesis, TU Dresden, Germany, Shaker Verlag, Aachen, ISSN 0945-0963, 2005.

28. A. M. Strydom, Z. Guo, S. Paschen, R. Viennois, and F. Steglich, *Physica* **B 359–361**, 293 (2005).

29. S. Paschen, M. Müller, J. Custers, M. Kriegisch, A. Prokofiev, G. Hilscher, W. Steiner, A. Pikul, F. Steglich, and A. M. Strydom, *J. Magn. Magn. Mater.* **316**, 90 (2007).

30. J. Kitagawa, N. Takeda, and M. Ishikawa, Phys. Rev. **B 53**, 5101 (1996).

31. S. Paschen, A. Bentien, S. Budnyk, A. M. Strydom, Y. Grin, and F. Steglich, *in Proc. 22nd Int. Conf. on Thermoelectrics* (168, Vienna, Austria, ISBN 1-4244-0810-5, 2006).

32. R. Venkatasubramanian, E. Siivola, T. Colpitts, and B. O'Quinn, Nature 413, 597 (2001).

33. D. Chung, T. Hogan, P. Brazis, M. Rocci-Lane, C. Kannewurf, M. Bastea, C. Uher, and M. G. Kanatzidis, *Science,* **287**, 1024 (2000).

34. C. D. W. Jones, K. A. Regan, and F. J. DiSalvo, *Phys. Rev.* **B 58**, 16057 (1998).

35. V. H. Tran, S. Paschen, A. Rabis, M. Baenitz, F. Steglich, P. de V. du Plessis, and A. M. Strydom, *Physica* **B 312-313**, 215 (2002).

36. B. C. Sales, B. C. Chakoumakos, and D. Mandrus, *Phys. Rev.* **B 61**, 2475 (2000).

37. S. Paschen, V. H. Tran, M. Baenitz, W. Carrillo-Cabrera, Yu. Grin, and F. Steglich, *Phys. Rev.* **B 65**, 134 435 (2002).

38. A. Bentien, S. Paschen, V. Pacheco, Y. Grin, and F. Steglich, in Proc. 22nd Int. Conf. on Thermoelectrics (p. 131, Vienna, Austria, ISBN 1-4244-0810-5, 2003).

Additional reference in Fig. 14: Cohn et al . *Phys. Rev. Lett.* **82**, 779 (1999).